Energy Systems

Series Editor:

Panos M. Pardalos, University of Florida, USA

For further volumes:
http://www.springer.com/series/8368

Soliman Abdel-Hady Soliman
Abdel-Aal Hassan Mantawy

Modern Optimization Techniques with Applications in Electric Power Systems

Soliman Abdel-Hady Soliman
Department of Electrical
Power and Machines
Misr University for Science
and Technology
6th of October City, Egypt

Abdel-Aal Hassan Mantawy
Department of Electrical
Power and Machines
Ain Shams University
Cairo, Egypt

ISSN 1867-8998 e-ISSN 1867-9005
ISBN 978-1-4614-1751-4 e-ISBN 978-1-4614-1752-1
DOI 10.1007/978-1-4614-1752-1
Springer New York Heidelberg Dordrecht London

Library of Congress Control Number: 2011942262

Mathematics Subject Classification (2010): T25015, T25031, T11014, T11006, T24043

© Springer Science+Business Media, LLC 2012
All rights reserved. This work may not be translated or copied in whole or in part without the written permission of the publisher (Springer Science+Business Media, LLC, 233 Spring Street, New York, NY 10013, USA), except for brief excerpts in connection with reviews or scholarly analysis. Use in connection with any form of information storage and retrieval, electronic adaptation, computer software, or by similar or dissimilar methodology now known or hereafter developed is forbidden.
The use in this publication of trade names, trademarks, service marks, and similar terms, even if they are not identified as such, is not to be taken as an expression of opinion as to whether or not they are subject to proprietary rights.

Printed on acid-free paper

Springer is part of Springer Science+Business Media (www.springer.com)

To the the Spirit of the martyrs of the EGYPTIAN 25th of January Revolution.
To them we say, "You did what other generations could not do". May GOD send your Spirit to Paradise.
"Think not of those who were killed in the way of Allah dead, but alive with their Lord they have provision"
(The Holy Quraan).
To my grandson Ali, the most beautiful flower in my life.
To my parents, I miss them.
To my wife, Laila, and my kids, Rasha, Shady, Samia, Hadier, and Ahmad, I love you all.
To my Great teacher G. S. Christensen

                                      (S.A. Soliman)

To the soul of my parents.
To my wife Mervat.
To my kids Sherouk, Omar, and Kareem.

                                      (A.H. Mantawy)

Preface

The growing interest in the application of artificial intelligence (AI) techniques to power system engineering has introduced the potentials of using this state-of-the-art technology. AI techniques, unlike strict mathematical methods, have the apparent ability to adapt to nonlinearities and discontinuities commonly found in power systems. The best-known algorithms in this class include evolution programming, genetic algorithms, simulated annealing, tabu search, and neural networks.

In the last three decades many papers on these applications have been published. Nowadays only a few books are available and they are limited to certain applications. The power engineering community is in need of a book containing most of these applications.

This book is unique in its subject, where it presents the application of some artificial intelligence optimization techniques in electric power system operation and control.

We present, with practical applications and examples, the application of functional analysis, simulated annealing, tabu search, genetic algorithms, and fuzzy systems on the optimization of power system operation and control.

Chapter 2 briefly explains the mathematical background behind optimization techniques used in this book including the minimum norm theorem and how it could be used as an optimization algorithm; it introduces fuzzy systems, the simulated annealing algorithm, tabu search algorithm, genetic algorithm, and the particle swarm as optimization techniques.

Chapter 3 explains the problem of economic operation of electric power systems, where the problem of short-term operation of a hydrothermal–nuclear power system is formulated as an optimal problem and using the minimum norm theory to solve this problem. The problem of fuzzy economic dispatch of all thermal power systems is also formulated and the algorithm suitable for solution is explained.

Chapter 4 explains the economic dispatch (ED) and unit commitment problems (UCP). The solution of the UCP problem using artificial intelligence techniques requires three major steps: a problem statement or system modeling, rules for generating trial solutions, and an efficient algorithm for solving the EDP. This chapter explains in detail the different algorithms used to solve the ED and UCP problems.

Chapter 5, "Optimal Power Flow," studies the load flow problem and presents the difference between the conventional load flow and the optimal load flow (OPF) problem and it introduces the different states used in formulating the OPF as a multiobjective problem. Furthermore this chapter introduces the particle swarm optimization algorithm as a tool to solve the optimal power flow problem.

Chapter 6, "Long-Term Operation of Hydroelectric Power Systems," formulates the problem of long-term operation of a multireservoir power system connected in cascade (series). The minimum norm approach, the simulated annealing algorithm, and the tabu search approach are implemented to solve the formulated problem.

Finally, in Chap. 7, "Electric Power Quality Analysis," presents applications of the simulated annealing optimization algorithm for measuring voltage flicker magnitude and frequency as well as the harmonic contents of the voltage signal. Furthermore, the implementation of SAA and tabu search to estimate the frequency, magnitude, and phase angle of a steady-state voltage signal, for a frequency relaying application is studied when the signal frequency is constant and is a variable with time. Two cases are studied: the linear variation of frequency with time and exponential variation. Effects of the critical parameters on the performance of these algorithms are studied in this book.

This book is useful for B. Sc. senior students in the electrical engineering discipline, MS and PhD students in the same discipline all over the world, electrical engineers working in utility companies, operation, control, and protection, as well as researchers working in operations research and water resources research.

Giza, Egypt	Soliman Abdel-Hady Soliman
Cairo, Egypt	Abdel-Aal Hassan Mantawy

Acknowledgments

I would like to acknowledge the support of the chancellor of Misr University for Science and Technology, Mr. Khalied Altokhy, and the president of the university during the course of writing this book. The help of the dean of engineering at Misr University for Science and Technology, Professor Hamdy Ashour is highly appreciated. Furthermore, I would like to acknowledge Dr. Jamal Madough, assistant professor at the College of Technological Studies, Kuwait, for allowing me to use some of the materials we coauthored in Chaps. 2 and 3. Finally, my appreciation goes to my best friend, Dr. Ahmad Al-Kandari, associate professor at the College of Technological Studies, Kuwait, for supporting me at every stage during the writing of this book. This book would not be possible without the understanding of my wife and children.

I would like to express my great appreciation to my wife, Mrs. Laila Mousa for her help and understanding, my kids Rasha, Shady, Samia, Hadeer, and Ahmad, and my grandchild, Ali, the most beautiful flower in my life. Ali, I love you so much, may God keep you healthy and wealthy and keep your beautiful smile for everyone in your coming life, Amen.

(S.A. Soliman)

I would like to express my deepest thanks to my PhD advisors, Professor Youssef L. Abdel-Magid and Professor Shokri Selim for their guidance, signs, and friendship, and allowing me to use some of the materials we coauthored in this book.

I am deeply grateful for the support of Ain Shams Uinversity and King Fahd Uinversity of Petroleum & Minerals, from which I graduated and continue my academic career.

Particular thanks go to my friend and coauthor of this book, Professor S. A. Soliman, for his encouragement and support of this work.

And last, but not least, I would like to thank my wife, Mervat, and my kids, Sherouk, Omar, and Kareem, for their love, patience, and understanding.

(A.H. Mantawy)

The authors of this book would like to acknowledge the effort done by Abiramasundari Mahalingam for reviewing this book many times and we appreciate her time, to her we say, you did a good job for us, you were sincere and honest in every stage of this book.

(The authors)

Contents

1 **Introduction** 1
   1.1 Introduction 1
   1.2 Optimization Techniques 2
       1.2.1 Conventional Techniques (Classic Methods) 3
       1.2.2 Evolutionary Techniques 7
   1.3 Outline of the Book 20
   References 21

2 **Mathematical Optimization Techniques** 23
   2.1 Introduction 23
   2.2 Quadratic Forms 24
   2.3 Some Static Optimization Techniques 26
       2.3.1 Unconstrained Optimization 27
       2.3.2 Constrained Optimization 30
   2.4 Pontryagin's Maximum Principle 37
   2.5 Functional Analytic Optimization Technique 42
       2.5.1 Norms 42
       2.5.2 Inner Product (Dot Product) 43
       2.5.3 Transformations 45
       2.5.4 The Minimum Norm Theorem 46
   2.6 Simulated Annealing Algorithm (SAA) 48
       2.6.1 Physical Concepts of Simulated Annealing 49
       2.6.2 Combinatorial Optimization Problems 50
       2.6.3 A General Simulated Annealing Algorithm 50
       2.6.4 Cooling Schedules 51
       2.6.5 Polynomial-Time Cooling Schedule 51
       2.6.6 Kirk's Cooling Schedule 53
   2.7 Tabu Search Algorithm 54
       2.7.1 Tabu List Restrictions 54
       2.7.2 Aspiration Criteria 55

		2.7.3	Stopping Criteria	55
		2.7.4	General Tabu Search Algorithm	56
	2.8	The Genetic Algorithm (GA)		57
		2.8.1	Solution Coding	58
		2.8.2	Fitness Function	58
		2.8.3	Genetic Algorithms Operators	59
		2.8.4	Constraint Handling (Repair Mechanism)	59
		2.8.5	A General Genetic Algorithm	60
	2.9	Fuzzy Systems		60
		2.9.1	Basic Terminology and Definition	64
		2.9.2	Support of Fuzzy Set	65
		2.9.3	Normality	66
		2.9.4	Convexity and Concavity	66
		2.9.5	Basic Operations	66
	2.10	Particle Swarm Optimization (PSO) Algorithm		71
	2.11	Basic Fundamentals of PSO Algorithm		74
		2.11.1	General PSO Algorithm	76
	References			78
3	**Economic Operation of Electric Power Systems**			**83**
	3.1	Introduction		83
	3.2	A Hydrothermal–Nuclear Power System		84
		3.2.1	Problem Formulation	84
		3.2.2	The Optimization Procedure	87
		3.2.3	The Optimal Solution Using Minimum Norm Technique	91
		3.2.4	A Feasible Multilevel Approach	94
		3.2.5	Conclusions and Comments	96
	3.3	All-Thermal Power Systems		96
		3.3.1	Conventional All-Thermal Power Systems; Problem Formulation	96
		3.3.2	Fuzzy All-Thermal Power Systems; Problem Formulation	97
		3.3.3	Solution Algorithm	105
		3.3.4	Examples	105
		3.3.5	Conclusion	111
	3.4	All-Thermal Power Systems with Fuzzy Load and Cost Function Parameters		112
		3.4.1	Problem Formulation	113
		3.4.2	Fuzzy Interval Arithmetic Representation on Triangular Fuzzy Numbers	123
		3.4.3	Fuzzy Arithmetic on Triangular L–R Representation of Fuzzy Numbers	128
		3.4.4	Example	129

3.5	Fuzzy Economical Dispatch Including Losses		145
	3.5.1	Problem Formulation	146
	3.5.2	Solution Algorithm	164
	3.5.3	Simulated Example	165
	3.5.4	Conclusion	167
References			183

4 Economic Dispatch (ED) and Unit Commitment Problems (UCP): Formulation and Solution Algorithms ... 185

4.1	Introduction		185
4.2	Problem Statement		186
4.3	Rules for Generating Trial Solutions		186
4.4	The Economic Dispatch Problem		186
4.5	The Objective Function		187
	4.5.1	The Production Cost	187
	4.5.2	The Start-Up Cost	187
4.6	The Constraints		188
	4.6.1	System Constraints	188
	4.6.2	Unit Constraints	189
4.7	Rules for Generating Trial Solutions		191
4.8	Generating an Initial Solution		193
4.9	An Algorithm for the Economic Dispatch Problem		193
	4.9.1	The Economic Dispatch Problem in a Linear Complementary Form	194
	4.9.2	Tableau Size for the Economic Dispatch Problem	196
4.10	The Simulated Annealing Algorithm (SAA) for Solving UCP		196
	4.10.1	Comparison with Other SAA in the Literature	197
	4.10.2	Numerical Examples	198
4.11	Summary and Conclusions		207
4.12	Tabu Search (TS) Algorithm		208
	4.12.1	Tabu List (TL) Restrictions	209
	4.12.2	Aspiration Level Criteria	212
	4.12.3	Stopping Criteria	213
	4.12.4	General Tabu Search Algorithm	213
	4.12.5	Tabu Search Algorithm for Unit Commitment	215
	4.12.6	Tabu List Types for UCP	216
	4.12.7	Tabu List Approach for UCP	216
	4.12.8	Comparison Among the Different Tabu Lists Approaches	217
	4.12.9	Tabu List Size for UCP	218
	4.12.10	Numerical Results of the STSA	218

4.13	Advanced Tabu Search (ATS) Techniques		220
	4.13.1	Intermediate-Term Memory	221
	4.13.2	Long-Term Memory	222
	4.13.3	Strategic Oscillation	222
	4.13.4	ATSA for UCP	223
	4.13.5	Intermediate-Term Memory Implementation	223
	4.13.6	Long-Term Memory Implementation	225
	4.13.7	Strategic Oscillation Implementation	226
	4.13.8	Numerical Results of the ATSA	226
4.14	Conclusions		230
4.15	Genetic Algorithms for Unit Commitment		231
	4.15.1	Solution Coding	232
	4.15.2	Fitness Function	232
	4.15.3	Genetic Algorithms Operators	233
	4.15.4	Constraint Handling (Repair Mechanism)	233
	4.15.5	A General Genetic Algorithm	234
	4.15.6	Implementation of a Genetic Algorithm to the UCP	234
	4.15.7	Solution Coding	235
	4.15.8	Fitness Function	236
	4.15.9	Selection of Chromosomes	237
	4.15.10	Crossover	237
	4.15.11	Mutation	237
	4.15.12	Adaptive GA Operators	239
	4.15.13	Numerical Examples	239
	4.15.14	Summary	244
4.16	Hybrid Algorithms for Unit Commitment		246
4.17	Hybrid of Simulated Annealing and Tabu Search (ST)		246
	4.17.1	Tabu Search Part in the ST Algorithm	247
	4.17.2	Simulated Annealing Part in the ST Algorithm	248
4.18	Numerical Results of the ST Algorithm		248
4.19	Hybrid of Genetic Algorithms and Tabu Search		251
	4.19.1	The Proposed Genetic Tabu (GT) Algorithm	251
	4.19.2	Genetic Algorithm as a Part of the GT Algorithm	251
	4.19.3	Tabu Search as a Part of the GT Algorithm	253
4.20	Numerical Results of the GT Algorithm		255
4.21	Hybrid of Genetic Algorithms, Simulated Annealing, and Tabu Search		259
	4.21.1	Genetic Algorithm as a Part of the GST Algorithm	261
	4.21.2	Tabu Search Part of the GST Algorithm	261
	4.21.3	Simulated Annealing as a Part of the GST Algorithm	263
4.22	Numerical Results of the GST Algorithm		263
4.23	Summary		268

	4.24	Comparisons of the Algorithms for the Unit Commitment Problem	269
		4.24.1 Results of Example 1	269
		4.24.2 Results of Example 2	271
		4.24.3 Results of Example 3	272
		4.24.4 Summary	274
	References		274
5	**Optimal Power Flow**		**281**
	5.1	Introduction	281
	5.2	Power Flow Equations	287
		5.2.1 Load Buses	288
		5.2.2 Voltage Controlled Buses	288
		5.2.3 Slack Bus	288
	5.3	General OPF Problem Formulations	291
		5.3.1 The Objective Functions	292
		5.3.2 The Constraints	295
		5.3.3 Optimization Algorithms for OPF	297
	5.4	Optimal Power Flow Algorithms for Single Objective Cases	299
		5.4.1 Particle Swarm Optimization (PSO) Algorithm for the OPF Problem	300
		5.4.2 The IEEE-30 Bus Power System	301
		5.4.3 Active Power Loss Minimization	301
		5.4.4 Minimization of Generation Fuel Cost	307
		5.4.5 Reactive Power Reserve Maximization	309
		5.4.6 Reactive Power Loss Minimization	310
		5.4.7 Emission Index Minimization	312
		5.4.8 Security Margin Maximization	317
	5.5	Comparisons of Different Single Objective Functions	319
	5.6	Multiobjective OPF Algorithm	327
	5.7	Basic Concept of Multiobjective Analysis	327
	5.8	The Proposed Multiobjective OPF Algorithm	329
		5.8.1 Multiobjective OPF Formulation	329
		5.8.2 General Steps for Solving Multi-Objective OPF Problem	330
	5.9	Generating Nondominated Set	330
		5.9.1 Generating techniques	330
		5.9.2 Weighting method	332
	5.10	Hierarchical Cluster Technique	333
	5.11	Conclusions	338
	Appendix		339
	References		342

6 Long-Term Operation of Hydroelectric Power Systems ... 347
- 6.1 Introduction ... 347
- 6.2 Problem Formulation ... 349
- 6.3 Problem Solution: A Minimum Norm Approach ... 350
  - 6.3.1 System Modeling ... 350
  - 6.3.2 Formulation ... 351
  - 6.3.3 Optimal Solution ... 354
  - 6.3.4 Practical Application ... 356
  - 6.3.5 Comments ... 356
  - 6.3.6 A Nonlinear Model ... 357
- 6.4 Simulated Annealing Algorithm (SAA) ... 366
  - 6.4.1 Generating Trial Solution (Neighbor) ... 367
  - 6.4.2 Details of the SAA for the LTHSP ... 368
  - 6.4.3 Practical Applications ... 370
  - 6.4.4 Conclusion ... 371
- 6.5 Tabu Search Algorithm ... 371
  - 6.5.1 Problem Statement ... 372
  - 6.5.2 TS Method ... 373
  - 6.5.3 Details of the TSA ... 373
  - 6.5.4 Step-Size Vector Adjustment ... 376
  - 6.5.5 Stopping Criteria ... 376
  - 6.5.6 Numerical Examples ... 376
  - 6.5.7 Conclusions ... 378
- References ... 378

7 Electric Power Quality Analysis ... 381
- 7.1 Introduction ... 381
- 7.2 Simulated Annealing Algorithm (SAA) ... 384
  - 7.2.1 Testing Simulated Annealing Algorithm ... 385
  - 7.2.2 Step-Size Vector Adjustment ... 385
  - 7.2.3 Cooling Schedule ... 386
- 7.3 Flicker Voltage Simulation ... 386
  - 7.3.1 Problem Formulation ... 386
  - 7.3.2 Testing the Algorithm for Voltage Flicker ... 387
  - 7.3.3 Effect of Number of Samples ... 388
  - 7.3.4 Effects of Sampling Frequency ... 388
- 7.4 Harmonics Problem Formulation ... 388
- 7.5 Testing the Algorithm for Harmonics ... 389
  - 7.5.1 Signal with Known Frequency ... 389
  - 7.5.2 Signal with Unknown Frequency ... 390
- 7.6 Conclusions ... 393
- 7.7 Steady-State Frequency Estimation ... 394
  - 7.7.1 A Constant Frequency Model, Problem Formulation ... 396
  - 7.7.2 Computer Simulation ... 397

		7.7.3 Harmonic-contaminated Signal	398
		7.7.4 Actual Recorded Data	400
7.8	Conclusions		401
		7.8.1 A Variable Frequency Model	401
		7.8.2 Simulated Example	402
		7.8.3 Exponential Decaying Frequency	405
7.9	Conclusions		407
References			407

Index 411

Chapter 1
Introduction

Objectives The primary objectives of this chapter are to

- Provide a broad overview of standard optimization techniques.
- Understand clearly where optimization fits into the problem.
- Be able to formulate a criterion for optimization.
- Know how to simplify a problem to the point at which formal optimization is a practical proposition.
- Have a sufficient understanding of the theory of optimization to select an appropriate optimization strategy, and to evaluate the results that it returns.

1.1 Introduction [1–11]

The goal of an optimization problem can be stated as follows. Find the combination of parameters (independent variables) that optimize a given quantity, possibly subject to some restrictions on the allowed parameter ranges. The quantity to be optimized (maximized or minimized) is termed the objective function; the parameters that may be changed in the quest for the optimum are called control or decision variables; the restrictions on allowed parameter values are known as constraints.

The problem formulation of any optimization problem can be thought of as a sequence of steps and they are:

1. Choosing design variables (control and state variables)
2. Formulating constraints
3. Formulating objective functions
4. Setting up variable limits
5. Choosing an algorithm to solve the problem
6. Solving the problem to obtain the optimal solution

Decision (control) variables are parameters that are deemed to affect the output in a significant manner. Selecting the best set of decision variables can sometimes be a challenge because it is difficult to ascertain which variables affect each specific behavior in a simulation. Logic determining control flow can also be classified as a decision variable. The domain of potential values for decision variables is typically restricted by constraints set by the user.

The optimization problem may have a single objective function or multiobjective functions. The multiobjective optimization problem (MOOP; also called the multicriteria optimization, multiperformance, or vector optimization problem) can be defined (in words) as the problem of finding a vector of decision variables that satisfies constraints and optimizes a vector function whose elements represent the objective functions. These functions form a mathematical description of performance criteria that are usually in conflict with each other. Hence, the term optimizes means finding such a solution that would give the values of all the objective functions acceptable to the decision maker.

Multiobjective optimization has created immense interest in the engineering field in the last two decades. Optimization methods are of great importance in practice, particularly in engineering design, scientific experiments, and business decision making. Most of the real-world problems involve more than one objective, making multiple conflicting objectives interesting to solve as multiobjective optimization problems.

1.2 Optimization Techniques

There are many optimization algorithms available to engineers with many methods appropriate only for certain type of problems. Thus, it is important to be able to recognize the characteristics of a problem in order to identify an appropriate solution technique. Within each class of problems there are different minimization methods, varying in computational requirements, convergence properties, and so on. Optimization problems are classified according to the mathematical characteristics of the objective function, the constraints, and the control variables.

Probably the most important characteristic is the nature of the objective function. These classifications are summarized in Table 1.1.

There are two basic classes of optimization methods according to the type of solution.

(a) Optimality Criteria

   Analytical methods: Once the conditions for an optimal solution are established, then either:

   - A candidate solution is tested to see if it meets the conditions.
   - The equations derived from the optimality criteria are solved analytically to determine the optimal solution.

1.2 Optimization Techniques

Table 1.1 Classification of the objective functions

Characteristic	Property	Classification
Number of control variables	One	Univariate
	More than one	Multivariate
Type of control variables	Continuous real numbers	Continuous
	Integers	Integer or discrete
	Both continuous real numbers and integers	Mixed integer
Problem functions	Linear functions of the control variables	Linear
	Quadratic functions of the control variables	Quadratic
	Other nonlinear functions of the control variables	Nonlinear
Problem formulation	Subject to constraints	Constrained
	Not subject to constraints	Unconmstrained

(b) Search Methods

Numerical methods: An initial trial solution is selected, either using common sense or at random, and the objective function is evaluated. A move is made to a new point (second trial solution) and the objective function is evaluated again. If it is smaller than the value for the first trial solution, it is retained and another move is made. The process is repeated until the minimum is found.

Search methods are used when:

- The number of variables and constraints is large.
- The problem functions (objective and constraint) are highly nonlinear.
- The problem functions (objective and constraint) are implicit in terms of the decision/control variables making the evaluation of derivative information difficult.

Other suggestions for classification of optimization methods are:

1. The first is based on *classic methods* such as the nonlinear programming technique, the weights method, and the ε-constraints method.
2. The second is based on the *evolutionary techniques* such as the NPGA method (niched Pareto genetic algorithm), NSGA (nondominated sorting genetic algorithm), SPEA (strength Pareto evolutionary algorithm), and SPEA2 (improving strength Pareto evolutionary algorithm).

1.2.1 Conventional Techniques (Classic Methods) [2, 3]

The classic methods present some inconveniences due to the danger of convergence, the long execution time, algorithmic complexity, and the generation of a weak number of nondominated solutions. Because of these inconveniences, evolutionary algorithms are more popular, thanks to their faculty to exploit vast amounts of research and the fact that they don't require prerecognition of the problem.

The two elements that most directly affect the success of an optimization technique are the quantity and domain of decision variables and the objective function. Identifying the decision variables and the objective function in an optimization problem often requires familiarity with the available optimization techniques and awareness of how these techniques interface with the system undergoing optimization.

The most appropriate method will depend on the type (classification) of problem to be solved. Some optimization techniques are more computationally expensive than others and thus the time required to complete an optimization is an important criterion. The setup time required of an optimization technique can vary by technique and is dependent on the degree of knowledge required about the problem. All optimization techniques possess their own internal parameters that must be tuned to achieve good performances. The time required to tweak these parameters is part of the setup cost.

Conventional optimization techniques broadly consist of calculus-based, enumerated, and random techniques. These techniques are based on well-established theories and work perfectly well for a case wherever applicable. But there are certain limitations to the above-mentioned methods. For example, the steepest descent method starts its search from a single point and finally ends up with an optimal solution. But this method does not ensure that this is the global optimum. Hence there is every possibility of these techniques getting trapped in local optima. Another great drawback of traditional methods is that they require complete information of the objective function, its dependence on each variable, and the nature of the function. They also make assumptions in realizing the function as a continuous one. All these characteristics of traditional methods make them inapplicable to many real-life problems where there is insufficient information on the mathematical model of the system, parameter dependence, and other such information. This calls for unconventional techniques to address many real-life problems.

The optimization methods that are incorporated in the optimal power flow tools can be classified based on optimization techniques such as

1. Linear programming (LP) based methods
2. Nonlinear programming (NLP) based methods
3. Integer programming (IP) based methods
4. Separable programming (SP) based methods
5. Mixed integer programming (MIP) based methods

Notably, linear programming is recognized as a reliable and robust technique for solving a wide range of specialized optimization problems characterized by linear objectives and linear constraints. Many commercially available power system optimization packages contain powerful linear programming algorithms for solving power system problems for both planning and operating engineers. Linear programming has extensions in the simplex method, revised simplex method, and interior point techniques.

1.2 Optimization Techniques

Interior point techniques are based on the Karmarkar algorithm and encompass variants such as the projection scaling method, dual affine method, primal affine method, and barrier algorithm.

In the case of nonlinear programming optimization methods, the following techniques are introduced.

- Sequential quadratic programming (SQP)
- Augmented Lagrangian method
- Generalized reduced gradient method
- Projected augmented Lagrangian
- Successive linear programming (SLP)
- Interior point methods

Sequential quadratic programming is a technique for the solution of nonlinearly constrained problems. The main idea is to obtain a search direction by solving a quadratic program, that is, a problem with a quadratic objective function and linear constraints. This approach is a generalization of Newton's method for unconstrained minimization. When solving optimization problems, SQP is not often used in its simple form. There are two major reasons for this: it is not guaranteed to converge to a local solution to the optimization problem, and it is expensive.

Gradient-based search methods are a category of optimization techniques that use the gradient of the objective function to find an optimal solution. Each iteration of the optimization algorithm adjusts the values of the decision variables so that the simulation behavior produces a lower objective function value. Each decision variable is changed by an amount proportionate to the reduction in objective function value. Gradient-based searches are prone to converging on local minima because they rely solely on the local values of the objective function in their search. They are best used on well-behaved systems where there is one clear optimum. Gradient-based methods will work well in high-dimensional spaces provided these spaces don't have local minima. Frequently, additional dimensions make it harder to guarantee that there are not local minima that could trap the search routine. As a result, as the dimensions (parameters) of the search space increase, the complexity of the optimization technique increases.

The benefits of traditional use of gradient-based search techniques are that computation and setup time are relatively low. However, the drawback is that global minima are likely to remain undiscovered. Nonlinear optimization problems with multiple nonlinear constraints are often difficult to solve, because although the available mathematical theory provides the basic principles for solution, it does not guarantee convergence to the optimal point. The straightforward application of augmented Lagrangian techniques to such problems typically results in slow (or lack of) convergence, and often in failure to achieve the optimal solution.

There are many factors that complicate the use of classical gradient-based methods including the presence of multiple local minima, the existence of regions in the design space where the functions are not defined, and the occurrence of an extremely large number of design variables.

All of these methods suffer from three main problems. First, they may not be able to provide an optimal solution and usually get stuck at a local optimal. Second, all these methods are based on the assumption of continuity and differentiability of the objective function which is not actually allowed in a practical system. Finally, all these methods cannot be applied with discrete variables.

Classical analytical methods include *Lagrangian methods* where necessary conditions known as the Karush–Kuhn–Tucker (KKT) conditions are used to identify candidate solutions. For n large, these classical methods, because of their combinatorial nature, become impractical, and solutions are obtained numerically instead by means of suitable numerical algorithms. The most important class of these methods is the so-called gradient-based methods. The most well known of these methods are various quasi-Newton and conjugate gradient methods for unconstrained problems, and the penalty function, gradient projection, augmented Lagrangian, and sequential quadratic programming methods for constrained problems.

Traditionally, different solution approaches have been developed to solve the different classes of the OPF problem. These methods are nonlinear programming techniques with very high accuracy, but their execution time is very long and they cannot be applied to real-time power system operations. Since the introduction of sequential or successive programming techniques, it has become widely accepted that *successive linear programming* algorithms can be used effectively to solve the optimization problem. In SLP, the original problem is solved by successively approximating the original problem using Taylor series expansion at the current operating point and then moving in an optimal direction until the solution converges.

Mixed integer programming is an integer programming used for optimizing linear functions that are constrained by linear bounds. Quite often, the variables that are being varied can have only integer value (e.g., in inventory problems where fractional values such as the number of cars in stock are meaningless). Hence, it is more appropriate to use integer programming. Mixed integer programming is a type of integer programming in which not all of the variables to be optimized have integer values. Due to the linear nature of the objective function it can be expressed mathematically as

$$\min \sum_{j,k=1}^{n} C_j X_k \qquad (1.1)$$

where C is the coefficient matrix and X is the attribute vector of attributes $x, \ldots, x_n$. Typically, MIP problems are solved by using branch-and-bound techniques to increase speed.

Mixed integer programming was found to have the widest application. It was preferred to routing airline crews and other similar problems that bore a close resemblance to the problem we had at hand. Furthermore, the mathematical rigor we were looking for was well established. However, as the nature of our problem is continuous and dynamic we preferred to use either simulated annealing or stochastic approximation (discussed later).

There is to date no universal method for solving all the optimization problems, even if restricted to cases where all the functions are analytically known, continuous, and smooth. Many inhibiting difficulties remain when these methods are applied to real-world problems. Typical optimization difficulties that arise are that the functions are often very expensive to evaluate. The existence of noise in the objective and constraint functions, as well as the presence of discontinuities in the functions, constitute further obstacles in the application of standard and established methods.

1.2.2 Evolutionary Techniques [4–11]

Recently the advances in computer engineering and the increased complexity of the power system optimization problem have led to a greater need for and application of specialized programming techniques for large-scale problems. These include dynamic programming, Lagrange multiplier methods, heuristic techniques, and *evolutionary techniques* such as genetic algorithms. These techniques are often hybridized with many other intelligent system techniques, including artificial neural networks (ANN), expert systems (ES), tabu search algorithms (TS), and fuzzy logic (FL).

Many researchers agree that first, having a population of initial solutions increases the possibility of converging to an optimum solution, and second, updating the current information of the search strategy from the previous history is a natural tendency. Accordingly, attempts have been made by researchers to restructure these standard optimization techniques in order to achieve the two goals mentioned.

To achieve these two goals, researchers have made concerted efforts in the last decade to invent novel optimization techniques for solving real-world problems, which have the attributes of memory update and population-based search solutions. Heuristic searches are one of these novel techniques.

1.2.2.1 Heuristic Search [3]

Several heuristic tools have evolved in the last decade that facilitate solving optimization problems that were previously difficult or impossible to solve. These tools include evolutionary computation, simulated annealing, tabu search, particle swarm, ant colony, and so on. Reports of applications of each of these tools have been widely published. Recently, these new heuristic tools have been combined among themselves and with knowledge elements, as well as with more traditional approaches such as statistical analysis, to solve extremely challenging problems. Developing solutions with these tools offers two major advantages:

1. Development time is much shorter than when using more traditional approaches.
2. The systems are very robust, being relatively insensitive to noisy and/or missing data.

Heuristic-based methods strike a balance between exploration and exploitation. This balance permits the identification of local minima, but encourages the discovery of a globally optimal solution. However, it is extremely difficult to fine-tune these methods to gain vital performance improvements. Because of their exploitation capabilities, heuristic methods may also be able to obtain the best value for decision variables. Heuristic techniques are good candidate solutions when the search space is large and nonlinear because of their exploration capabilities.

1.2.2.2 Evolutionary Computation [8]

The mainstream algorithms for evolutionary computation are genetic algorithms (GAs), evolutionary programming (EP), evolution strategies (EvS), and genetic programming (GP). These algorithms have been recognized as important mathematical tools in solving continuous optimization problems. Evolutionary algorithms possess the following salient characteristics.

(a) Genetic variation is largely a chance phenomenon and stochastic processes play a significant role in evolution.
(b) A population of agents with nondeterministic recruitment is used.
(c) Inherent parallel search mechanisms are used during evolution.
(d) Evolution is a change in adaptation and diversity, not merely a change in gene frequencies.
(e) Evolutionary algorithms operate with a mechanism of competition–cooperation.

These algorithms simulate the principle of evolution (a two-step process of variation and selection), and maintain a population of potential solutions (individuals) through repeated application of some evolutionary operators such as mutation and crossover. They yield individuals with successively improved fitness, and converge, it is hoped, to the fittest individuals representing the optimum solutions. The evolutionary algorithms can avoid premature entrapment in local optima because of the stochastic search mechanism. Genetic algorithms and evolutionary programming are among the two most widely used evolutionary computation algorithms.

Evolutionary algorithms are robust and powerful global optimization techniques for solving large-scale problems that have many local optima. However, they require high CPU times, and they are very poor in terms of convergence performance. On the other hand, local search algorithms can converge in a few iterations but lack a global perspective. The combination of global and local search procedures should offer the advantages of both optimization methods while offsetting their disadvantages.

Evolutionary algorithms seem particularly suitable to solve multiobjective optimization problems because they deal simultaneously with a set of possible solutions (the so-called population). This allows us to find several members of the Pareto optimal set in a single run of the algorithm, instead of having to perform a series of separate runs as in the case of traditional mathematical programming techniques. In addition, evolutionary algorithms are less susceptible to the shape or continuity of the Pareto front (e.g., they can easily deal with discontinuous or concave Pareto fronts), whereas these two issues are a real concern with mathematical programming techniques.

What Do You Mean by Pareto Optimal Set?

We say that a vector of decision variables $\vec{x}^* \in f$ is Pareto optimal if there does not exist another $\vec{x} \in f$ such that $f_i(\vec{x}) \leq f_i(\vec{x}^*)$ for all $i = 1,\ldots,k$ and $f_j(\vec{x}) \leq f_j(\vec{x}^*)$ for at least one j. In words, this definition says that $\vec{x}^*$ is Pareto optimal if there is no feasible vector of decision variables $\vec{x} \in f$ that would decrease some criterion without causing a simultaneous increase in at least one other criterion. Unfortunately, this concept almost always does not give a single solution, but rather a set of solutions called the Pareto optimal set. The vectors $\vec{x}^*$ corresponding to the solutions included in the Pareto optimal set are called *nondominated*. The plot of the objective functions whose nondominated vectors are in the Pareto optimal set is called the Pareto front.

The advantage of evolutionary algorithms is that they have minimum requirements regarding the problem formulation: objectives can be easily added, removed, or modified. Moreover, because they operate on a set of solution candidates, evolutionary algorithms are well suited to generate Pareto set approximations. This is reflected by the rapidly increasing interest in the field of evolutionary multiobjective optimization. Finally, it has been demonstrated in various applications that evolutionary algorithms are able to tackle highly complex problems and therefore they can be seen as a complementary approach to traditional methods such as integer linear programming.

Evolutionary computation paradigms generally differ from traditional search and optimization paradigms in three main ways:

1. Evolutionary computation paradigms utilize a population of points in their search.
2. Evolutionary computation paradigms use direct "fitness" information instead of function derivatives.
3. Evolutionary computation paradigms use direct "fitness" information instead of other related knowledge.
4. Evolutionary computation paradigms use probabilistic, rather than deterministic, transition rules.

1.2.2.3 Genetic Algorithm [7]

The genetic algorithm is a search algorithm based on the conjunction of natural selection and genetics. The features of the genetic algorithm are different from other search techniques in several aspects. The algorithm is multipath, searching many peaks in parallel, and hence reducing the possibility of local minimum trapping. In addition, GA works with a coding of parameters instead of the parameters themselves. The parameter coding will help the genetic operator to evolve the current state into the next state with minimum computations. The genetic algorithm evaluates the fitness of each string to guide its search instead of the optimization function. The GA only needs to evaluate the objective function (fitness) to guide its search. There is no requirement for derivatives or other auxiliary knowledge. Hence, there is no need for computation of derivatives or other auxiliary functions. Finally, GA explores the search space where the probability of finding improved performance is high.

At the start of a genetic algorithm optimization, a set of decision variable solutions is encoded as members of a population. There are multiple ways to encode elements of solutions including binary, value, and tree encodings. Crossover and mutation operators based on reproduction are used to create the next generation of the population. *Crossover* combines elements of solutions in the current generation to create a member of the next generation. *Mutation* systematically changes elements of a solution from the current generation in order to create a member of the next generation. Crossover and mutation accomplish exploration of the search space by creating diversity in the members of the next generation. Traditional uses of GAs leverage the fact that these algorithms explore multiple areas of the search space to find a global minimum. Through the use of the crossover operator, these algorithms are particularly strong at combining the best features from different solutions to find one global solution. Genetic algorithms are also well suited for searching complex, highly nonlinear spaces because they avoid becoming trapped in a local minimum. Genetic algorithms explore multiple solutions simultaneously. These sets of solutions make it possible for a user to gain, from only one iteration, multiple types of insight of the algorithm.

The genetic algorithm approach is quite simple.

1. Randomly generate an initial solution population.
2. Evaluate these solutions for fitness.
3. If time or iteration constraints are not yet satisfied, then
4. Select parents (best solutions so far).
5. Recombine parents using portions of original solutions.
6. Add possible random solution "mutations."
7. Evaluate new solutions for fitness.
8. Return to Step 3.

So a genetic algorithm for any problem should have the following five components:

1. A genetic representation of potential solutions to the problem
2. A way to create an initial population of potential solutions
3. An evaluation function that plays the role of the environment and rates solutions in terms of their fitness
4. Genetic operators to alter the composition of a string

1.2 Optimization Techniques

5. Values for various parameters that the genetic algorithm uses (population size, probabilities of applying genetic operators, etc.)

Genetic algorithms use probablistic transition rules rather than deterministic procedures. Hence, the search will be multidirectional. GAs generally require a great number of iterations and they converge slowly, especially in the neighborhood of the global optimum. It thus makes sense to incorporate a faster local optimization algorithm into a GA in order to overcome this lack of efficiency while retaining the advantages of both optimization methods.

Genetic algorithms seem to be the most popular algorithms at present. Their advantage lies in the ease of coding them and their inherent parallelism. The use of genotypes instead of phenotypes to travel in the search space makes them less likely to get stuck in local minima. There are, however, certain drawbacks to them. Genetic algorithms require very intensive computation time and hence they are slow. They have been shown to be useful in the optimization of multimodal functions in highly complex landscapes, especially when the function does not have an analytic description and is noisy or discontinuous. The usefulness of genetic algorithms for such problems comes from their evolutionary and adaptive capabilities. When given a measure of the fitness (performance) of a particular solution and a population of adequately coded feasible solutions, the GA is able to search many regions of the parameter space simultaneously. In the GA, better than average solutions are sampled more frequently and thus, through the genetic operations of crossover and mutation, new promising solutions are generated and the average fitness of the whole population improves over time. Although these algorithms are not guaranteed to find the global optimum, they tend to converge toward good regions of high fitness.

There are many algorithms in GA such as the vector-evaluated genetic algorithm (VEGA). Some of the most recent ones are the nondominated sorting genetic algorithm-II (NSGA-II), strength Pareto evolutionary algorithm-II (SPEA-II), and Pareto envelope-based selection-II (PESA-II). Most of these approaches propose the use of a generational GA. But the elitist steady-state multiobjective evolutionary algorithm (MOEA) attempts to maintain spread while attempting to converge to the true Pareto-optimal front. This algorithm requires sorting of the population for every new solution formed, thereby increasing its time complexity. Very high time complexity makes the elitist steady-state MOEA impractical for some problems. The area of steady-state multiobjective GAs has not been widely explored. Also constrained multiobjective optimization, which is very important for real-world application problems, has not received its deserved exposure.

1.2.2.4 Evolution Strategies and Evolutionary Programming

Evolution strategies employ real-coded variables and, in their original form, relied on mutation as the search operator, and a population size of one. Since then they have evolved to share many features with GAs. The major similarity between these two types of algorithms is that they both maintain populations of potential solutions and use a selection mechanism for choosing the best individuals from the population. The main differences are: EvSs operate directly on floating point vectors whereas classical GAs operate on binary strings; GAs rely mainly on recombination to explore the search space, and EvSs use mutation as the dominant operator; and EvS is an abstraction of evolution at the individual behavior level, stressing the behavioral link between an individual and its offspring, whereas GAs maintain the genetic link.

Evolutionary programming is a stochastic optimization strategy similar to GA, which places emphasis on the behavioral linkage between parents and their offspring, rather than seeking to emulate specific genetic operators as observed in nature. EP is similar to evolutionary strategies, although the two approaches developed independently. Like both EvS and GAs, EP is a useful method of optimization when other techniques such as gradient descent or direct analytical discovery are not possible. Combinatorial and real-valued function optimizations, in which the optimization surface or fitness landscape is "rugged," possessing many locally optimal solutions, are well suited for evolutionary programming.

EP was initially developed as different from the basic GA in two main aspects:

1. The authors of EP felt that their representations (whether real or binary) represented phenotypic behavior whereas the authors of GA felt that their representations represented genotypic traits.
2. Evolutionary programming depends more on mutation and selection operations whereas GA mainly relies on crossover.

It is noted that, given the wide availability and development in encoding/decoding techniques for GA, the first difference between the two algorithms is diminishing. However, the inherent characteristics of EP have made it a widely practiced evolutionary computation algorithm in many applications, especially where search diversity is a key concern in the optimization process.

As branches of evolutionary algorithms, EvS and EP share many common features, including the real-valued representation of search points, emphasis on the utilization of normally distributed random mutations as the main search operator, and most important, the concept of self-adaptation of strategy parameters online during the search. There exist, however, some striking differences, such as the specific representation of mutation, most notably the missing recombination operator in EP and the softer, probabilistic selection mechanism used in EP. The combination of these properties seems to have some negative impact on the performance of EP.

As a powerful and general global optimization tool, EP seeks the optimal solution by evolving a population of candidate solutions over a number of generations or iterations. A new population is generated from an existing population through the use of a mutation operator. This operator perturbs each component of every solution in the population by a Gaussian random variable δ with zero mean and preselects variance σ^2 to produce new ones. A mutation operator with high efficiency should fully reflect the principle of organic evolution in nature, that is,

1.2 Optimization Techniques

the lower the fitness score is, the higher the mutation possibility is, and vice versa. Through the use of a competition scheme, the individuals in each population compete with each other. The winning individuals will form a resultant population that is regarded as the next generation. For optimization to occur, the competition scheme must ensure that the more optimal solutions have a greater chance of survival than the poorer solutions. Through this process, the population is expected to evolve toward the global optimum. It is known that there is more research needed in the mathematical foundation for the EP or its variants with regard to experimental and empirical research. The state-of-the-art of EP mainly focuses on the application of solving optimization problems, especially for the application to real-valued function optimization. So far, to the best of the authors' knowledge, there has been very little theoretical research available explaining the mechanisms of the successful search capabilities of EP or its variants even though some convergence proofs with certain assumptions for the EP have been carried out with varying degrees of success in the past few years.

EP is similar to GAs in principle. It works on a population of trial solutions, imposes random changes to those solutions to create offspring, and incorporates the use of selection to determine which solutions to maintain into future generations and which to remove from the pool of trials. However, in contrast to GAs, the individual component of a trial solution in EP is viewed as a behavioral trait, not as a gene. In other words, EP emphasizes the behavioral link between parents and offspring rather than the genetic link. It is assumed whatever genetic transformations occur, the resulting change in each behavioral trait will follow a Gaussian distribution with zero mean difference and some standard deviation.

1.2.2.5 Differential Evolutions

Differential evolution (DE) is an improved version of the genetic algorithm for faster optimization. Unlike a simple GA that uses binary coding for representing problem parameters, differential evolution uses real coding of floating point numbers. Among the DE's advantages are its simple structure, ease of use, speed, and robustness.

Differential strategies can be adopted in a DE algorithm depending upon the type of problem to which DE is applied. The strategies can vary based on the vector to be perturbed, number of difference vectors considered for perturbation, and finally the type of crossover used. The general convention used in these strategies is *DE/x/y/z*. *DE* stands for differential evolution, x represents a string denoting the vector to be perturbed, y is the number of difference vectors considered for the perturbation of x, and z stands for the type of crossover being used (exp:exponential; bin:binomial).

Differential evolution has been successfully applied in various fields. Some of the successful applications include digital filter design, batch fermentation process, estimation of heat transfer parameters in a trickle bed reactor, optimal design of heat exchangers, synthesis and optimization of a heat-integrated distillation system, scenario-integrated optimization of dynamic systems, optimization of nonlinear functions, optimization of thermal cracker operation, optimization of nonlinear chemical processes, global optimization of nonlinear chemical engineering processes, optimization of water pumping systems, and optimization of biomass pyrolysis, among others. Applications of DE to multiobjective optimization are scarce.

1.2.2.6 Particle Swarm [9]

Particle swarm optimization is an exciting new methodology in evolutionary computation that is somewhat similar to a genetic algorithm in that the system is initialized with a population of random solutions. Unlike other algorithms, however, each potential solution (called a particle) is also assigned a randomized velocity and then flown through the problem hyperspace. Particle swarm optimization has been found to be extremely effective in solving a wide range of engineering problems. It is very simple to implement (the algorithm comprises two lines of computer code) and solves problems very quickly. In a PSO system, the group is a community composed of all particles, and all particles fly around in a multidimensional search space. During flight, each particle adjusts its position according to its own experience and the experience of neighboring particles, making use of the best position encountered by itself and its neighbors. The swarm direction of each particle is defined by the set of particles neighboring the particle and its historical experience.

Particle swarm optimization shares many similarities with evolutionary computation techniques in general and GAs in particular. All three techniques begin with a group of a randomly generated population and all utilize a fitness value to evaluate the population. They all update the population and search for the optimum with random techniques. A large inertia weight facilitates global exploration (search in new areas), whereas a small one tends to assist local exploration. The main difference between the PSO approach compared to EC and GA is that PSO does not have genetic operators such as crossover and mutation. Particles update themselves with internal velocity; they also have a memory that is important to the algorithm. Compared with EC algorithms (such as evolutionary programming, evolutionary strategy, and genetic programming), the information-sharing mechanism in PSO is significantly different. In EC approaches, chromosomes share information with each other, thus the whole population moves as one group toward an optimal area. In PSO, only the "best" particle gives out the information to others. It is a one-way information-sharing mechanism; the evolution only looks for the best solution. Compared with ECs, all the particles tend to converge to the best solution quickly even in the local version in most cases. Compared to GAs, the advantages are that PSO is easy to implement and there are few parameters to adjust.

There is an improvement of the PSO method to facilitate a multiobjective approach and called multiobjective PSO (MOPSO). The important part in multiobjective particle swarm optimization is to determine the best global particle for each particle i of the population. In the single-objective PSO, the global best particle is determined easily by selecting the particle with the best position. Because

1.2 Optimization Techniques

multiobjective optimization problems have a set of Pareto-optimal solutions as the optimum solutions, each particle of the population should use Pareto-optimal solutions as the basis for selecting one of its global best particles. The detailed computational flow of the MOPSO technique for the economic load dispatch problem can be described in the following steps.

Step 1: Input parameters of the system, and specify the lower and upper boundaries of each variable.

Step 2: Randomly initialize the speed and position of each particle and maintain the particles within the search space.

Step 3: For each particle of the population, employ the Newton–Raphson power flow analysis method to calculate the power flow and system transmission loss, and evaluate each of the particles in the population.

Step 4: Store the positions of the particles that represent nondominated vectors in the repository NOD.

Step 5: Generate hypercubes of the search space explored so far, and locate the particles using these hypercubes as a co-ordinate system where each particle's co-ordinates are defined according to the values of its objective function.

Step 6: Initialize the memory of each particle in which a single local best for each particle is contained (this memory serves as a guide to travel through the search space. This memory is stored in the other repository P_{BEST}).

Step 7: Update the time counter $t = t + 1$.

Step 8: Determine the best global particle g_{best} for each particle i from the repository NOD. First, those hypercubes containing more than one particle are assigned a fitness value equal to the result of dividing any number $x > 1$ by the number of particles that they contain. Then, we apply roulette wheel selection using these fitness values to select the hypercube from which we will take the corresponding particle. Once the hypercube has been selected, we randomly select a particle as the best global particle g_{best} for particle i within such a hypercube.

Step 9: Compute the speed and the new position of each particle and maintain the particles within the search space in case they go beyond its boundaries.

Step 10: Evaluate each particle in the population by the Newton–Raphson power flow analysis method.

Step 11: Update the contents of the repository NOD together with the geographical representation of the particles within the hypercubes. This update consists of inserting all the currently nondominated locations into the repository. Any dominated locations from the repository are eliminated in the process. Because the size of the repository is limited, whenever it gets full a secondary criterion for retention is applied: those particles located in less-populated areas of objective space are given priority over those lying in highly populated regions.

Step 12: Update the contents of the repository P_{BEST}. If the current position of the particle is dominated by the position in the repository P_{BEST}, then the position in the repository P_{BEST} is kept; otherwise, the current position replaces the one in memory; if neither of them is dominated by the other, one of them is randomly selected.

Step 13: If the maximum iterations $iter_{max}$ are satisfied then go to Step 14. Otherwise, go to Step 7.

Step 14: Input a set of the Pareto-optimal solutions from the repository NOD.

The MOPSO approach is efficient for solving multiobjective optimization problems where multiple Pareto-optimal solutions can be found in one simulation run. In addition, the nondominated solutions in the obtained Pareto-optimal set are well distributed and have satisfactory diversity characteristics.

1.2.2.7 Tabu Search [8–12]

Tabu search is basically a gradient descent search with memory. The memory preserves a number of previously visited states along with a number of states that might be considered unwanted. This information is stored in a tabu list. The definition of a state, the area around it and the length of the tabu list are critical design parameters. In addition to these tabu parameters, two extra parameters are often used: aspiration and diversification. *Aspiration* is used when all the neighboring states of the current state are also included in the tabu list. In that case, the tabu obstacle is overridden by selecting a new state. *Diversification* adds randomness to this otherwise deterministic search. If the tabu search does not converge, the search is reset randomly.

Tabu search has the advantage of not using hill-climbing strategies. Its performance can also be enhanced by branch-and-bound techniques. However, the mathematics behind this technique are not as strong as those behind neural networks or simulated annealing. Furthermore, a solution space must be generated. Hence, tabu search requires knowledge of the entire operation at a more detailed level.

1.2.2.8 Simulated Annealing [8–12]

In statistical mechanics, a physical process called annealing is often performed in order to relax the system to a state with minimum free energy. In the annealing process, a solid in a heat bath is heated up by increasing the temperature of the bath until the solid is melted into liquid, then the temperature is lowered slowly. In the liquid phase all particles of the solid arrange themselves randomly. In the ground state the particles are arranged in a highly structured lattice and the energy of the system is a minimum. The ground state of the solid is obtained only if the maximum temperature is sufficiently high and the cooling is done sufficiently slowly. Based on the annealing process in statistical mechanics, simulated annealing was introduced for solving complicated combinatorial optimization.

1.2 Optimization Techniques

The name "simulated annealing" originates from the analogy with the physical process of solids, and the analogy between the physical system and simulated annealing is that the cost function and the solution (configuration) in the optimization process correspond to the energy function and the state of statistical physics, respectively. In a large combinatorial optimization problem, an appropriate perturbation mechanism, cost function, solution space, and cooling schedule are required in order to find an optimal solution with simulated annealing. The process is effective in network reconfiguration problems for large-scale distribution systems, and its search capability becomes more significant as the system size increases. Moreover, the cost function with a smoothing strategy enables simulated annealing to escape more easily from local minima and to reach the vicinity of an optimal solution rapidly.

The major strengths of simulated annealing are that it can optimize functions with arbitrary degrees on nonlinearity, stochasticity, boundary conditions, and constraints. It is also statistically guaranteed to find an optimal solution. However, it has its disadvantages too. Like GAs it is very slow; its efficiency is dependent on the nature of the surface it is trying to optimize and it must be adapted to specific problems. The availability of supercomputing resources, however, mitigates these drawbacks and makes simulated annealing a good candidate.

1.2.2.9 Stochastic Approximation

Some nonclassical optimization techniques are able to optimize on discontinuous objective functions, however, they are unable to do so when the complexity of the data becomes very large. In this case the complexity of the system requires that the objective function be estimated. Furthermore, the models that are used to estimate the objective function may be stochastic due to the dynamic and random nature of the system and processes.

The basic idea behind the stochastic approximation method is the gradient descent method. Here the variable that the objective function is to be optimized upon is varied in small increments and the impact of this variation (measured by the gradient) is used to determine the direction of the next step. The magnitude of the step is controlled to have larger steps when the perturbations in the system are small and vice versa.

Stochastic approximation algorithms based on various techniques have been developed recently. They have been applied to both continuous and discrete objective functions. Recently, their convergence has been proved for the degenerate case as well.

Stochastic approximation did not have as many applications reported as the other techniques. This could have been because of various factors such as the lack of a metaphorical concept to facilitate understanding and proofs that are complex. It has recently shown great promise, however, especially in optimizing nondiscrete problems. The stochastic nature of our model along with the complexity of the application domain makes this an attractive candidate.

1.2.2.10 Fuzzy [13]

Fuzzy systems are knowledge-based or rule-based systems. The heart of a fuzzy system is a knowledge base consisting of the so-called fuzzy IF–THEN rules. A fuzzy IF–THEN rule is an IF–THEN statement in which some words are characterized by continuous membership functions. For example, the following is a fuzzy IF–THEN rule.

IF *the speed of a car is high*, THEN *apply less force to the accelerator*,

where the words *high* and *less* are characterized by the membership functions.

A fuzzy set is a set of ordered pairs with each containing an element and the degree of membership for that element. A higher membership value indicates that an element more closely matches the characteristic feature of the set. By fuzzy theory we mean all theories that use the basic concept of fuzzy sets or continuous membership function. Fuzzy theory can be roughly classified into five major branches:

- *Fuzzy mathematics*: Where classical mathematical concepts are extended by replacing classical sets with fuzzy sets
- *Fuzzy logic and artificial intelligence*: Where approximations to classical logic are introduced and expert systems are developed based on fuzzy information and approximate reasoning
- *Fuzzy systems*: Which include fuzzy control and fuzzy approaches in signal processing and communications
- *Uncertainty and information*: Where different kinds of uncertainties are analyzed
- *Fuzzy decision making*: Which considers optimization problems with soft constraints

Of course, these five branches are not independent and there are strong interconnections among them. For example, fuzzy control uses concepts from fuzzy mathematics and fuzzy logic. Fuzzy mathematics provides the starting point and basic language for fuzzy systems and fuzzy control. Understandably, only a small portion of fuzzy mathematics has found applications in engineering.

Fuzzy logic implements experience and preferences through membership functions. The membership functions have different shapes depending on the designer's preference and experience. Fuzzy rules may be formed that describe relationships linguistically as antecedent-consequent pairs of IF–THEN statements. Basically, there are four approaches to the derivation of fuzzy rules: (1) from expert experience and knowledge, (2) from the behavior of human operators, (3) from the fuzzy model of a process, and (4) from learning. Linguistic variables allow a system to be more comprehensible to a nonexpert operator. In this way, fuzzy logic can be used as a general methodology to incorporate knowledge, heuristics, or theory into controllers and decision making.

1.2 Optimization Techniques

Here is a list of general observations about fuzzy logic.

- Fuzzy logic is conceptually easy to understand.
- Fuzzy logic is flexible.
- Fuzzy logic is tolerant of imprecise data.
- Fuzzy logic can model nonlinear functions of arbitrary complexity.
- Fuzzy logic can be built on top of the experience of experts.
- Fuzzy logic can be blended with conventional control techniques.
- Fuzzy logic is based on natural language.

The last statement is perhaps the most important one and deserves more discussion. Natural language, that which is used by ordinary people on a daily basis, has been shaped by thousands of years of human history to be convenient and efficient. Sentences written in ordinary language represent a triumph of efficient communication. We are generally unaware of this because ordinary language is, of course, something we use every day. Because fuzzy logic is built atop the structures of qualitative description used in everyday language, fuzzy logic is easy to use.

Fuzzy logic is not a cure-all. When should you not use fuzzy logic? The safest statement is the first one made in this introduction: fuzzy logic is a convenient way to map an input space to an output space. If you find it's not convenient, try something else. If a simpler solution already exists, use it. Fuzzy logic is the codification of common sense: use common sense when you implement it and you will probably make the right decision. Many controllers, for example, do a fine job without using fuzzy logic. However, if you take the time to become familiar with fuzzy logic, you'll see it can be a very powerful tool for dealing quickly and efficiently with imprecision and nonlinearity.

Fuzzy systems have been applied to a wide variety of fields ranging from control, signal processing, communications, integrated circuit manufacuring, and expert systems to business, medicine, pychology, and so on. However, the most significant applications have concentrated on control problems. There are essentially three groups of applications: rule-based systems with fuzzy logic, fuzzy logic controllers, and fuzzy decision systems.

The broadest class of problems within power system planning and operation is decision making and optimization, which include transmission planning, security analysis, optimal power flow, state estimation, and unit commitment, among others. These general areas have received great attention in the research community with some notable successes; however, most utilities still rely more heavily on experts than on sophisticated optimization algorithms. The problem arises from attempting to fit practical problems into rigid models of the system that can be optimized. This results in reduction in information either in the form of simplified constraints or objectives. The simplifications of the system model and subjectivity of the objectives may often be represented as uncertainties in the fuzzy model.

Consider optimal power flow. Objectives could be cost minimization, minimal control adjustments, and minimal emission of pollutants or maximization of adequate security margins. Physical constraints must include generator and load bus voltage levels, line flow limits, and reserve margins. In practice, none of these constraints or objectives is well defined. Still, a compromise is needed among these various considerations in order to achieve an acceptable solution. Fuzzy mathematics provides a mathematical framework for these considerations. The applications in this category are an attempt to model such compromises.

1.3 Outline of the Book

This book consists of seven chapters including this chapter. The objectives of Chapter 2, "Mathematical Optimization Techniques," are:

- Explaining some of optimization techniques
- Explaining the minimum norm theorem and how it could be used as an optimization algorithm, where a set of equations can be obtained
- Introducing the fuzzy system as an optimization technique
- Introducing the simulated annealing algorithm as an optimization technique
- Introducing the tabu search algorithm as an optimization technique
- Introducing the genetic algorithm as an optimization technique
- Introducing the particle swarm as an optimization technique

The purpose of Chap. 3, "Economic Operation of Electric Power Systems," is:

- To formulate the problem of optimal short-term operation of hydrothermal–nuclear systems
- To obtain the solution by using a functional analytical optimization technique that employs the minimum norm formulation
- To propose an algorithm suitable for implementing the optimal solution
- To present and formulate the fuzzy economic dispatch of all thermal power systems and explaining the algorithm that is suitable for solution
- To formulate the fuzzy economic dispatch problem of hydrothermal power systems and its solution

The objectives of Chap. 4, "Economic Dispatch (ED) and Unit Commitment Problems (UCP): Formulation and Solution Algorithms," are:

- Formulating the objectives function for ED and UCP
- Studying the system and unit constraints
- Proposing rules for generating solutions
- Generating an initial solution
- Explaining an algorithm for the economic dispatch problem
- Applying the simulated annealing algorithm to solve the problems
- Comparing simulated annealing with other simulated annealing algorithms
- Offering numerical results for the simulated annealing algorithm

The objective of Chap. 5, "Optimal Power Flow," is:

- Studying the load flow problem and representing the difference between conventional and optimal load flows (OPF)
- Introducing the different states used in formulating the OPF
- Studying the multiobjective optimal power flow
- Introducing the particle swarm optimization algorithm to solve the optimal power flow

The objective of Chap. 6, "Long-Term Operation of Hydroelectric Power System," is:

- Formulating the problem of long-term operation problem of a multireservoir power system connected in cascade (series)
- Implementing the minimum norm approach to solving the formulated problem
- Implementing the simulated annealing algorithm to solve the long-term hydro scheduling problem (LTHSP)
- Introducing an algorithm enhancement for randomly generating feasible trial solutions
- Implementing an adaptive cooling schedule and a method for variable discretization to enhance the speed and convergence of the original SAA
- Using the short-term memory of the tabu search approach to solve the nonlinear optimization problem in continuous variables of the LTHSP

The objectives of Chap. 7, "Electric Power Quality Analysis," are:

- Applications to a simulated annealing optimization algorithm for measuring voltage flicker magnitude and frequency as well as the harmonic content of the voltage signal, for power quality analysis.
- Moreover, the power system voltage magnitude, frequency, and phase angle of the fundamental component are estimated by the same technique.
- The nonlinear optimization problem in continuous variables is solved using a SA algorithm with an adaptive cooling schedule.
- A method for variable discretization is implemented.
- The algorithm minimizes the sum of the absolute value of the error in the estimated voltage signal.
- The algorithm is tested on simulated and actual recorded data.
- Effects of sampling frequency as well as the number of samples on the estimated parameters are discussed. It is shown that the proposed algorithm is able to identify the parameters of the voltage signal.

References

1. Maust, R.S.: Optimal power flow using a genetic algorithm and linear algebra. Doctor of Philosophy in Engineering, West Virginia University, Morgantown (1999)
2. Sbalzariniy, I.F., Müllery, S., Koumoutsakosyz, P.: Multiobjective optimization using evolutionary algorithms. In: Proceedings of the Summer Program, Center for Turbulence Research, Stanford University, Stanford, 2000, pp. 63–74

3. Lee, K.Y., El-Sharkawi, M.A.: Modern Heuristic Optimization Techniques with Applications to Power Systems. IEEE Power Engineering Society, New York (2002)
4. Ueda, T., Koga, N., Okamoto, M.: Efficient numerical optimization technique based on real-coded genetic algorithm. Genome Inform **12**, 451–453 (2001)
5. Rangel-Merino, A., Lpez-Bonilla, J.L., Linares y Miranda, R.: Optimization method based on genetic algorithms. Apeiron **12**(4), 393–406 (2005)
6. Zhao, B., Cao, Y-j: Multiple objective particle swarm optimization technique for economic load dispatch. J Zhejiang Univ Sci **6**(5), 420–427 (2005)
7. Konak, A., Coit, D.W., Smith, A.E.: Multi-objective optimization using genetic algorithms: a tutorial. Reliab. Eng. Syst. Saf. **91**, 992–1007 (2006)
8. Shi, L., Dong, Z.Y., Hao, J., Wong, K.P.: Mathematical analysis of the heuristic optimisation mechanism of evolutionary programming. Int. J. Comput. Intell. Res. **2**(4), 357–366 (2006)
9. Jones, K.O.: Comparison of genetic algorithms and particle swarm optimization for fermentation feed profile determination. In: International Conference on Computer Systems and Technologies – CompSysTech, University of Veliko Tarnovo, Bulgaria, 15–16 June 2006, pp. 1–7
10. Sayah, S., Zehar, K.: Economic load dispatch with security constraints of the algerian power system using successive linear programming method. Leonardo J. Sci. **9**, 73–86 (2006)
11. Ben Aribia, H., Hadj Abdallah, H.: Multi objectives reactive dispatch optimization of an electrical network. Leonardo J. Sci. **10**, 101–114 (2007)
12. Aarts, E., Korst, J.: Simulated Annealing and Boltzman Machines: A Stochastic Approach to Combinatorial Optimization and Neural Computing. Wiley, New York (1989)
13. El-Hawary, M.E.: Electric Power Applications of Fuzzy Systems. IEEE Press, New York (1998)

Chapter 2
Mathematical Optimization Techniques

Objectives The objectives of this chapter are:

- Explaining some optimization techniques.
- Explaining the minimum norm theorem and how it could be used as an optimization algorithm, where a set of equations can be obtained.
- Introducing the fuzzy system as an optimization technique.
- Introducing the simulated annealing algorithm (SAA) as an optimization technique.
- Introducing the tabu search algorithm (TSA) as an optimization technique.
- Introducing the genetic algorithm (GA) as an optimization technique.
- Introducing the particle swarm (PS) as an optimization technique.

2.1 Introduction

Growing interest in the application of Artificial Intelligence (AI) techniques to power system engineering has introduced the potential of using this state-of-the-art technology. AI techniques, unlike strict mathematical methods, have the apparent ability to adapt to nonlinearities and discontinuities commonly found in power systems. The best-known algorithms in this class include evolution programming, genetic algorithms, simulated annealing, tabu search, and neural networks.

In the last three decades many optimization techniques have been invented and successfully applied to the operation and control of electric power systems.

This chapter introduces the mathematical background behind the algorithms used in this book, without going deeply into the mathematical proofs of these algorithms, to help the reader understand the application of these algorithms. Different examples are offered, where they are needed, to help the reader understand a specific optimization algorithm.

2.2 Quadratic Forms [1]

An algebraic expression of the form

$$f_{(x,y)} = ax^2 + bxy + cy^2$$

is said to be a quadratic form. If we let

$$X = \begin{bmatrix} x \\ y \end{bmatrix}$$

Then we obtain

$$f_{(x,y)} = \begin{bmatrix} x & y \end{bmatrix} \begin{bmatrix} a & \frac{b}{2} \\ \frac{b}{2} & c \end{bmatrix} \begin{bmatrix} x \\ y \end{bmatrix}, \text{ or}$$

$$f_{(X)} = X^T A X$$

The above equation is in quadratic form. The matrix A in this form is a symmetrical matrix.

A more general form for the quadratic function can be written in matrix form as

$$F_{(X)} = X^T A X + B^T X + C$$

where X is an $n \times 1$ vector, A is an $n \times n$ matrix, B is an $n \times 1$ vector, and C is a 1×1 vector.

Example (2.1)

Given the function

$$f_{(x,y)} = 2x^2 + 4xy - y^2 = 0$$

it is necessary to write this function in a quadratic form.

Define the vector

$$X = \begin{bmatrix} x \\ y \end{bmatrix}$$

Then

$$f_{(x,y)} = \begin{bmatrix} x & y \end{bmatrix} \begin{bmatrix} 2 & 2 \\ 2 & -1 \end{bmatrix} \begin{bmatrix} x \\ y \end{bmatrix}$$

$$F_{(x)} = X^T A X$$

2.2 Quadratic Forms

where

$$A = \begin{bmatrix} 2 & 2 \\ 2 & -2 \end{bmatrix}$$

Example (2.2)

Obtain the quadratic form for the function

$$f_{(x_1,x_2)} = 3x_1^2 + 4x_1x_2 - 4x_2^2$$

Define the vector X as

$$X = \begin{bmatrix} x_1 & x_2 \end{bmatrix}^T$$

Then

$$f_{(x_1,x_2)} = \begin{bmatrix} x_1 & x_2 \end{bmatrix}^T \begin{bmatrix} 3 & 2 \\ 2 & -4 \end{bmatrix} \begin{bmatrix} x_1 \\ x_2 \end{bmatrix}$$

Then

$$f_{(X)} = X^T A X$$

where

$$A = \begin{bmatrix} 3 & 2 \\ 2 & -4 \end{bmatrix}$$

Let A be an $n \times n$ matrix and let X be an $n \times 1$ vector. Then, irrespective of whether A is symmetric:

$$X^T A X = \sum_{i=1}^{n} \sum_{j=1}^{n} x_i a_{ij} x_j$$

$$= \sum_{i=1}^{n} \sum_{j=1}^{n} x_i a_{ji} x_j$$

- An $n \times n$ matrix A is positive definite if and only if

$$X^T A X > 0 \quad (\forall X \in R^n, X \neq 0)$$

- and is positive semidefinite if and only if

$$X^T A X \geq 0 \quad (\forall X \in R^n)$$

- Similarly, A is negative definite if, and only if,

$$X^T AX < 0 \quad (\forall X \in R^n, X \neq 0)$$

- and A is negative semidefinite if and only if

$$X^T AX \leq 0 \quad (\forall X \in R^n)$$

2.3 Some Static Optimization Techniques [1–10]

In this section we discuss the general optimization problem without going into details of mathematical analysis. The first part of the section introduces unconstrained optimization, which has many applications throughout this book, and the second part introduces the constrained optimization problem. Generally speaking, the optimization problem has the form

$$\text{Minimize} \\ f(x_1, \ldots, x_n) \tag{2.1}$$

$$\text{Subject to} \\ \phi_i(x_1, \ldots, x_n) = 0, (i = 1, \ldots, \ell) \tag{2.2}$$

$$\Psi_j(x_1, \ldots, x_n) \leq 0, (j = 1, \ldots, m) \tag{2.3}$$

Equation 2.2 represents ℓ, $\ell < n$, equality constraints, whereas Eq. 2.3 represents m inequality constraints. By using vector notation, we may express the general constrained optimization problem as follows.

$$\text{Minimize} \\ f(X) \tag{2.4}$$

$$\text{Subject to} \\ \phi(X) = 0 \tag{2.5}$$

$$\Psi(X) \leq 0, X \in R^n \tag{2.6}$$

The problem formulated in Eqs. 2.4, 2.5, and 2.6 is usually referred to as the general nonlinear programming problem. Any point X that satisfies these equations is called a feasible point.

2.3 Some Static Optimization Techniques

2.3.1 Unconstrained Optimization

The calculus of variations is concerned with the determination of extrema (maxima and minima) or stationary values of functionals. A functional can be defined as a function of several other functions. The calculus of variations is a powerful method for the solution of problems in optimal economic operation of power systems. In this section we introduce the subject of variational calculus through a derivation of the Euler equations and associated transversality conditions.

In the unconstrained optimization problem, we need to find the value of the vector $X = [x_1, \ldots, x_n]^T$ that minimizes the function

$$f(x_1, \ldots, x_n) \tag{2.7}$$

provided that the function f is continuous and has a first-order derivative.

To obtain the minimum and/or maximum of the function f we set its first derivative with respect to x to zero

$$\frac{\partial f(x_1, \ldots, x_n)}{\partial x_1} = 0 \tag{2.8}$$

$$\frac{\partial f(x_1, \ldots, x_n)}{\partial x_2} = 0 \tag{2.9}$$

$$\vdots$$

$$\frac{\partial f(x_1, \ldots, x_n)}{\partial x_1} = 0 \tag{2.10}$$

Equations 2.8, 2.9, and 2.10 represent n equations in n unknowns. The solution of these equations produces candidate solution points. If the function f has second partial derivatives, then we calculate the Hessian matrix,

$$H = \frac{\partial^2 f(x_1, \ldots, x_n)}{\partial x_i^2}.$$

If the matrix H is positive definite, then the function f is a minimum at the candidate points, but if the matrix H is negative definite then f is a maximum at the candidate points. The following examples illustrate these steps.

Example (2.3)

Minimize
$$f(x_1, x_2) = x_1^2 + x_1 x_2 + x_2^2 \; (x \in R^2)$$

To obtain the candidate solution points, we have

$$\frac{\partial f(x_1, x_2)}{\partial x_1} = 2x_1 + x_2 = 0$$

and

$$\frac{\partial f(x_1, x_2)}{\partial x_2} = x_1 + 2x_2 = 0$$

Solving the above equations yields the candidate solution point as

$$[x_1^*, x_2^*]^T = [0, 0]^T$$

Next, we calculate the Hessian matrix using

$$H = \begin{bmatrix} \frac{\partial^2 f}{\partial x_1^2} & \frac{\partial^2 f}{\partial x_1 \partial x_2} \\ \frac{\partial^2 f}{\partial x_2 \partial x_1} & \frac{\partial^2 f}{\partial x_2^2} \end{bmatrix}$$

to obtain

$$H = \begin{bmatrix} 2 & 1 \\ 1 & 2 \end{bmatrix}$$

so

$$X^T H X = \begin{bmatrix} x_1 & x_2 \end{bmatrix} \begin{bmatrix} 2 & 1 \\ 1 & 2 \end{bmatrix} \begin{bmatrix} x_1 \\ x_2 \end{bmatrix}$$
$$= 2(x_1^2 + x_1 x_2 + x_2^2)$$
$$= 2\left\{ \left(x_1 + \frac{1}{2}x_2\right)^2 + \frac{3}{4}x_2^2 \right\}$$

Therefore

$$X^T H X > 0 \quad (\forall X \neq 0)$$

and so H is positive definite, and the function f is a minimum at the candidate point.

Note that the positive definiteness of H can also be verified just by calculating the values of the different determinants, produced from H as

$$\Delta_1(H) = 2 = h_{11}$$
$$\Delta_2(H) = (4 - 1) = 3$$

Because all Δs are positive, H is a positive definite matrix.

2.3 Some Static Optimization Techniques

Example (2.4)

Minimize
$$f(x_1, x_2) = 34x_1^2 - 24x_1x_2 + 41x_2^2$$

Set the first derivatives to zero to obtain

$$\frac{\partial f(x_1, x_2)}{\partial x_1} = 68x_1 - 24x_2 = 0$$

$$\frac{\partial f(x_1, x_2)}{\partial x_2} = -24x_1 + 82x_2 = 0$$

The solution to the above equation gives

$$\begin{bmatrix} x_1^* \\ x_2^* \end{bmatrix} = \begin{bmatrix} 0 \\ 0 \end{bmatrix}$$

Calculate the Hessian matrix as

$$H = \begin{bmatrix} 68 & -24 \\ -24 & 82 \end{bmatrix}$$

Check the definiteness for the Hessian matrix as

$$\Delta_1(H) = h_{11} = 68 > 0$$

$$\Delta_2(H) = 68 \times 82 - 24 \times 24 = 1328 > 0$$

Hence H is a positive definite matrix; or calculate the quadratic form:

$$X^T H X = \begin{bmatrix} x_1 & x_2 \end{bmatrix} \begin{bmatrix} 68 & -24 \\ -24 & 82 \end{bmatrix} \begin{bmatrix} x_1 \\ x_2 \end{bmatrix}$$

$$= 68x_1^2 - 48x_1x_2 + 82x_2^2$$

$$= 2(4x_1 - 3x_2)^2 + 32x_1^2 + 64x_2^2$$

so

$$X^T H X > 0 \quad (\forall X \neq 0)$$

hence H is positive definite and f is a minimum at the feasible points.

Example (2.5)

Minimize
$$f(x_1, x_2) = x_1^3 - 2x_1^2 x_2 + x_2^2$$

We have

$$\frac{\partial f(x_1, x_2)}{\partial x_1} = 3x_1^2 - 4x_1 x_2 = 0$$

$$\frac{\partial f(x_1, x_2)}{\partial x_2} = -2x_1^2 + 2x_2 = 0$$

Solving the above two equations yields the critical points to be

$$x^* = \begin{bmatrix} x_1^* \\ x_2^* \end{bmatrix} = \begin{bmatrix} \frac{3}{4} \\ \frac{9}{16} \end{bmatrix}$$

The Hessian matrix is calculated as

$$H = \begin{bmatrix} (6x_1 - 4x_2) & -4x_1 \\ -4x_1 & 2 \end{bmatrix}$$

At the solution points, we calculate H as

$$H(x_1^*, x_2^*) = \begin{bmatrix} \frac{9}{4} & -3 \\ -3 & 2 \end{bmatrix}$$

$$\Delta_1(H) = \frac{9}{4} > 0$$

$$\Delta_2(H) = \frac{18}{4} - 9 = -\frac{18}{4} < 0$$

Hence $H(x_1^*, x_2^*)$ is positive semidefinite and so nothing can be concluded about the nature of the solution point x^*. The solution point in this case is called a saddle point.

2.3.2 Constrained Optimization

The problem examined in Sect. 2.3.1 excluded consideration of optimal control problems having constraint relationships between the scalar elements of the state trajectory, which occurs in many physical problems. The problem including such a

2.3 Some Static Optimization Techniques

constraint can be formulated as follows. Find the function $x(t)$ that minimizes the following cost functional.

$$J = \int_{t_0}^{t_f} L[x(t), \dot{x}(t), t] dt \tag{2.11}$$

Subject to satisfying the following constraint:

$$\int_{t_0}^{t_f} g[x(t), \dot{x}(t), t] dt = 0 \tag{2.12}$$

An example of this problem that occurs in the power system area is the minimization of the fuel cost of a power plant subject to satisfying the active power balance equation for the system.

We can form an augmented cost functional by augmenting the cost functional of Eq. 2.11 by Eq. 2.12 via Lagrange's multiplier λ,

$$\tilde{J} = \int_{t_0}^{t_f} \tilde{L}[x(t)\dot{x}(t), t] dt \tag{2.13}$$

where

$$\tilde{L}(.) = L(.) + \lambda g(.) \tag{2.14}$$

As a result we obtain the following modified Euler equation.

$$\tilde{L}_{x(t)} - \frac{d}{dt}\left(\tilde{L}_{\dot{x}(t)}\right) = 0 \tag{2.15}$$

or

$$\left(L_{x(t)} + g_{x(t)}\right) - \frac{d}{dt}\left(L_{\dot{x}(t)} + \lambda g_{\dot{x}(t)}\right) = 0 \tag{2.16}$$

Example (2.6)

We wish to maximize

$$J = \int_{-1}^{1} x(t) dt$$

subject to satisfying

$$\int_{-1}^{1} [1+\dot{x}^2(t)]^{1/2} dt = 1$$

The augmented cost functional is given by

$$\tilde{J} = \int_{-1}^{1} \left\{ x(t) + \lambda [1+\dot{x}^2(t)]^{1/2} \right\} dt$$

$$\tilde{L}(x,\dot{x},t) = x(t) + \lambda [1+\dot{x}^2(t)]^{1/2}$$

$$\tilde{L}_{x(t)} = \frac{\partial \tilde{L}(.)}{\partial x(t)} = 1$$

$$\tilde{L}_{\dot{x}(t)} = \frac{\partial \tilde{L}(.)}{\partial \dot{x}(t)} = \frac{\lambda \dot{x}(t)}{[1+\dot{x}^2(t)]^{1/2}}$$

Substituting into Eq. 2.14, one obtains the following Euler equation,

$$1 - \lambda \frac{d}{dt} \frac{\dot{x}(t)}{[1+\dot{x}^2(t)]^{1/2}} = 0$$

or

$$\frac{\dot{x}(t)}{[1+\dot{x}(t)]^{1/2}} = \frac{1}{\lambda} t + C$$

The solution of the above equation is given by

$$(x-x_1)2 + (t-t_1)2 = r^2$$

where the parameters x_1, t_1, and r are chosen to satisfy the boundary conditions.
Another mathematical form of the constrained optimization problem has the form

Minimize

$$f(x_1,\ldots,x_n) \qquad (2.17)$$

subject to satisfying

$$\phi_i(x_1,\ldots,x_n) = 0, (i=1,\ldots,\ell) \qquad (2.18)$$

2.3 Some Static Optimization Techniques

and

$$\Psi_j(x_1,\ldots,x_n) \leq 0, (j=1,\ldots,m) \tag{2.19}$$

Let us consider, for instance, the case when the objective function is subject only to equality constraints. We form the augmented objective function by adjoining the equality constraints to the function via Lagrange multipliers to obtain the alternative form.

Minimize

$$\tilde{f}(x_1,\ldots,x_n,\lambda_i) = f(x_1,\ldots x_n) + \sum_{i=1}^{l} \lambda_i \phi_i(x_1,\ldots,x_n) \tag{2.20}$$

or, in vector form,

$$\tilde{f}(X,\lambda) = f(X) + \lambda^T \phi(x) \tag{2.21}$$

Putting the first derivative to zero, we obtain

$$\frac{\partial \tilde{f}(X,\lambda)}{\partial x_i} = \frac{\partial f(x)}{\partial x_i} + \sum_{j=1}^{\ell} \lambda_j \frac{\partial \phi_j}{\partial x_i} = 0 \tag{2.22}$$

Equation 2.22 is a set of n equations in $(n+\ell)$ unknowns (x_i; $i=1,\ldots,n : \lambda_j$; $j=1,\ldots,\ell$). To obtain the solution, the equality constraints must be satisfied; that is,

$$\phi_i(x_1,\ldots,x_n) = 0 \quad i=1,\ldots,\ell \tag{2.23}$$

Solving Eqs. 2.22 and 2.23, we obtain x_i^* and λ_j^*. This is illustrated in the following examples.

Example (2.7)

Minimize
$$f(x_1,x_2) = x_1^2 + x_2^2$$
Subject to
$$\phi[x_1,x_2] = x_1 + 2x_2 + 1 = 0$$

For this problem $n=2$, $\ell=1$, $(n+\ell=3)$. The augmented cost function is given by

$$\tilde{f}(x_1,x_2,\lambda) = x_1^2 + x_2^2 + \lambda(x_1 + 2x_2 + 1)$$

Putting the first derivatives to zero gives

$$\frac{\partial \tilde{f}}{\partial x_1} = 0 = 2x_1^* + \lambda$$

$$\frac{\partial \tilde{f}}{\partial x_2} = 0 = 2x_2^* + 2\lambda$$

and

$$\frac{\partial \tilde{f}}{\partial \lambda} = 0 = x_1^* + 2x_2^* + 1 \quad (\text{equality constraint})$$

Solving the above three equations gives

$$x_1^* = -\frac{1}{5}, x_2^* = -\frac{2}{5}, \lambda = \frac{2}{5}$$

Example (2.8)

Minimize
$$\tilde{f}(x_1, x_2, \lambda) = \left(10 + 5x_1 + 0.2x_1^2\right) + \left(20 + 3x_2 + 0.1x_2^2\right)$$
Subject to
$$x_1 + x_2 = 10$$

The augmented cost function is

$$\tilde{f}(x_1, x_2, \lambda) = \left(30 + 5x_1 + 0.2x_1^2 + 3x_2 + 0.1x_2^2\right) + \lambda(10 - x_1 - x_2)$$

Putting the first derivatives to zero we obtain

$$\frac{\partial \tilde{f}}{\partial x_1} = 0 = 5 + 0.4x_1^* - \lambda$$

$$\frac{\partial \tilde{f}}{\partial x_2} = 0 = 3 + 0.2x_2^* - \lambda$$

$$\frac{\partial \tilde{f}}{\partial \lambda} = 0 = 10 - x_1^* - x_2^*$$

Solving the above three equations gives

$$x_1^* = 0, x_2^* = 10 \text{ and } \lambda = 5$$

and the minimum of the function is

$$f(0, 10) = 30 + 30 + 10 = 70$$

2.3 Some Static Optimization Techniques

If there are inequality constraints, then the augmented function is obtained by adjoining these inequality constraints via Kuhn–Tucker multipliers, to obtain

$$\tilde{f}(X, \lambda, \mu) = f(X) + \lambda^T \phi(X) + \mu^T \Psi(X) \tag{2.24}$$

Putting the first derivative to zero, we obtain

$$\frac{\partial \tilde{f}}{\partial X} = 0 = \frac{\partial f(X^*)}{\partial X} + \lambda^T \frac{\partial \phi(X^*)}{\partial X} + \mu^T \frac{\partial \Psi(X^*)}{\partial X} \tag{2.25}$$

and

$$\frac{\partial \tilde{f}}{\partial \lambda} = 0 = \phi(X^*) \tag{2.26}$$

with

$$\mu^T \Psi(X^*) = 0 \tag{2.27}$$

If $\Psi(X^*) > 0$, then $\mu = 0$.

Solving the above equations gives the candidate solution (X^*, λ, μ).

Example (2.9)

Recall the previous example; we have

Minimize
$$f(x_1, x_2) = 0.1x_2^2 + 0.2x_1^2 + 3x_2 + 5x_1 + 30$$
Subject to the following constraints
$$x_1 + x_2 = 10$$

with
$$x_1 \geq 0$$
$$0 \leq x_2 \leq 15$$

We form the augmented function as

$$\tilde{f}(x_1, x_2, \lambda, \mu_1, \mu_2, \mu_3) = f(x_1, x_2) + \lambda(10 - x_1 - x_2) + \mu_1 x_1 + \mu_2 x_2 + \mu_3(15 - x_2)$$

Putting the first derivatives to zero leads to

$$\frac{\partial \tilde{f}}{\partial x_1} = 0 = \frac{\partial f}{\partial x_1} - \lambda + \mu_1 + \mu_2$$

$$\frac{\partial \tilde{f}}{\partial x_2} = 0 = \frac{\partial f}{\partial x_2} - \lambda + \mu_2 - \mu_3$$

$$\frac{\partial \tilde{f}}{\partial \lambda} = 0 = 10 - x_1 - x_2$$

with

$$\mu_1 x_1 = 0;$$
$$\mu_2 x_2 = 0;$$
$$\mu_3 (15 - x_2) = 0$$

Now we have six equations for six unknowns, however, solving these equations is very difficult. We assume that none of the variables violates its limits; thus we obtain

$$\mu_1 = 0$$
$$\mu_2 = 0$$
$$\mu_3 = 0$$

and we must check the solution obtained for these conditions. The solution in this case is:

$$x_1^* = 0, \quad x_2^* = 10, \quad \lambda = 5$$

Indeed, as we see, the variables do not violate their limits, and the optimal solution in this case is

$$x_1^* = 0, \quad x_2^* = 10, \quad \lambda^* = 5$$
$$\mu_1^* = 0, \quad \mu_2^* = 0, \quad \mu_3^* = 0$$

However, if we change the second inequality constraint to be

$$0 \leq x_2 \leq 8$$

then we can see that for the solution above, $x_2^* = 10$ violates the upper limit. In this case we put

$$x_2^* = 8; \quad \text{with } \mu_3 (8 - x_2^*) = 0$$

and recalculate x_1^* as

$$x_1^* = 10 - x_2^* = 10 - 8$$
$$x_1^* = 2, \quad (x_1^* > 0)$$

Under this solution $\mu_3 \neq 0$, but $\mu_1^* = 0$ and $\mu_2^* = 0$. To calculate λ^* and μ_3^* we use the first two equations as

$$0 = 0.4 x_1^* + 5 - \lambda^*$$

or

$$\lambda^* = 0.4(2) + 5$$
$$= 5.8$$

and

$$0 = 0.2x_2^* + 3 - \lambda^* - \mu_3^*$$

or

$$\mu_3^* = 1.6 + 3 - 5.8$$
$$= -1.2$$

2.4 Pontryagin's Maximum Principle [11–14]

Let $u(t)$ be an admissible control and $x(t)$ be the corresponding trajectory of the system described by

$$\dot{x}(t) = (x(t), u(t), t) \qquad (2.28)$$

Let $x(t_0)$, t_0, and t_f be specified and $x(t_f)$ be free. The necessary conditions for $u(t)$ to be an optimal control, that is, to be the control that takes $x(t)$ from $x(0)$ to some state $x(tf)$ while minimizing the functional J

$$J = G\left[x(t_f, t_f) + \int_{t_0}^{t_f} L[x(t), u(t), t] dt\right] \qquad (2.29)$$

are as follows.

1. There exists a function or vector $\lambda(t)$ such that $x(t)$ and $\lambda(t)$ are the solutions of the following equations.

$$\dot{x}(t) = \frac{\partial H}{\partial \lambda(t)} \qquad (2.30)$$

$$\dot{\lambda}(t) = -\frac{\partial H}{\partial x(t)} \qquad (2.31)$$

subject to the boundary conditions given by

$$x(t_0) = x(0) \qquad (2.32)$$

$$\lambda(t_f) = \left.\frac{\partial G(.)}{\partial x(t)}\right|_{t=t_f} \quad \text{at} \quad x(t) = x(t_f) \tag{2.33}$$

where the function H is a scalar function, which is called the Hamiltonian and is given by

$$H[x(t), u(t), \lambda(t), t] = L[x(t), u(t), t] + \lambda^T(t)f[x(t), u(t), t] \tag{2.34}$$

2. The functional $H[x(t), u(t), \lambda(t), t]$ has a local minimum at

$$\frac{\partial H}{\partial u(t)} = 0 \tag{2.35}$$

In many practical problems, there are inequality constraints on the admissible control and states and these constraints must be taken into account. As a result Eq. is no longer applicable. The basic contribution of the maximum principle addresses this difficulty. In place of Eq. the necessary condition is that the Hamiltonian function H $[x(t), u(t), \lambda(t), t]$ attains an absolute minimum as a function of $u(t)$ over the admissible region Ω for all t in the interval (t_0, t_f). In other words, the optimal $u\varepsilon$ satisfies

$$H[x(t), u_\varepsilon(t), \lambda(t), t] \leq H[x(t), u(t), \lambda(t), t] \quad u_\varepsilon \in \Omega \tag{2.36}$$

Example (2.10)

A linear differential system is described by

$$\dot{x} = Ax + Bu$$

where

$$A = \begin{bmatrix} 0 & 1 \\ 0 & 0 \end{bmatrix}, \quad B = \begin{bmatrix} 1 & 0 \\ 0 & 1 \end{bmatrix}, \quad x^T = [x_1, x_1], \quad u^T = [u_1, u_2]$$

Find $u(t)$ such that

$$J = \frac{1}{2}\int_0^2 \|u\|^2 dt$$

is minimum, given $xT(0) = [1, 1]$ and $x1(2) = 0$.

Define the Hamiltonian H as

$$H = \tfrac{1}{2}u^T u + \lambda^T(Ax + Bu)$$

2.4 Pontryagin's Maximum Principle

Equation 2.24 leads to the following costate equations.

$$\dot{\lambda}(t) = -\frac{\partial H}{\partial x}$$

$$\dot{\lambda}(t) = A^T \lambda$$

or

$$\dot{\lambda}_1(t) = 0 \qquad \text{(a)}$$

$$\dot{\lambda}(t) = -\lambda_1(t) \qquad \text{(b)}$$

with

$$\lambda_1(2) = \lambda_2(2) = 0 \quad (\text{because } G = 0) \qquad \text{(c)}$$

Integration of Eqs. (a) and (b) with the boundary Eq. (c) gives

$$\lambda_1(t) = C_1 \qquad \text{(d)}$$

$$\lambda_2(t) = C_1(2-t) \qquad \text{(e)}$$

H has a local minimum at

$$\frac{\partial H}{\partial u(t)} = 0 \qquad \text{(f)}$$

or

$$u(t) = -BT \, \lambda(t) \qquad \text{(g)}$$

We have

$$\dot{x} = Ax + Bu$$

Substituting for $u(t)$ from Eq. (g) in the above equation, we obtain

$$\dot{x} = Ax - BB^T \lambda \qquad \text{(h)}$$

The solution of the above equation with Eqs. (d) and (e) gives

$$x_t(t) = -C_1 t^2 + \tfrac{1}{6} C_1 t^3 + t - C_1 t + C_2$$

$$x_2(t) = -2C_1 t + \tfrac{1}{2} C_1 t^2 + C_3$$

Using the boundary conditions at $t = 0$, $x_T(0) = [1, 1]$ gives

$$1 = C_2$$

$$1 = C_3$$

The state $x_1(t)$ now is given by

$$x_1(t) = C_1 t^2 + \tfrac{1}{6} C_1 t^3 + t - C_1 t + 1$$

By using the boundary condition at $t = 2$, $x1(2) = 0$,

$$0 = -4C_1 + \tfrac{8}{6} C_1 + 2 - 2C_1 + 1$$

$$C_1 = \tfrac{9}{14}$$

We are now in a position to write the system equations that satisfied the boundary conditions as

$$\lambda_1(t) = \tfrac{9}{14}$$

$$\lambda_2(t) = \tfrac{9}{14}(2 - t)$$

$$x_1(t) = -\tfrac{9}{14}t^2 + \tfrac{3}{28}t^2 + \tfrac{5}{14}t + 1$$

$$x_2(t) = -\tfrac{9}{7}t + \tfrac{9}{28}t^2 + 1$$

$$u_1(t) = -\tfrac{9}{14}$$

$$u_2(t) = -\tfrac{9}{14}(2 - t)$$

Example (2.11)

For the fixed plant dynamics given by

$$\dot{x}(t) = u(t), x(0) = x_0,$$

determine the optimal closed-loop control that minimizes for fixed t_f,

$$J = \tfrac{1}{2}Sx^2(t_f) + \tfrac{1}{2}\int_0^{t_f} u(t)^2 dt$$

where S is an arbitrary constant. Do this by first determining the optimum open-loop control and trajectory and then let $u(t) = k(t) x(t)$.

2.4 Pontryagin's Maximum Principle

Define the Hamiltonian as

$$H = \tfrac{1}{2}u^2(t) + \lambda(t)u(t)$$

We have

$$\dot\lambda(t) = -\frac{\partial H}{\partial x} = 0$$

Integrating directly we obtain

$$\lambda(t) = C_1 \tag{a}$$

From the boundary term $G = \tfrac{1}{2}Sx^2(t_f)$, we have

$$\lambda(t_f) = \left.\frac{\partial G}{\partial x(t)}\right|_{t=t_f}$$
$$\lambda(t_f) = Sx(t_f) \tag{b}$$

Substituting from Eq. b into Eq. 2.31a at $t = tf$, $\lambda(tf) = Sx(tf)$,

$$C_1 = Sx(t_f)$$
$$\lambda(t) = Sx(t_f) \tag{c}$$

H has a local minimum at

$$\frac{\partial H}{\partial u(t)} = 0$$

or

$$0 = u(t) + \lambda(t)$$

Hence, the open-loop control is given by

$$u(t) = -Sx(t_f) \tag{d}$$

Also, we have

$$\dot x(t) = u(t)$$

or

$$\dot x(t) = -Sx(t_f)$$

Integrating the above equation directly and by using at $t = 0$, $x(t) = x(0)$, we obtain

$$x(t) = -Sx(t_f)t + x(0) \qquad (e)$$

Now, the closed-loop control is given by

$$u(t) = k(t)[-Sx(t_f)t + x(0)]$$

or

$$u(t) = -Sk(t)x(t_f)t + k(t)x(0)$$

2.5 Functional Analytic Optimization Technique [6]

The aim of this section is to discuss the application of one important minimum norm result as an optimization technique that has been used as a powerful tool in the solution of problems treated in this book [6, 15]. Before we do this, a brief discussion of relevant concepts from functional analysis is given.

2.5.1 Norms

A norm, commonly denoted by $\|\cdot\|$, is a real-valued and positive definite scalar. The norm satisfies the following axioms.

(1) $\|x\| \geq 0$ for all $x \in X$, $\|x\| = 0 \leftrightarrow x = 0$

(2) $\|x + y\| \leq \|x\| + \|y\|$ for each $x, y \in X$

(3) $\|\alpha x\| = |\alpha| \cdot \|x\|$ for all scalars α and each $x \in X$

A normed linear (vector) space X is a linear space in which every vector x has a norm (length). The norm functional is used to define a distance and a convergence measure

$$d(x,y) = \|x - y\|$$

For example, let $[0, T]$ be a closed bounded interval. The space of continuous functions $x(t)$ on $[0, T]$ can have one of the following norms.

$$\|x\|_1 = \int_0^T |x(t)| dt$$

2.5 Functional Analytic Optimization Technique

$$\|x\|_2 = \left[\int_0^T |x(t)|^2 dt\right]^{1/2}$$

The normed linear space becomes a metric space under the (usual) metric [5].

2.5.2 Inner Product (Dot Product)

Consider the real three-dimensional Euclidean space E_3. A vector or element in E_3 is an ordered real triple $x = (x_1, x_2, x_3)$ in which the norm is defined by

$$\|x\| = \left(|x_1|^2 + |x_2|^2 + |x_3|^2\right)^{1/2}$$

From linear algebra in a Euclidean space E, if x is multiplied with another vector $y = (y_1, y_2, y_3)$, the result is (dot product)

$$(x, y) = x_1 y_1 + x_2 y_2 + x_3 y_3$$

or

$$<x, y> = \sum_{i=1}^{3} x_i y_i$$

In the space E_3 the angle θ between x and y is also defined and is related to the norm by the equation

$$\langle x, y \rangle = \|x\| \cdot \|y\| \cos \theta$$

If the two vectors are orthogonal, $\theta = 90°$, then their inner product is zero

$$\langle x, y \rangle = 0$$

and they are collinear, $\theta = 0$, if their inner product is equal to

$$\langle x, y \rangle = \pm \|x\| \cdot \|y\|$$

In the complex space E_3 the inner product between any two vectors $x = (x_1, x_2, x_3)$ and $y = (y_1, y_2, y_3)$ can be defined by

$$\langle x, y \rangle = x_1 y_1^* + x_2 y_2^* + x_3 y_3^*$$

Using these ideas as a background, let us now formulate the basic definition of an inner product in an abstract linear space.

Let X be a linear space. A rule that assigns a scalar $\langle x,y \rangle$ to every pair of elements $x, y \in X$ is called an inner product function if the following conditions are satisfied.

1. $\langle x,y \rangle = \langle y,x \rangle$

2. $\langle \alpha x + \beta y, z \rangle = \alpha \langle x,z \rangle + \beta \langle y,z \rangle$

3. $\langle \lambda x, y \rangle = \lambda \langle x, y \rangle$

4. $\langle x, y \rangle \geq 0, \quad \langle x, x \rangle = 0 \leftrightarrow x = 0$

5. $\langle x, x \rangle = \|x\|^2$

A linear space X is called a Hilbert space if X is an inner product space that is complete with respect to the norm induced by the inner product.

Equivalently, a Hilbert space is a Banach space whose norm is induced by an inner product. Let us now consider some specific examples of Hilbert spaces. The space En is a Hilbert space with inner product as defined by

$$\langle x, y \rangle = x^T y$$

or

$$\langle x, y \rangle = \sum_{i=1}^{n} x_i y_i$$

The space $L_2[0, T]$ is a Hilbert space with inner product

$$\langle x, y \rangle = \int_0^T x(t)y(t)dt$$

We have in this book a very useful Hilbert space. The elements of the space are vectors whose components are functions of time such as active power generation by the system units over the interval $[0, T_f]$. Given a positive definite matrix $B(t)$ whose elements are functions of time as well, we can define the Hilbert space $L_{2B}^n(0, T_f)$. The inner product in this case is given by

$$\langle V(t), (t) \rangle = \int_0^{T_f} V^T(t) B(t) U(t) dt$$

for every $V(t)$ and $U(t)$ in the space.

2.5.3 Transformations

Let X and Y be linear vector spaces and let D be a subset of X. A rule that associates with every element $x \in D$ and element $y \in Y$ is said to be a transformation from X to Y with domain D. If y corresponds to x under T, we write $y = T(x)$ and y is referred to as the image of x under T. Alternatively, a transformation is referred to as an operator.

The transformation $T: X \to Y$ is said to be linear if

$$T(\alpha_1 x_1 + \alpha_2 x_2) = \alpha_1 T(x_1) + \alpha_2 T(x_2)$$

for all $\alpha 1, \alpha 2 \in R$ (the real line) and for every $x1, x2 \in X$.

The linear operator T from a normed space X to a normed space Y is said to be bounded if there is a constant M such that $\|Tx\| \leq M\|x\|$ for all $x \in X$. The normed space of all bounded linear operators from the normed space X into the normed space Y is denoted by $B(X, Y)$. Examples of bounded linear operators include one transformation useful for our purposes. This is $T: L_{1B}^n(0, T_f) \to R^m$ defined by $b = T[U(t)]$

$$b = \int_0^{T_f} M^T I(t) dt$$

In practical applications there are many transformations; among these transformations we have [6] the following.

1. If two power plants supply a demand $PD(t)$ such that the active power balance equation is satisfied

$$P_D(t) = P_1(t) + P_2(t)$$

2. If time is not included as a parameter the above equation can be written as

$$P_D(t) = M^T P(t)$$

where

$$M = \text{col}(1, 1)$$

$$P(t) = \text{col}[P_1(t), P_2(t)]$$

This defines a transformation $T: L_2^2(0, T_f) \to L_2(0, T_f)$ sending functions $[P_1(t), P_2(t)]$ into their image $PD(t)$. Observe that $T = MT$.

A functional is a transformation from a linear space into the space of real or complex scalars. A typical functional is the objective functional of optimal economy operation of m thermal plants given by

$$J(P_s) = \int_0^{T_f} \sum_{i=1}^m \left[\alpha_1 + \beta_1 P_{si}(t) + \gamma_1 P_{si}^2(t)\right] dt$$

In the above equation the space $L_2^m(0, T_f)$ of thermal power generation vector functions is mapped into the real scalar space. In a linear space X, a functional f is linear if for any two vectors $x_1, x_2 \in X$, and any two scalars $\alpha 1$ and $\alpha 2$ we have

$$f(\alpha_1 x_1 + \alpha_1 x_2) = \alpha_1 f(x_1) + \alpha_2 f(x_2)$$

On a normed space a linear functional f is bounded if there is a constant M such that $|f(x)| \leq M\|x\|$ for every $x \in X$. The space of these linear functionals is a normed linear space X^*. X^* is the normed dual space of X and is a Banach space. If X is a Hilbert space then $X = X^*$. Thus Hilbert spaces are self-dual. For the normed linear space to be a reflexive, $X = X^{**}$. Any Hilbert space is reflexive.

Let X and Y be normed spaces and let $T \in B(X, Y)$. The adjoint (conjugate) operator $T^*: Y^* \to X^*$ is defined by

$$\langle x, T^* y \rangle = \langle Tx, y^* \rangle$$

An important special case is that of a linear operator $T: H \to G$ where H and G are Hilbert spaces. If G and H are real, then they are their own duals and the operator T^* can be regarded as mapping G into H. In this case the adjoint relation becomes

$$\langle Tx, y \rangle = \langle x, T^* y \rangle$$

Note that the left-hand side inner product is taken in G whereas the right-hand side inner product is taken in H.

Composite transformations can be formed as follows. Suppose T and G are transformations $T: X \to Y$ and $G: Y \to Z$. We define the transformation $GT: X \to Z$ by

$$(GT)(x) = G(T(x))$$

We then say that GT is a composite of G and T, respectively.

2.5.4 The Minimum Norm Theorem

With the above-outlined definitions in mind, we can now introduce one powerful result in optimization theory. The theorem described here is only one of a wealth of results that utilize functional analytic concepts to solve optimization problems effectively.

2.5 Functional Analytic Optimization Technique

Theorem. *Let B and D be Banach spaces. Let T be a bounded linear transformation defined on B with values in D. Let û be a given vector in B. For each ξ in the range of T, there exists a unique element $u\xi \in B$ that satisfies*

$$\xi = Tu$$

while minimizing the objective functional

$$J(u) = \|u - \hat{u}\|$$

The unique optimal $u\xi \in B$ is given by

$$u_\xi = T^+[\xi - T\hat{u}] + \hat{u}$$

where the pseudoinverse operator $T+$ in the case of Hilbert spaces is given by

$$T+ \ \xi = T^*[TT^*]-1 \ \xi$$

provided that the inverse of TT^* exists.

The theorem as stated is an extension of the fundamental minimum norm problem where the objective functional is

$$J(u) = \|u\|$$

The optimal solution for this case is

$$u \ \xi = T+ \ \xi$$

with $T+$ being the pseudoinverse associated with T.

The above equations for the optimal control vector u can be obtained by using the Lagrange multiplier argument. The augmented cost functional can be obtained by adjoining to the cost functional the equality constraint via the Lagrange multiplier as follows.

$$\tilde{J}(u) = \|u - \hat{u}\|^2 + \langle \lambda, (\xi - Tu) \rangle$$

where λ is a multiplier (in fact $\lambda \in D$) to be determined so that the constraint $\xi = Tu$ is satisfied. By utilizing properties of the inner product we can write

$$\tilde{J}(u) = \|u - \hat{u} - T^*(\lambda/2)\|^2 - \|T^*(\lambda/2)\|^2 + \langle \lambda, \xi \rangle$$

Only the first norm of the above equation depends explicitly on the control vector u. To minimize $\tilde{J}$ we consider only

$$\tilde{J}(u) = \|u - \hat{u} - T^*(\lambda/2)\|$$

The minimum of the above equation is clearly achieved when

$$u_\xi = \hat{u} + T^*(\lambda/2)$$

To find the value of $(\lambda/2)$ we use the equality constraint

$$\xi = Tu\ \xi$$

which gives

$$(\lambda/2) = [TT^*]^{-1}[\xi - T\hat{u}]$$

It is therefore clear that with an invertible TT^* we write

$$u_\xi = T^*[TT^*]^{-1}[\xi - T\hat{u}]$$

which is the required result. In the above equation if $\hat{u} = 0$ we obtain

$$u\ \xi = T^*[TT^*] - 1\ \xi$$

which is the same result obtained for the fundamental minimum norm problem.

In applying this result to our physical problem we need to recall two important concepts from ordinary constrained optimization. These are the Lagrange multiplier rule and the Kuhn–Tucker multipliers. An augmented objective functional is obtained by adjoining to the original cost functional terms corresponding to the constraints using the necessary multipliers. The object in this case is to ensure that the augmented functional can indeed be cast as a norm in the chosen space. A set of linear constraints on the control vector is singled out; this set, under appropriate conditions, defines the bounded linear transformation T.

2.6 Simulated Annealing Algorithm (SAA) [16–26]

Annealing is the physical process of heating up a solid until it melts, followed by cooling it down until it crystallizes into a state with a perfect lattice. During this process, the free energy of the solid is minimized. Practice shows that the cooling must be done carefully in order not to get trapped in a locally optimal lattice structure with crystal imperfections. In combinatorial optimization, we can define a similar process. This process can be formulated as the problem of finding (among a potentially very large number of solutions) a solution with minimal cost. Now, by establishing a correspondence between the cost function and the free energy, and between the solutions and physical states, we can introduce a solution method in the field of combinatorial optimization based on a simulation of the physical annealing

2.6 Simulated Annealing Algorithm (SAA)

process. The resulting method is called simulated annealing (SA). The salient features of the SA method may be summarized as follows.

- It could find a high-quality solution that does not strongly depend on the choice of the initial solution.
- It does not need a complicated mathematical model of the problem under study.
- It can start with any given solution and try to improve it. This feature could be utilized to improve a solution obtained from other suboptimal or heuristic methods.
- It has been theoretically proved to converge to the optimum solution [16].
- It does not need large computer memory.

In this chapter we propose an implementation of a Simulated Annealing Algorithm (SAA) to solve the Unit Commitment Problem (UCP). The combinatorial optimization subproblem of the UCP is solved using the proposed SAA and the Economic Dispatch Problem (EDP) is solved via a quadratic programming routine. Two different cooling schedules are implemented and compared. Three examples are solved to test the developed computer model.

2.6.1 Physical Concepts of Simulated Annealing [79]

Simulated annealing was independently introduced by Kirkpatrick, Gela, and Vecchi in 1982 and 1983 [18] and Cerny in 1985 [19]. The slow cooling in annealing is achieved by decreasing the temperature of the environment in steps. At each step the temperature is maintained constant for a period of time sufficient for the solid to reach thermal equilibrium. At equilibrium, the solid could have many configurations, each corresponding to different spins of the electrons and to a specific energy level.

At equilibrium the probability of a given configuration, P_{confg}, is given by the Boltzman distribution,

$$P_{confg} = K.e^{\left(-\frac{E_{confg}}{Cp}\right)},$$

where E_{confg} is the energy of the given configuration and K is a constant [20].

Reference [21] proposed a Monte Carlo method to simulate the process of reaching thermal equilibrium at a fixed temperature Cp. In this method, a randomly generated perturbation of the current configuration of the solid is applied so that a trial configuration is obtained. Let E_c and E_t denote the energy level of the current and trial configurations, respectively. If $E_c > E_t$, then a lower energy level has been reached, and the trial configuration is accepted and becomes the current configuration. On the other hand, if $E_c \leq E_t$ then the trial configuration is accepted as the current configuration with probability $e^{[(E_c-E_t)/Cp]}$. The process continues where a transition to a higher energy level configuration is not necessarily rejected.

Eventually thermal equilibrium is achieved after a large number of perturbations, where the probability of a configuration approaches a Boltzman distribution. By gradually decreasing Cp and repeating Metropolis simulation, new lower energy levels become achievable. As Cp approaches zero, the least energy configurations will have a positive probability of occurring.

2.6.2 Combinatorial Optimization Problems

By making an analogy between the annealing process and the optimization problem, a large class of combinatorial optimization problems can be solved following the same procedure of transition from an equilibrium state to another, reaching the minimum energy level of the system. This analogy can be set as follows [20].

- Solutions in the combinatorial optimization problem are equivalent to states (configurations) of the physical system.
- The cost of a solution is equivalent to the energy of a state.
- A control parameter Cp is introduced to play the role of the temperature in the annealing process.

In applying the SAA to solve the combinatorial optimization problem, the basic idea is to choose a feasible solution at random and then get a neighbor to this solution. A move to this neighbor is performed if either it has a lower objective function value or, in the case of a higher objective function value, if $\exp(-\Delta E/Cp) \geq U(0,1)$, where ΔE is the increase in the objective function value if we moved to the neighbor. The effect of decreasing Cp is that the probability of accepting an increase in the objective function value is decreased during the search.

The most important part in using the SAA is to have good rules for finding a diversified and intensified neighborhood so that a large amount of the solution space is explored. Another important issue is how to select the initial value of Cp and how it should be decreased during the search.

2.6.3 A General Simulated Annealing Algorithm [16–26]

A general SAA can be described as follows.

Step (0): Initialize the iteration count $k = 0$ and select the temperature $Cp = Cp_o$ to be sufficiently high such that the probability of accepting any solution is close to 1.

Step (1): Set an initial feasible solution = current solution X_i with corresponding objective function value E_i.

Step (2): If the equilibrium condition is satisfied, go to Step (5); else execute Steps (3) and (4).

2.6 Simulated Annealing Algorithm (SAA)

Step (3): Generate a trial solution X_j, as a neighbor to X_i. Let E_j be the corresponding objective function value.

Step (4): Acceptance test: If $E_j \leq E_i$: accept the trial solution, set $X_i = X_j$, and go to Step (2). Otherwise: if $\exp[(E_i - E_j)/Cp] \geq U(0,1)$ set $X_i = X_j$ and go to Step (2); else go to Step (2).

Step (5): If the stopping criterion is satisfied then stop; else decrease the temperature Cp^k and go to Step (2).

2.6.4 Cooling Schedules

A finite-time implementation of the SAA can be realized by generating homogeneous Markov chains of a finite length for a finite sequence of descending values of the control parameter Cp. To achieve this, one must specify a set of parameters that governs the convergence of the algorithm. These parameters form a cooling schedule. The parameters of the cooling schedules are as follows.

- An initial value of the control parameter.
- A decrement function for decreasing the control parameter.
- A final value of the control parameter specified by the stopping criterion.
- A finite length of each homogeneous Markov chain.

The search for adequate cooling schedules has been the subject of study in many papers [16, 23–26].

In this chapter, two cooling schedules are implemented, namely the polynomial-time and Kirk's cooling schedules. The description of these cooling schedules is presented in the following sections.

2.6.5 Polynomial-Time Cooling Schedule

This cooling schedule leads to a polynomial-time execution of the SAA, but it does not guarantee the convergence of the final cost, as obtained by the algorithm, to the optimal value. The different parameters of the cooling schedule are determined based on the statistics calculated during the search. In the following we describe these parameters [16, 23, 24].

2.6.5.1 Initial Value of the Control Parameter

The initial value of Cp, is obtained from the requirement that virtually all proposed trial solutions should be accepted. Assume that a sequence of m trials is generated at a certain value of Cp. Let m_1 denote the number of trials for which the objective function value does not exceed the respective current solution. Thus, $m_2 = m - m_1$ is the number of trials that result in an increasing cost.

It can be shown that the acceptance ratio X can be approximated by [16]:

$$X \approx (m_1 + m_2 \cdot \exp(-\overset{(+)}{\Delta f}/Cp))/(m_1 + m_2) \qquad (2.37)$$

where, $\overset{(+)}{\Delta f}$ is the average difference in cost over the m_2 cost-increasing trials. From which the new temperature Cp is

$$Cp = \overset{(+)}{\Delta f}/\ln(m_2/(m_2 \cdot X - m_1(1-X)) \qquad (2.38)$$

2.6.5.2 Decrement of the Control Parameter

The next value of the control parameter, Cp^{k+1}, is related to the current value Cp^k by the function [16]:

$$Cp^{k+1} = Cp^k/(1 + (Cp^k \cdot \ln(1+\delta)/3\sigma Cp^k) \qquad (2.39)$$

where σ is calculated during the search. Small values of δ lead to small decrements in Cp. Typical values of δ are between 0.1 and 0.5.

2.6.5.3 Final Value of the Control Parameter

Termination in the polynomial-time cooling schedule is based on an extrapolation of the expected average cost at the final value of the control parameter. Hence, the algorithm is terminated if for some value of k we have [16, 23, 24]

$$\left. \frac{Cp^k}{\langle f \rangle_\infty} \cdot \frac{\partial \langle f \rangle_{Cp}}{\partial Cp} \right|_{Cp=Cp_k} < \varepsilon \qquad (2.40)$$

where
$\langle f \rangle_\infty \approx \langle f \rangle_{Cp_o}$ is the average cost at initial value of control parameter Cp_o.
$\langle f \rangle_{Cp}$ is the average cost at the kth Markov chain.
$\left. \frac{\partial \langle f \rangle_{Cp}}{\partial Cp} \right|_{Cp=Cp_k}$ is the rate of change in the average cost at Cp^k.
ε is some small positive number. In our implementation $\varepsilon = 0.00001$.

2.6.5.4 The Length of Markov Chains

In [16] it is concluded that the decrement function of the control parameter, as given in Eq. 2.39, requires only a "small" number of trial solutions to rapidly approach the stationary distribution for a given next value of the control parameter. The word "small" can be specified as the number of transitions for which the algorithm has a sufficiently large probability of visiting at least a major part of the neighborhood of a given solution.

In general, a chain length of more than 100 transitions is reasonable [16]. In our implementation good results have been reached at a chain length of 150.

2.6.6 Kirk's Cooling Schedule

This cooling schedule was originally proposed by Kirkpatrick, Gelatt, and Vecchi [17]. It has been used in many applications of the SAA and is based on a number of conceptually simple empirical rules. The parameters of this cooling schedule are described in the following subsections [16, 18].

2.6.6.1 Initial Value of the Control Parameter

It is recommended to start with an arbitrary control parameter Cp [16]. If the percentage of the accepted trials solutions is close to 1, then this temperature is a satisfactory starting Cp. On the other hand, if this acceptance ratio is not close to 1, then Cp has to be increased iteratively until the required acceptance ratio is reached.

This can be achieved by starting off at a small positive value of Cp and multiplying it with a constant factor, larger than 1, until the corresponding value of the acceptance ratio, calculated from the generated transitions, is close to 1. In the physical system analogy, this corresponds to heating up the solid until all particles are randomly arranged in the liquid phase.

In our implementation, this procedure is accelerated by multiplying Cp by the reciprocal of the acceptance ratio.

2.6.6.2 Decrement of the Control Parameter

It is important to make "small" decrements in the values of the control parameter, to allow for a very slow cooling and consequently reaching an equilibrium at each value of the control parameter Cp. A frequently used decrement function is given by

$$Cp^{k+1} = \alpha.Cp^k, \quad k = 1, 2, \tag{2.41}$$

where α is a constant smaller than but close to 1. Typical values lie among 0.8 and 0.99.

2.6.6.3 Final Value of the Control Parameter

Execution of the algorithm is terminated if the value of the cost function of the solution obtained in the last trial of the Markov chain remains unchanged for a number of consecutive chains (Lm). In our implementation, Lm is taken as 500 chains.

2.6.6.4 Length of the Markov Chain

The length of Markov chains L^k is based on the requirement that equilibrium is to be restored at each value of Cp. This is achieved after the acceptance of at least some fixed number of transitions. However, because the transitions are accepted with decreasing probability, one would obtain $L^k \to \infty$ as $Cp^k \to 0$. Consequently, L^k is bounded by some constant L_{max} to avoid extremely long Markov chains for small values of Cp^k. In this work, the chain length is guided by the changes of the best solution that has been obtained thus far. The chain length is assumed equal to 150 unless the best solution changes. If so, the chain length is extended by another 150 iterations. In the next section, two algorithms based on tabu search methods are described.

2.7 Tabu Search Algorithm

Tabu search is a powerful optimization procedure that has been successfully applied to a number of combinatorial optimization problems [22–46]. It has the ability to avoid entrapment in local minima. TS employs a flexible memory system (in contrast to "memoryless" systems, such as SA and GAs, and rigid memory systems as in branch-and-bound). Specific attention is given to the short-term memory (STM) component of TS, which has provided solutions superior to the best obtained by other methods for a variety of problems [30]. Advanced TS procedures are also used for sophisticated problems. These procedures include, in addition to the STM, intermediate-term memory (ITM), long-term memory (LTM), and strategic oscillations (SO). In this section, two algorithms based on the TS method are discussed. The first algorithm uses the STM procedure, and the second algorithm is based on advanced TS procedures.

In general terms, TS is an iterative improvement procedure that starts from some initial feasible solution and attempts to determine a better solution in the manner of a greatest descent algorithm. However, TS is characterized by an ability to escape local optima (which usually cause simple descent algorithms to terminate) by using a short-term memory of recent solutions. Moreover, TS permits backtracking to previous solutions, which may ultimately lead, via a different direction, to better solutions [31].

The main two components of a TSA are the TL restrictions and the aspiration level (AV) of the solution associated with these restrictions. Discussion of these terms is presented in the following sections.

2.7.1 Tabu List Restrictions

TS may be viewed as a "metaheuristic" superimposed on another heuristic. The approach undertakes to surpass local optimality by a strategy of forbidding (or, more broadly, penalizing) certain moves. The purpose of classifying certain moves as forbidden – that is, "tabu" – is basically to prevent cycling. Moves that hold tabu status are generally a small fraction of those available, and a move loses its tabu status to become once again accessible after a relatively short time.

The choice of appropriate types of the tabu restrictions "list" depends on the problem under study. The elements of the TL are determined by a function that utilizes historical information from the search process, extending up to Z iterations in the past, where Z (TL size) can be fixed or variable depending on the application or the stage of the search.

The TL restrictions could be stated directly as a given change of variables (moves) or indirectly as a set of logical relationships or linear inequalities. Usage of these two approaches depends on the size of the TL for the problem under study.

A TL is managed by recording moves in the order in which they are made. Each time a new element is added to the "bottom" of a list, the oldest element on the list is dropped from the "top." The TL is designed to ensure the elimination of cycles of length equal to the TL size. Empirically [30], TL sizes that provide good results often grow with the size of the problem and stronger restrictions are generally coupled with smaller lists.

The way to identify a good TL size for a given problem class and choice of tabu restrictions is simply to watch for the occurrence of cycling when the size is too small and the deterioration in solution quality when the size is too large (caused by forbidding too many moves). The best sizes lie in an intermediate range between these extremes. In some applications a simple choice of Z in a range centered around 7 seems to be quite effective [28].

2.7.2 Aspiration Criteria

Another key issue of TS arises when the move under consideration has been found to be tabu. Associated with each entry in the TL there is a certain value for the evaluation function called the aspiration level. If the appropriate aspiration criteria are satisfied, the move will still be considered admissible in spite of the tabu classification. Roughly speaking, AV criteria are designed to override tabu status if a move is "good enough" with the compatibility of the goal of preventing the solution process from cycling [28]. Different forms of aspiration criteria are available. The one we use in this study is to override the tabu status if the tabu moves yield a solution that has a better evaluation function than the one obtained earlier for the same move.

2.7.3 Stopping Criteria

There may be several possible stopping conditions for the search. In our implementation we stop the search if either of the following two conditions is satisfied.

- The number of iterations performed since the best solution last changed is greater than a prespecified maximum number of iterations.
- The maximum allowable number of iterations is reached.

2.7.4 General Tabu Search Algorithm

In applying the TSA to solve a combinatorial optimization problem, the basic idea is to choose a feasible solution at random and then get a neighbor to this solution. A move to this neighbor is performed if either it does not belong to the TL or, in the case of being in the TL, it passes the AV test. During these search procedures the best solution is always updated and stored aside until the stopping criteria are satisfied.

A general TSA, based on the STM, for combinatorial optimization problems is described below, with this notation used in the algorithm:

X: The set of feasible solutions for a given problem
x: Current solution, $x \in X$
x'': Best solution reached
x': Best solution among a sample of trial solutions
$E(x)$: Evaluation function of solution x
$N(x)$: Set of neighborhood of $x \in X$ (trial solutions)
$S(x)$: Sample of neighborhood, of x, $S(x) \in N(x)$
$SS(x)$: Sorted sample in ascending order according to their evaluation functions, $E(x)$

Step (0): Set the TL as empty and the AV to be zero.
Step (1): Set iteration counter $K = 0$. Select an initial solution $x \in X$, and set $x'' = x$.
Step (2): Generate randomly a set of trial solutions $S(x) \in N(x)$ (neighbor to the current solution x) and sort them in an ascending order, to obtain $SS(x)$. Let x' be the best trial solution in the sorted set $SS(x)$ (the first in the sorted set).
Step (3): If $E(x') > E(x'')$, go to Step (4); else set the best solution $x'' = x'$ and go to Step (4).
Step (4): Perform the tabu test. If x' is not in the TL, then accept it as a current solution, set $x = x'$, update the TL and AV, and go to Step (6); else go to Step (5).
Step (5): Perform the AV test. If satisfied, then override the tabu state, set $x = x'$, update the AV, and go to Step (7); else go to Step (6).
Step (6): If the end of the $SS(x)$ is reached, go to Step (7); otherwise, let x' be the next solution in the $SS(x)$ and go to Step (3).
Step (7): Perform the termination test. If the stopping criterion is satisfied then stop; else set $K = K + 1$ and go to Step (2).

The main steps of the TSA are also shown in the flowchart of Fig. 2.1.
In the following section we describe the details of the general TSA.

2.8 The Genetic Algorithm (GA)

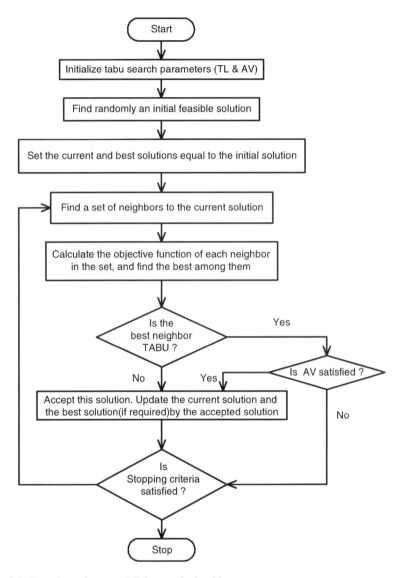

Fig. 2.1 Flowchart of a general Tabu search algorithm

2.8 The Genetic Algorithm (GA)

GAs are general-purpose search techniques based on principles inspired by the genetic and evolution mechanisms observed in natural systems and populations of living beings. Their basic principle is the maintenance of a population of solutions to a problem (genotypes) in the form of encoded individual information that evolves in time. A GA for a particular problem must have the following five components [39, 40, 47].

- A genetic representation for a potential solution to the problem.
- A way to create an initial population of potential solutions.
- An evaluation function that plays the role of the environment, rating solutions in terms of their "fitness."
- Genetic operators that alter the composition of children.
- Values for various parameters that the GA uses (population size, probabilities of applying genetic operators, etc.)

A genetic search starts with a randomly generated initial population within which each individual is evaluated by means of a fitness function. Individuals in this and subsequent generations are duplicated or eliminated according to their fitness values. Further generations are created by applying GA operators. This eventually leads to a generation of high-performing individuals [44].

2.8.1 Solution Coding

GAs require the natural parameter set of the optimization problem to be coded as a finite-length string over some finite alphabet. Coding is the most important point in applying the GA to solve any optimization problem. Coding could be in a real or binary form. Coded strings of solutions are called "chromosomes." A group of these solutions (chromosomes) is called a population.

2.8.2 Fitness Function

The fitness function is the second important issue in solving optimization problems using GAs. It is often necessary to map the underlying natural objective function to a fitness function through one or more mappings. The first mapping is done to transform the objective function into a maximization problem rather than minimization to suit the GA concepts of selecting the fittest chromosome that has the highest objective function.

A second important mapping is the scaling of the fitness function values. Scaling is an important step during the search procedures of the GA. This is done to keep appropriate levels of competition throughout a simulation. Without scaling, there is a tendency early on for a few superindividuals to dominate the selection process. Later, when the population has largely converged, competition among population members is weaker and simulation tends to wander. Thus, scaling is a useful process to prevent both the premature convergence of the algorithm and the random improvement that may occur in the late iterations of the algorithm. There are many methods for scaling such as linear, sigma truncation, and power law scaling [42]. Linear scaling is the most commonly used. In the sigma truncation method, population variance information to preprocess raw fitness values prior to scaling is used. It is called sigma (σ) truncation because of the use of population standard deviation information; a constant is subtracted from raw fitness values as follows:

$$\mathbf{f}' = \mathbf{f} - (\mathbf{f}' - c.\sigma) \qquad (2.41a)$$

In Eq. 2.41a the constant c is chosen as a reasonable multiple of the population standard deviation and negative results ($\mathbf{f}'<0$) are arbitrarily set to 0. Following sigma truncation, fitness scaling can proceed as described without the danger of negative results.

2.8.3 Genetic Algorithms Operators

There are usually three operators in a typical GA [44]. The first is the production operator which makes one or more copies of any individual that possesses a high fitness value; otherwise, the individual is eliminated from the solution pool.

The second operator is the recombination (also known as the "crossover") operator. This operator selects two individuals within the generation and a crossover site and performs a swapping operation of the string bits to the right-hand side of the crossover site of both individuals. The crossover operator serves two complementary search functions. First, it provides new points for further testing within the hyperplanes already represented in the population. Second, crossover introduces representatives of new hyperplanes into the population, which are not represented by either parent structure. Thus, the probability of a better performing offspring is greatly enhanced.

The third operator is the "mutation" operator. This operator acts as a background operator and is used to explore some of the unvisited points in the search space by randomly flipping a "bit" in a population of strings. Frequent application of this operator would lead to a completely random search, therefore a very low probability is usually assigned to its activation.

2.8.4 Constraint Handling (Repair Mechanism)

Constraint-handling techniques for the GAs can be grouped into a few categories [40]. One way is to generate a solution without considering the constraints but to include them with penalty factors in the fitness function. This method has been used previously [48–52].

Another category is based on the application of a special repair algorithm to correct any infeasible solution so generated.

The third approach concentrates on the use of special representation mappings (decoders) that guarantee (or at least increase the probability of) the generation of a feasible solution or the use of problem-specific operators that preserve feasibility of the solutions.

In our implementation, we are always generating solutions that satisfy the minimum up/down constraints. However, due to applying the crossover and mutation operations the load demand and/or the reserve constraints might be violated. A mechanism to restore feasibility is applied by randomly committing more units at the violated time periods and keeping the feasibility of the minimum up/down time constraints.

2.8.5 A General Genetic Algorithm

In applying the GAs to optimization problems, certain steps for simulating evolution must be performed. These are described as follows [39]:

Step (1): Initialize a population of chromosomes.
Step (2): Evaluate each chromosome in the population.
Step (3): Create new chromosomes by mating current chromosomes; apply mutation and recombination as the parent chromosomes mate.
Step (4): Delete members of the population to make room for the new chromosomes.
Step (5): Evaluate the new chromosomes and insert them into the population.
Step (6): If the termination criterion is satisfied, stop and return the best chromosomes; otherwise, go to Step (3).

2.9 Fuzzy Systems [78]

Human beings make tools for their use and also think to control the tools as they desire [53–60]. A feedback concept is very important to achieve control of the tools. As modern plants with many inputs and outputs become more and more complex, the description of a modern control system requires a large number of equations. Since about 1960 modern control theory has been developed to cope with the increased complexity of modern plants. The most recent developments may be said to be in the direction of optimal control of both deterministic and stochastic systems as well as the adaptive and learning control of time-variant complex systems. These developments have been accelerated by the digital computer.

Modern plants are designed for efficient analysis and production by human beings. We are now confronted by control of living cells, which are nonlinear, complex, time-variant, and "mysterious." They cannot easily be mastered by classical or control theory and even modern artificial intelligence (AI) employing

2.9 Fuzzy Systems

a powerful digital computer. Thus our problems are seen in terms of decision, management, and predictions. Solutions are seen in terms of faster access to more information and of increased aid in analyzing, understanding, and utilizing that information to discern its usefulness. These two elements, a large amount of information coupled with a large amount of uncertainty, taken together constitute the ground of many of our problems today: complexity. How do we manage to cope with complexity as well as we do and how could we manage to cope better? These are the reasons for introducing fuzzy notations, because the fuzzy sets method is very useful for handling uncertainties and essential for a human expert's knowledge acquisitions. First we have to know the meaning of fuzzy, which is vague or imprecise information.

Everyday language is one example of the ways in which vagueness is used and propagated such as driving a car or describing the weather and classifying a person's age, and so on. Therefore fuzzy is one method engineers use to describe the operation of a system by means of fuzzy variables and terms. To solve any control problem you might have a variable; this variable is a crisp set in the conventional control method (i.e., it has a definite value and a certain boundary). We define two groups as follows.

1. *Members:* Those that certainly belong in the set inside the boundary.
2. *Nonmembers:* Those that certainly don't.

But sometimes we have collections and categories with boundaries that seem vague and the transition from member to nonmember appears gradual rather than abrupt. These are what we call fuzzy sets. Thus fuzzy sets are a generalization of conventional set theory. Every fuzzy set can be represented by a membership function, and there is no unique membership. A membership function for any fuzzy set exhibits a continuous curve changing from 0 to 1 or vice versa, and this transition region represents a fuzzy boundary of the term.

In computer language we define fuzzy logic as a method of easily representing analog processes with continuous phenomena that are not easily broken down into discrete segments; the concepts involved are often difficult to model. In conclusion, we can use fuzzy when:

1. One or more of the control variables is continuous.
2. When a mathematical model of the process does not exist, or exists but is too difficult to encode.
3. When a mathematical model is too complex to be evaluated fast enough for real-time operation.
4. When a mathematical model involves too much memory on the designated chip architecture.
5. When an expert is available who can specify the rules underlying the system behavior and the fuzzy sets that represent the characteristics of each variable.
6. When a system has uncertainties in either the input or definition.

On the other hand, for systems where conventional control equations and methods are already optimal or entirely equal we should avoid using fuzzy logic.

One of the advantages of fuzzy logic is that we can implement systems too complex, too nonlinear, or with too much uncertainty to implement using traditional techniques. We can also implement and modify systems more quickly and squeeze additional capability from existing designs. Finally it is simple to describe and verify. Before we introduce fuzzy models we need some definitions.

- *Singletons*:
  A deterministic term or value, for example: male and female, dead and alive, 80°C, 30 Kg. These deterministic words and numerical values have neither flexibility nor intervals. So a numerical value to be substituted into a mathematical equation representing a scientific law is a singleton.
- *Fuzzy number*:
  A fuzzy linguistic term that includes an imprecise numerical value, for example, around 80°C, bigger than 25.
- *Fuzzy set*:
  A fuzzy linguistic term that can be regarded as a set of singletons where the grades are not only [1] but also range from [0 to 1]. It can also be a set that allows partial membership states. Ordinary or crisp sets have only two membership states: inclusion and exclusion (member or nonmember). Fuzzy sets allow degrees of membership as well. Fuzzy sets are defined by labels and membership functions, and every fuzzy set has an infinite number of membership functions (μFs) that may represent it.
- *Fuzzy linguistic terms*:
  Elements of which are ordered, are fuzzy intervals, and the membership function is a bandwidth of this fuzzy linguistic term. Elements of fuzzy linguistic terms such as "robust gentleman" or "beautiful lady" are discrete and also disordered. This type of term cannot be defined by a continuous membership function, but defined by vectors.
- *A characteristic function of*:
  Singletons, an interval and a fuzzy linguistic term are given by:

  (a) Singleton.
  (b) An interval.
  (c) Fuzzy linguistic term.

- *Control variable*:
  A variable that appears in the premise of a rule and controls the state of the solution variables.
- *Defuzzification*:
  The process of converting an output fuzzy set for a solution variable into a single value that can be used as an output.
- *Overlap*:
  The degree to which the domain of one fuzzy set overlaps that of another.
- *Solution fuzzy set*:
  A temporary fuzzy set created by the fuzzy model to resolve the value of a corresponding solution variable. When all the rules have been fired the solution fuzzy set is defuzzified into the actual solution variable.

2.9 Fuzzy Systems

- *Solution variable*:
  The variable whose value the fuzzy logic system is meant to find.
- *Fuzzy model*:
  The components of conventional and fuzzy systems are quite alike, differing mainly in that fuzzy systems contain "fuzzifers" which convert inputs into their fuzzy representations and "defuzzifiers" which convert the output of the fuzzy process logic into "crisp" (numerically precise) solution variables.

In a fuzzy system, the values of a fuzzified input execute all the values in the knowledge repository that have the fuzzified input as part of their premise. This process generates a new fuzzy set representing each output or solution variable. Defuzzification creates a value for the output variable from that new fuzzy set. For physical systems, the output value is often used to adjust the setting of an actuator that in turn adjusts the states of the physical systems. The change is picked up by the sensors, and the entire process starts again. Finally we can say that there are four steps to follow to design a fuzzy model.

1. First step: "Define the model function and operational characteristics"
   The goal is to establish the architectural characteristics of the system and also to define the specific operating properties of the proposed fuzzy system. The fuzzy system designer's task lies in defining what information (datapoint) flows into the system, what basic information is performed on the data, and what data elements are output from the system. Even if the designer lacks a mathematical model of the system process, it is essential that she have a deep understanding of these three phenomena. This step is also the time to define exactly where the fuzzy subsystem fits into the total system architecture, which provides a clear picture of how inputs and outputs flow to and from the subsystem. The designer can then estimate the number and ranges of input and output that will be required. It also reinforces the input-process-output design step.
2. The second step: "Define the control surfaces"
   Each control and solution variable in the fuzzy model is decomposed into a set of a fuzzy regions. These regions are given a unique name, called labels, within the domain of the variable. Finally a fuzzy set that semantically represents the concept associated with the label is created. Some rules of thumb help in defining fuzzy sets.

   (a) First, the number of labels associated with a variable should generally be an odd number from [56–58].
   (b) Second, each label should overlap somewhat with its neighbors. To get a smooth stable surface fuzzy controller, the overlap should be between 10% and 50% of the neighboring space. And the sum of vertical points of the overlap should always be less than one.
   (c) Third, the density of the fuzzy sets should be the highest around the optimal control point of the system, and this should be out as the distance from that point increases.

3. The third step: "Define the behavior of the control surfaces"
   This step involves writing the rules that tie the input values to the output model properties. These rules are expressed in natural language with syntax such as

   IF<fuzzy proposition>, then<fuzzy proposition>

   that is, IF, THEN rule, where fuzzy proposition are "x is y" or "x is not y" x is a scalar variable, and y is a fuzzy set associated with that variable. Generally the number of rules a system requires is simply related to the number of control variables.

4. The fourth step: "Select a method of defuzzification"
   It is a way to convert an output fuzzy set into a crisp solution variable. The two most common ways are:

   - The composite maximum.
   - Calculation of the concentration.

   Once the fuzzy model has been constructed, the process of solution and protocycling begins. The model is compared against known test cases to validate the results. When the results are not as desired, changes are made either to the fuzzy set descriptions or to the mappings encoded in the rules.

5. Fuzzy Sets and Membership
   Fuzzy set theory was developed to improve the oversimplified model, thereby developing a more robust and flexible model in order solve real-world complex systems involving human aspects. Furthermore, it helps the decision maker not only to consider the existing alternatives under given constraints (optimize a given system), but also to develop new alternatives (design a system). Fuzzy set theory has been applied in many fields, such as operations research, management science, control theory, artificial intelligence/expert systems, and human behavior, among others.

6. Membership Functions
   A classical (crisp or hard) set is a collection of distinct objects, defined in such a manner as to separate the elements of a given universe of discourse into two groups: those that belong (members), and those that do not belong (nonmembers). The transition of an element between membership and nonmembership in a given set in the universe is abrupt and well defined. The crisp set can be defined by the so-called characteristic function, for example, let U be a universe of discourse, the characteristic function of a crisp set.

2.9.1 Basic Terminology and Definition

Let X be a classical set of objects, called the universe, whose generic elements are denoted by x. The membership in a crisp subset of X is often viewed as the characteristic function μ_A from X to $\{0, 1\}$ such that:

$$\mu_A(x) = 1 \text{ if and only if } x \in A$$
$$= 0 \text{ otherwise} \quad (2.42)$$

where $\{0, 1\}$ is called a valuation set.

2.9 Fuzzy Systems

If the valuation set is allowed to be the real interval [0, 1], $\tilde{A}$ is called a fuzzy set as proposed by Zadeh. $\mu_A(x)$ is the degree of membership of x in A. The closer the value of $\mu_A(x)$ is to 1, the more x belongs to A. Therefore, $\tilde{A}$ is completely characterized by the set of ordered pairs:

$$\tilde{A} = \{(x, \mu_A(x)) | x \in X\} \tag{2.43}$$

It is worth noting that the characteristic function can be either a membership function or a possibility distribution. In this study, if the membership function is preferred, then the characteristic function is denoted as $\mu_A(x)$. On the other hand, if the possibility distribution is preferred, the characteristic function is specified as $\pi(x)$. When X is a finite set $\{x_1, x_2, ..., x_n\}$, a fuzzy set A is then expressed as

$$\tilde{A} = \mu_A(x_1)/x_1 + ... + \mu_A(x_n)/x_n = \sum_i \mu_A(x_i)/x_i \tag{2.44}$$

When X is not a finite set, A then can be written as

$$A = \int_X \mu_A(x)/x \tag{2.45}$$

Sometimes, we might only need objects of a fuzzy set (but not its characteristic function), in order to transfer a fuzzy set. To do so, we need two concepts: support and α-level cut.

2.9.2 Support of Fuzzy Set

The support of a fuzzy set A is the crisp set of all $x \in U$ such that $(x) > 0$. That is,

$$\text{supp}(A) = \{x \in U | \mu_A > 0\} \tag{2.46}$$

- **α-Level Set (α-Cut)**

  The α-level set (α-cut) of a fuzzy set A is a crisp subset of X and is shown in Fig. 2.2. An α-cut of a fuzzy set $\tilde{A}$ is a crisp set A which contains all the elements of the universe U that have a membership grade in $\tilde{A}$ greater than or equal to α. That is,

$$A_\alpha = \{x | \mu_A(x) \geq \alpha \text{ and } x \in X\} \tag{2.47}$$

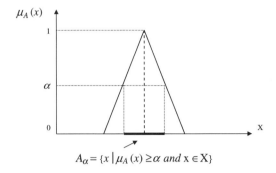

Fig. 2.2 Membership and α-cut

If $A_\alpha = \{x | \mu_A(x) > \alpha\}$, then A_α is called a strong α-cut of a given fuzzy set A is called a level set of A. That is,

$$\prod_A = \{\alpha | \mu_A(x) = \alpha, \text{ for some } x \in \cup\} \tag{2.48}$$

2.9.3 Normality

A fuzzy set A is normal if and only if $\text{Sup}_x \mu_A(x) = 1$; that is, the supreme of $\mu_A(x)$ over X is unity. A fuzzy set is subnormal if it is not normal. A nonempty subnormal fuzzy set can be normalized by dividing each $\mu_A(x)$ by the factor $\text{Sup}_x \mu_A(x)$. (A fuzzy set is empty if and only if $\mu_A(x) = 0$ for $\forall x \in X) \forall x \in X$.

2.9.4 Convexity and Concavity

A fuzzy set A in X is convex if and only if for every pair of point x^1 and x^2 in X, the membership function of A satisfies the inequality:

$$\mu_A(\partial x^1 + (1 - \partial) x^2) \geq \min(\mu_A(x^1), \mu_A(x^2)) \tag{2.49}$$

where $\partial \in [0, 1]$ (see Fig. 2.3). Alternatively, a fuzzy set is convex if all α-level sets are convex.

Dually, A is concave if its complement A^c is convex. It is easy to show that if A and B are convex, so is $A \cap B$. Dually, if A and B are concave, so is $A \cup B$.

2.9.5 Basic Operations [53]

This section is a summary of some basic set-theoretic operations useful in fuzzy mathematical programming and fuzzy multiple objective decision making. These operations are based on the definitions from Bellman and Zadeh.

2.9 Fuzzy Systems

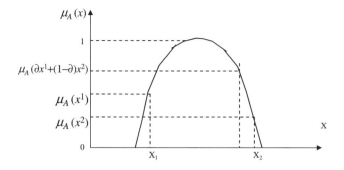

Fig. 2.3 A convex fuzzy set

2.9.5.1 Inclusion

Let A and B be two fuzzy subsets of X; then A is included in B if and only if:

$$\mu_A(x) \leq \mu_B(x) \text{ for } \forall x \in X \tag{2.50}$$

2.9.5.2 Equality

A and B are called equal if and only if:

$$\mu_A(x) = \mu_B(x) \text{ for } \forall x \in X \tag{2.51}$$

2.9.5.3 Complementation

A and B are complementary if and only if:

$$\mu_A(x) = 1 - \mu_B(x) \text{ for } \forall x \in X \tag{2.52}$$

2.9.5.4 Intersection

The intersection of A and B may be denoted by $A \cap B$ which is the largest fuzzy subset contained in both fuzzy subsets A and B. When the min operator is used to express the logic "and," its corresponding membership is then characterized by

$$\begin{aligned}\mu_{A \cap B}(x) &= \min(\mu_A(x), \mu_B(x)) \text{ for } \forall x \in X \\ &= \mu_A(x) \wedge \mu_B(x)\end{aligned} \tag{2.53}$$

where $\wedge$ is a conjunction.

2.9.5.5 Union

The union $(A \cup B)$ of A and B is dual to the notion of intersection. Thus, the union of A and B is defined as the smallest fuzzy set containing both A and B.

The membership function of $A \cup B$ is given by

$$\mu_{A \cup B}(x) = \max(\mu_A(x), \mu_B(x)) \text{ for } \forall x \in X$$
$$= \mu_A(x) \vee \mu_B(x) \qquad (2.54)$$

2.9.5.6 Algebraic Product

The algebraic product AB of A and B is characterized by the following membership function,

$$\mu_{AB}(x) = \mu_A(x) \mu_B(x) \text{ for } \forall x \in X \qquad (2.55)$$

2.9.5.7 Algebraic Sum

The algebraic sum $A \oplus B$ of A and B is characterized by the following membership function,

$$\mu_{A \oplus B}(x) = \mu_A(x) + \mu_B(x) - \mu_A(x) \mu_B(x) \qquad (2.56)$$

2.9.5.8 Difference

The difference $A - B$ of A and B is characterized by

$$\mu_{A \cap B^c}(x) = \min(\mu_A(x), \mu_{B^c}(x)) \qquad (2.57)$$

2.9.5.9 Fuzzy Arithmetic

(a) **Addition of Fuzzy Number**
The addition of X and Y can be calculated by using the α-level cut and max–min convolution.

- **α-level cut.** Using the concept of confidence intervals, the α-level sets of X and Y are $X_\alpha = [X_\alpha^L, X_\alpha^U]$ and $Y_\alpha = [Y_\alpha^L, Y_\alpha^U]$ where the result Z of the addition is:

$$Z_\alpha = X_\alpha(+)Y_\alpha = [X_\alpha^L + Y_\alpha^L, X_\alpha^U + Y_\alpha^U] \qquad (2.58)$$

2.9 Fuzzy Systems

for every $\alpha \in [0, 1]$.
- **Max–Min Convolution.** The addition of the fuzzy number X and Y is represented as

$$Z(z) = \max_{z=x+y} \left[\min[\mu_X(x), \mu_Y(y)]\right] \tag{2.59}$$

(b) Subtraction of Fuzzy Numbers

- **α-level cut.** The subtraction of the fuzzy numbers X and Y in the α-level cut representation is:

$$Z_\alpha = X_\alpha(-)Y_\alpha = \left[X_\alpha^L - Y_\alpha^U, X_\alpha^U - Y_\alpha^L\right] \text{ for every } \alpha \in [0, 1]. \tag{2.60}$$

- **Max–Min Convolution.** The subtraction of the fuzzy number X and Y is represented as

$$\mu_Z(Z) = \max_{z=x-y} \{[\mu_x(x), \mu_Y(y)]\}$$
$$\max_{z=x+y} \{[\mu_x(x), \mu_Y(-y)]\}$$
$$\max_{z=x+y} \{[\mu_x(x), \mu_{-Y}(y)]\} \tag{2.61}$$

(c) Multiplication of Fuzzy Numbers

- **α-level cut.** The multiplication of the fuzzy numbers X and Y in the α-level cut representation is

$$Z_\alpha = X_\alpha(.)Y_\alpha = [[X_\alpha^L y_\alpha^L X_\alpha^U Y_\alpha^U]] \tag{2.62}$$

- for every $\alpha \in [0, 1]$.
- **Max–Min Convolution.** The multiplication of the fuzzy number X and Y is represented by Kaufmann and Gupta in the following procedure as

  1. First, find Z1 (the peak of the fuzzy number Z) such that $\mu_Z(z^1) = 1$; then we calculate the left and right legs.
  2. The left leg of $\mu_Z(z)$ is defined as

$$\mu_z(z) = \max_{xy \leq z} \{\min[\mu_x(x), \mu_Y(y)]\} \tag{2.63}$$

  3. The right leg of $\mu_Z(z)$ is defined as

$$\mu_z(z) = \max_{xy \geq z} \{\min[\mu_x(x), \mu_Y(y)]\} \tag{2.64}$$

(d) **Division of Fuzzy Numbers**

$$\alpha - \text{level cut.} \ Z_\alpha = X_\alpha(:)Y_\alpha = [[x_\alpha^L/y_\alpha^U, x_\alpha^U/y_\alpha^L]] \quad (2.65)$$

- **Max–Min Convolution.** As defined earlier we must find the peak, then the left and right legs.
    1. The peak $Z = X \ (:) \ Y$ is used.
    2. The left leg is presented as

$$\mu_z(z) = \max_{x/y \leq z} \{\min[\mu_x(x), \mu_Y(y)]\}$$

$$\max_{xy \leq z} \{\min[\mu_x(x), \mu_Y(1/y)]\}$$

$$\max_{xy \leq z} \left\{\min[\mu_x(x), \mu_{1/Y}(y)]\right\} \quad (2.66)$$

    3. The right leg is presented as

$$\mu_z(z) = \max_{x/y \geq z} \{\min[\mu_x(x), \mu_Y(y)]\}$$

$$\max_{xy \geq z} \{\min[\mu_x(x), \mu_Y(1/y)]\}$$

$$\max_{xy \geq z} \left\{\min[\mu_x(x), \mu_{1/Y}(y)]\right\} \quad (2.67)$$

2.9.5.10 *LR*-Type Fuzzy Number

A fuzzy number is defined to be of the *LR* type if there are reference functions *L* and *R* and positive scalars as shown in Fig. 2.4: α(left spread), β(right spread), and *m* (mean) such that [53]:

$$\mu_M(x) = \begin{cases} L\left(\dfrac{m-x}{\alpha}\right) & \text{for} \quad x \leq m \\ R\left(\dfrac{x-m}{\beta}\right) & \text{for} \quad x \geq m \end{cases} \quad (2.68)$$

As the spread increases, *M* becomes fuzzier and fuzzier. Symbolically we write:

$$M = (m, \alpha, \beta)_{LR} \quad (2.69)$$

2.10 Particle Swarm Optimization (PSO) Algorithm

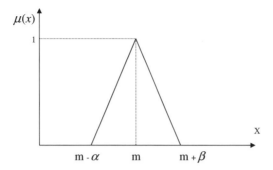

Fig. 2.4 Triangular membership

2.9.5.11 Interval Arithmetic

Interval arithmetic is normally used with uncertain data obtained from different instruments if we enclose those values obtained in a closed interval on the real line R; that is, this uncertain value is inside an interval of confidence $R, x \in [a_1, a_2]$, where $a_1 \leq a_2$.

2.9.5.12 Triangular and Trapezoidal Fuzzy Numbers

The triangular and trapezoidal fuzzy number is considered one of the most important and useful tools in solving possibility mathematical programming problems. Tables 2.1 and 2.2 show all the formulae used in the *L–R* representation of the fuzzy number and interval arithmetic methods.

2.10 Particle Swarm Optimization (PSO) Algorithm

One of the most difficult parts encountered in practical engineering design optimization is handling constraints [61–76]. Real-world limitations frequently introduce multiple nonlinear and nontrivial constraints in engineering design problems [77]. Constraints often limit the feasible solutions to a small subset of the design space. A general engineering optimization problem can be defined as

Minimize $f(X)$, $X = \{x_1, x_2, \ldots, x_n\} \in R$
Subject to $g_i(X) \leq 0, i = 1, 2, \ldots, p$

$$h_i(X) = 0, \quad i = 1, 2, \ldots, m$$

Where

$$x_i^{(L)} \leq x_i \leq x_i^{(U)}, \quad i = 1, 2, \ldots, n$$

Table 2.1 Fuzzy arithmetic on triangular L–R representation of fuzzy numbers $X = (x, \alpha, \beta)$ & $Y = (y, r, \delta)$

Image of Y: $\quad -Y = (-y, \delta, r) - Y = (-y, \delta, r)$
Inverse of Y: $\quad Y^{-1} = (y^{-1}, \delta y^{-2}, r y^{-2})$
Addition: $\quad X(+)Y = (x + y, \alpha + r, \beta + \delta)$
Subtraction: $\quad X(-)Y = X(+) - Y = (x - y, \alpha + \delta, \beta + r)$
Multiplication:
$\quad X>0, Y>0 : X(\bullet)Y = (xy, xr + y\alpha, x\delta + y\beta)$
$\quad X<0, Y>0 : X(\bullet)Y = (xy, y\alpha - x\delta, y\beta - xr)$
$\quad X<0, Y<0 : X(\bullet)Y = (xy, -x\delta - y\beta, -xr - y\alpha)$
Scalar multiplication:
$\quad a>0, a \in R : a(\bullet)X = (ax, a\alpha, a\beta)$
$\quad a<0, a \in R : a(\bullet)X = (ax, -a\beta, -a\alpha)$
Division:
$\quad X>0, Y>0 : X(:)Y = \left(x/y, (x\delta + y\alpha)/y^2, (xr + y\beta)/y^2\right)$
$\quad X<0, Y>0 : X(:)Y = \left(x/y, (y\alpha - xr)/y^2, (y\beta - x\delta)/y^2\right)$
$\quad X<0, Y<0 : X(:)Y = \left(x/y, (-xr - y\beta)/y^2, (-x\delta - y\alpha)/y^2\right)$

Table 2.2 Fuzzy interval arithmetic on triangular fuzzy numbers $X = (x^m, x^p, x^o)$ & $Y = (y^m, y^p, y^o)$

Image of Y: $-Y = (-y^m, -y^o, -y^p)$
Inverse of Y: $Y^{-1} = (1/y^m, 1/y^o, 1/y^p)$
Addition: $X(+)Y = (x^m + y^m, x^p + y^p, x^o + y^o)$
Subtraction: $X(-)Y = X(+) - Y = (x^m - y^m, x^p - y^o, x^o - y^p)$
Multiplication:
$\quad X>0, Y>0 : X(\bullet)Y = (x^m y^m, x^p y^p, x^o y^o)$
$\quad X<0, Y>0 : X(\bullet)Y = (x^m y^m, x^p y^o, x^o y^p)$
$\quad X<0, Y<0 : X(\bullet)Y = (x^m y^m, x^o y^o, x^p y^p)$
Scalar multiplication:
$\quad a>0, a \in R : a(\bullet)X = (ax^m, ax^p, ax^o)$
$\quad a<0, a \in R : a(\bullet)X = (ax^m, ax^o, ax^p)$
Division:
$\quad X>0, Y>0 : X(:)Y = (x^m/y^m, x^p/y^o, x^o/y^p)$
$\quad X<0, Y>0 : X(:)Y = (x^m/y^m, x^o/y^o, x^p/y^p)$
$\quad X<0, Y<0 : X(:)Y = (x^m/y^m, x^o/y^p, x^p/y^o)$

Due to the complexity and unpredictability of constraints, a general deterministic solution is difficult to find. In recent years, several evolutionary algorithms have been proposed for constrained engineering optimization problems and many methods have been proposed for handling constraints, the key point of the optimization process. Recently a new evolutionary computational technique, called particle swarm optimization (PSO) has been proposed and introduced [61–64].

2.10 Particle Swarm Optimization (PSO) Algorithm

Particle swarm optimization is a population-based stochastic optimization technique developed in [65], inspired by the social behavior of flocks of birds or schools of fish [65]. PSO shares many similarities with evolutionary computation techniques such as genetic algorithms. The system is initialized with a population of random feasible solutions and searches for optima by updating generations. However, unlike GA, PSO has no evolution operators such as crossover and mutation. The PSO algorithm has also been demonstrated to perform well on genetic algorithm test functions [66].

In PSO, the potential solutions, called particles, fly through the problem space by following the current optimum particles [67]. The particles change their positions by flying around in a multidimensional search space until a relatively unchanged position has been encountered, or until computational limitations are exceeded [68]. In a social science context, a PSO system combines a social-only model and a cognition-only model. The social-only component suggests that individuals ignore their own experience and fine-tune their behavior according to the successful beliefs of the individual in the neighborhood. On the other hand, the cognition-only component treats individuals as isolated beings. A particle changes its position using these models.

Each particle keeps track of its co-ordinates in the problem space, which is associated with the best solution, fitness, it has achieved so far. The fitness value is also stored. This value is called $p$best. Another best value that is tracked by the particle swarm optimizer is the best value obtained thus far by any particle in the neighbors of the particle. This location is called $l$best. When a particle takes the whole population as its topological neighbors, the best value is a global best and is called $g$best.

The concept of the PSO consists of, at each timestep, changing the velocity of (accelerating) each particle toward its $p$best and $l$best locations (local version of PSO). Acceleration is weighted by a random term, with separate random numbers being generated for acceleration toward $p$best and $l$best locations. In the past few years, PSO has been successfully applied in many research and application areas. It has been demonstrated that PSO gets better results in a faster and cheaper way compared with other methods. As shown in the literature, the PSO algorithm has been successfully applied to various problems [68–75].

Another reason that PSO is attractive is that there are few parameters to adjust. One version, with slight variations, works well in a wide variety of applications. Particle swarm optimization has been used for approaches that can be used across a wide range of applications, as well as for specific applications focused on a specific requirement. Many advantages of PSO over other traditional optimization techniques can be summarized as follows [68].

(a) PSO is a population-based search algorithm (i.e., PSO has implicit parallelism). This property ensures that PSO is less susceptible to being trapped on local minima.
(b) PSO uses payoff (performance index or objective function) information to guide the search in the problem space. Therefore, PSO can easily deal with

nondifferentiable objective functions. In addition, this property relieves PSO of assumptions and approximations, which are often required by traditional optimization models.
(c) PSO uses probabilistic transition rules and not deterministic rules. Hence, PSO is a kind of stochastic optimization algorithm that can search a complicated and uncertain area. This makes PSO more flexible and robust than conventional methods.
(d) Unlike the genetic and other heuristic algorithms, PSO has the flexibility to control the balance between global and local exploration of the search space. This unique feature of a PSO overcomes the premature convergence problem and enhances search capability.
(e) Unlike traditional methods, the solution quality of the proposed approach does not depend on the initial population. Starting anywhere in the search space, the algorithm ensures convergence to the optimal solution.

2.11 Basic Fundamentals of PSO Algorithm

The basic fundamentals of the PSO technique are stated and defined as follows [68].

1. **Particle $X(i)$**: A candidate solution represented by a k-dimensional real-valued vector, where k is the number of optimized parameters; at iteration i, the jth particle $X(i, j)$ can be described as

$$X_j(i) = [x_{j,1}(i); x_{j,2}(i); \ldots; x_{j,k}(i); \ldots \ldots; x_{j,d}(i)] \quad (2.70)$$

Where: xs are the optimized parameters
$x_k(i,j)$ is the kth optimized parameter in the jth candidate solution
d represents the number of control variables

2. **Population**: This is a set of n particles at iteration i.

$$\text{pop}(i) = [X_1(i), X_2(i), \ldots \ldots \ldots X_n(i)]^T \quad (2.71)$$

where n represents the number of candidate solutions.

3. **Swarm**: This is an apparently disorganized population of moving particles that tend to cluster together and each particle seems to be moving in a random direction.

4. **Particle velocity $V(i)$**: The velocity of the moving particles represented by a d-dimensional real-valued vector; at iteration i, the jth particle $V_j(i)$ can be described as

$$V_j(i) = [v_{j,1}(i); v_{j,2}(i); \ldots; v_{j,k}(i); \ldots \ldots; v_{j,d}(i)] \quad (2.72)$$

where: $v_{j,k}(i)$ is the velocity component of the jth particle with respect to the kth dimension.

5. **Inertia weight $w(i)$**: This is a control parameter, used to control the impact of the previous velocity on the current velocity. Hence, it influences the tradeoff between the global and local exploration abilities of the particles. For initial stages of the search process, a large inertia weight to enhance global exploration is recommended whereas it should be reduced at the last stages for better local exploration. Therefore, the inertia factor decreases linearly from about 0.9 to 0.4 during a run. In general, this factor is set according to Eq. (2.73):

$$W = W_{\max} - \frac{(W_{\max} - W_{\min})}{iter_{\max}} * iter \quad (2.73)$$

where $iter_{\max}$ is the maximum number of iterations and $iter$ is the current number of iterations.

6. **Individual best $X^*(i)$**: During the movement of a particle through the search space, it compares its fitness value at the current position to the best fitness value it has ever reached at any iteration up to the current iteration. The best position that is associated with the best fitness encountered thus far is called the individual best $X^*(i)$. For each particle in the swarm, $X^*(i)$ can be determined and updated during the search. For the jth particle, individual best can be expressed as

$$X_j^*(i) = [x_{j1}^*(i), x_{j,2}^*(i), \ldots, x_{j,d}^*(i)]^T \quad (2.74)$$

In a minimization problem with only one objective function f, the individual best of the jth particle $X_j^*(i)$ is updated whenever $f(X_j^*(i)) < f(X_j^*(i-1))$. Otherwise, the individual best solution of the jth particle will be kept as in the previous iteration.

7. **Global best $X^{**}(t)$**: This is the best position among all of the individual best positions achieved thus far.

8. **Stopping criteria**: The search process will be terminated whenever one of the following criteria is satisfied.

   - The number of iterations since the last change of the best solution is greater than a prespecified number.
   - The number of iterations reaches the maximum allowable number.

The particle velocity in the kth dimension is limited by some maximum value, $v_k^{\max} v_k^{\max}$. This limit enhances local exploration of the problem space and it realistically simulates the incremental changes of human learning. The maximum velocity in the kth dimension is characterized by the range of the kth optimized parameter and given by

$$V_k^{\max} = \frac{(x_k^{\max} - x_k^{\max})}{N} \quad (2.75)$$

where N is a chosen number of intervals in the kth dimension.

2.11.1 General PSO Algorithm

In a PSO algorithm, the population has n particles that represent candidate solutions. Each particle is a k-dimensional real-valued vector, where k is the number of the optimized parameters [69]. Therefore, each optimized parameter represents a dimension of the problem space. The PSO technique steps can be described as follows.

Step 1: **Initialization**: Set $i = 0$ and generate random n particles $\{X_j(0), j = 1, 2, \ldots, n\}$. Each particle is considered to be a solution for the problem and it can be described as $X_j(0) = [x_{i,1}(0); x_{i,2}(0); \ldots, x_{i,k}(0)]$. Each control variable has a range $[x_{\min}, x_{\max}]$. Each particle in the initial population is evaluated using the objective function f. If the candidate solution is a feasible solution (i.e., all problem constraints have been met), then go to Step 2; else repeat this step.

Step 2: **Counter updating**: Update the counter $i = i + 1$.

Step 3: **Compute the objective function.**

Step 4: **Velocity updating**: Using the global best and individual best, the jth particle velocity in the kth dimension in this study (integer problem) is updated according to the following equation.

$$V(k, j, i+1) = w^* V(k, j, i) + C_1^* rand^* (pbestx(j, k) - x(k, j, i)) \\ + C_2^* rand^* (gbestx(k) - x(k, j, i)) \quad (2.76)$$

Where i is the iteration number
j is the particle number
k is the kth control variable
w is the inertia weighting factor
c_1, c_2 are acceleration constants
$rand\,()$ is a uniform random value in the range of [0,1]
$V(k,j,i)$ is the velocity of particle j at iteration i
$x(k,j,i)$ is the current position of particle j at iteration j

Then, check the velocity limits. If the velocity violated its limit, set it at its proper limit. The second term of the above equation represents the cognitive part of the PSO where the particle changes its velocity based on its own thinking and memory. The third term represents the social part of PSO where the particle changes its velocity based on the social–psychological adaptation of knowledge.

Step 5: **Position updating**: Based on the updated velocity, each particle changes its position according to Eq. 2.77 (Fig. 2.5).

$$x(k, j, i+1) = x(k, j-1, i) + v(k, j, i) \quad (2.77)$$

2.11 Basic Fundamentals of PSO Algorithm

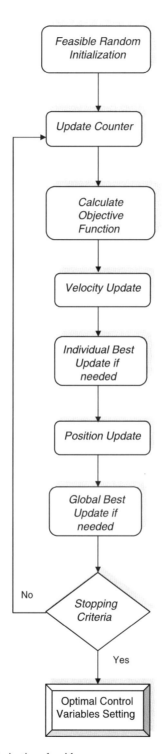

Fig. 2.5 Particle swarm optimization algorithm

Step 6: **Individual best updating**: Each particle is evaluated and updated according to the update position.
Step 7: Search for the minimum value in the individual best where its solution has ever been reached in every iteration and considered it as the minimum.
Step 8: **Stopping criteria**: If one of the stopping criteria is satisfied, then stop; otherwise go to Step 2.

References

1. El-Hawary, M.E., Christensen, G.S.: Optimal Economic Operation of Electric Power Systems. Academic, New York (1979)
2. Horst, R., Pardalos, P.M. (eds.): Handbook of Global Optimization. Kluwer, Netherlands (1995)
3. Kuo, B.C.: Automatic Control Systems, 4th edn. Prentice-Hall, Englewood Cliffs (1982)
4. Nemhauser, G.L., Rinnooy Kan, A.H.G., Todd, M.J. (eds.): Optimization. Elsevier Science, Netherlands (1989)
5. Wolfe, M.A.: Numerical Methods for Unconstrained Optimization: An Introduction. Van Nostrand Reinhold, New York (1978)
6. Zill, D.G., Cullen, M.R.: Advanced Engineering Mathematics. PWS, Boston (1992)
7. Porter, W.A.: Modern Foundations of Systems Engineering. Macmillan, New York (1966)
8. Luenberger, D.G.: Optimization by Vector Space Methods. Wiley, New York (1969)
9. Sage, A.: Optimum System Controls. Prentice-Hall, Englewood Cliffs (1968)
10. Sage, A.P., White, C.C.: Optimum Systems Control. Prentice-Hall, Englewood Cliffs (1977)
11. Bryson, A.E., Ho, Y.C.: Applied Optimal Control. Wiley, New York (1975)
12. Rao, S.S.: Optimization Theory and Applications. Wiley Eastern, New Delhi (1979)
13. Leitmann, G.: The Calculus of Variations and Optimal Control. Plenum Press, New York (1981)
14. Kirk, D.E.: Optimal Control Theory: An Introduction. Prentice-Hall, Englewood Cliffs (1970)
15. Narici, B.: Functional Analysis. Academic, New York (1966)
16. Aarts, E., Korst, J.: Simulated Annealing and Boltzman Machines: A Stochastic Approach to Combinatorial Optimization and Neural Computing. Wiley, New York (1989)
17. Kirkpatrick, S., Gelatt, C.D., Vecchi, M.P.: Optimization by simulated annealing. Science **220**, 671–680 (1983)
18. Cerny, V.: Thermodynamical approach to the traveling salesman problem: an efficient simulation algorithm. J. Optim. Theor. Appl. **45**(1), 41–51 (1985)
19. Selim, S.Z., Alsultan, K.: A simulated annealing algorithm for the clustering problem. Pattern Recogn. **24**(10), 1003–1008 (1991)
20. Metropolis, N., Rosenbluth, A., Rosenbluth, M., Teller, A., Teller, E.: Equations of state calculations by fast computing machines. J. Chem. Phys. **21**, 1087–1092 (1953)
21. Tado, M., Kubo, R., Saito, N.: Statistical Physics. Springer, Berlin (1983)
22. Aarts, E.H.L., van Laarhoven, P.J.M.: Statistical cooling: a general approach to combinatorial optimization problems. Philips J. Res. **40**, 193–226 (1985)
23. Aarts, E.H.L., van Laarhoven, P.J.M.: A new polynomial time cooling schedule. In: Proceedings of the IEEE International Conference on Computer-Aided Design, pp. 206–208. Santa Clara (1985)
24. Aarts, E.H.L., van Laarhoven, P.J.M.: Simulated annealing: a pedestrian review of the theory and some applications. In: Devijver, P.A., Kittler, J. (eds.) Pattern Recognition Theory and Applications. NASI Series on Computer and Systems Sciences 30, pp. 179–192. Springer, Berlin (1987)

References

25. Glover, F., Greenberg, H.J.: New approach for heuristic search: a bilateral linkage with artificial intelligence. Eur. J. Oper. Res. **39**, 119–130 (1989)
26. Glover, F.: Future paths for integer programming and links to artificial intelligence. Comput. Oper. Res. **13**(5), 533–549 (1986)
27. Glover, F.: Tabu search-part I. Orsa J. Comput. **1**(3), 190–206 (1989). Summer
28. Glover, F.: Artificial intelligence, heuristic frameworks and tabu search. Manage. Decis. Econ. **11**, 365–375 (1990)
29. Glover, F.: Tabu search-part II. Orsa J. Comput. **2**(1), 4–32 (1990). Winter
30. Bland, J.A., Dawson, G.P.: Tabu search and design optimization. Comput. Aided Des. **23**(3), 195–201 (1991). April
31. Glover, F.: A user's guide to tabu search. Ann. Oper. Res. **41**, 3–28 (1993)
32. Laguna, M., Glover, F.: Integrating target analysis and tabu search for improved scheduling systems. Expert Syst. Appl. **6**, 287–297 (1993)
33. Kelly, J.P., Olden, B.L., Assad, A.A.: Large-scale controlled rounding using tabu search with strategic oscillation. Ann. Oper. Res. **41**, 69–84 (1993)
34. Barnes, J.W., Laguna, M.: A tabu search experience in production scheduling. Ann. Oper. Res. **41**, 141–156 (1993)
35. Charest, M., Ferland, J.A.: Preventive maintenance scheduling of power generating units. Ann. Oper. Res. **41**, 185–206 (1993)
36. Daniels, R.L., Mazzola, J.B.: A tabu search heuristic for the flexible-resource flow shop scheduling problem. Ann. Oper. Res. **41**, 207–230 (1993)
37. Amico, M.D., Trubian, M.: Applying tabu search to the jop-shop scheduling problem. Ann. Oper. Res. **41**, 231–252 (1993)
38. Mooney, E.L., Rardin, R.L.: Tabu search for a class of scheduling problems. Ann. Oper. Res. **41**, 253–278 (1993)
39. Davis, L. (ed.): Handbook of Genetic Algorithms. Van Nostrand, New York (1991)
40. Michalewicz, Z.: Genetic Algorithms + Data Structures = Evolution Programs. Springer, Berlin/Heidelberg/New York (1992)
41. Grefenstette, J.J.: Optimization of control parameters for genetic algorithms. IEEE Trans. Syst. Man Cybern. **16**(1), 122–128 (1986)
42. Grefenstette, J.J., Baker, J.E.: How genetic algorithm work: a critical look at implicit parallelism. In: The Proceedings of the Third International Conference on Genetic Algorithms. Morgan Kaufmann, San Mateo (1989)
43. Buckles, B.P., Petry, F.E., Kuester, R.L.: Schema survival rates and heuristic search in genetic algorithms. In: Proceedings of Tools for AI, pp. 322–327. Washington, DC (1990)
44. Awadh, B., Sepehri, N., Hawaleshka, O.: A computer-aided process planning model based on genetic algorithms. Comput. Oper. Res. **22**(8), 841–856 (1995)
45. Goldberg, D.E., Deb, K., Clark, J.H.: Genetic algorithms, noise, and the sizing of populations. Complex Syst. **6**, 333–362 (1992)
46. Homaifar, A., Guan, S., Liepins, G.E.: Schema analysis of the traveling salesman problem using genetic algorithms. Complex Syst. **6**, 533–552 (1992)
47. Goldberg, D.E.: Genetic Algorithms in Search, Optimization and Machine Learning. Addison Wesely, Reading Mass (1989)
48. Mantawy, H., Abdel-Magid, Y.L., Selim, S.Z.: A simulated annealing algorithm for unit commitment. IEEE Trans. Power Syst. **13**(1), 197–204 (1998)
49. Dasgupta, D., Mcgregor, D.R.: Thermal unit commitment using genetic algorithms. IEE Proc. Gener. Transm. Distrib. **141**(5), 459–465 (1994). September
50. Ma, X., El-Keib, A.A., Smith, R.E., Ma, H.: A genetic algorithm based approach to thermal unit commitment of electric power systems. Electr. Power Syst. Res. **34**, 29–36 (1995)
51. Kazarlis, S.A., Bakirtzis, A.G., Petridis, V.: A genetic algorithm solution to the unit commitment problem. IEEE Trans. Power Syst. **11**(1), 83–91 (1996). February

52. Yang, P.-C., Yang, H.-T., Huang, C.-L.: Solving the unit commitment problem with a genetic algorithm through a constraint satisfaction technique. Electr. Power Syst. Res. **37**, 55–65 (1996)
53. Ross, T.J.: Fuzzy Logic with Engineering Applications. McGraw-Hill, New York (1995)
54. Nazarka, J., Zalewski, W.: An application of the fuzzy regression analysis to the electrical load estimation. Electrotechnical Conference: MELECON'96, Bari, Italy, vol. 3, pp. 1563–1566. IEEE Catalog #96CH35884, 13–16 May 1996
55. Tanaka, H., Uejima, S., Asai, K.: Linear regression analysis with fuzzy model. IEEE Trans. Syst. Man Cybern. **12**(6), 903–907 (1983)
56. Chang, P.T., Lee, E.S.: Fuzzy least absolute deviations regression based on the ranking of fuzzy numbers. IEEE World Congress on Fuzzy Systems, Orlando, FL, USA, IEEE Proceeding, vol. 2, pp. 1365–1369 (1994)
57. Watada, J., Yabuchi, Y.: Fuzzy robust regression analysis. In: IEEE World Congress on Fuzzy Systems, Orlando, FL, USA, IEEE Proceeding, vol. 2, pp. 1370–1376 (1994)
58. Alex, R., Wang, P.Z.: A new resolution of fuzzy regression analysis. In: IEEE International Conference on Systems, Man, and Cybernetics, San Diego, California, USA, vol. 2, pp. 2019–2021. (1998)
59. Ishibuchi, H., Nii, M.: Fuzzy regression analysis by neural networks with non-symmetric fuzzy number weights. In: Proceedings of IEEE International Conference on Neural Networks, Washington, DC, USA, vol. 2, pp. 1191–1196 (1996)
60. Ghoshray, S.: Fuzzy linear regression analysis by symmetric triangular fuzzy number coefficients. In: Proceedings of IEEE International Conference on Intelligent Engineering Systems, Budapest, Hungary, pp. 307–313 (1997)
61. Hu, X., Eberhart, R.C., Shi, Y.: Engineering optimization with particle swarm. In: IEEE International Conference on Evolutionary Computation, pp. 53–57 (2003)
62. Kennedy, J.: The particle swarm: social adaptation of knowledge. In: Proceedings of 1997 IEEE International Conference on Evolutionary Computation (ICEC '97), pp. 303–8. Indianapolis (1997)
63. Angeline, P.: Evolutionary optimization versus particle swarm optimization: philosophy and performance differences. In: Proceedings of the 7th Annual Conference on Evolutionary Programming, San Diego, California, USA, pp. 601–10 (1998)
64. Shi, Y., Eberhart, R.: Parameter selection in particle swarm optimization. In: Proceedings of the 7th Annual Conference on Evolutionary Programming, San Diego, California, USA, pp. 591–600 (1998)
65. Stott, B., Hobson, E.: Power system security control calculation using linear programming. IEEE Transactions on Power Apparatus and Systems, vol. PAS-97, pp. 1713–1731 (1978)
66. Ozcan, E., Mohan, C.: Analysis of a simple particle swarm optimization system. Intell. Eng. Syst. Artif. Neural Networks **8**, 253–258 (1998)
67. Kennedy, J., Eberhart, R.: Particle swarm optimization. In: IEEE International Conference on Evolutionary Computation, Perth, WA, Australia, pp. 1942–1948 (1995)
68. Eberhart, R.C., Shi, Y.: Comparing inertia weights and constriction factors in particle swarm optimization. In: IEEE International Conference on Evolutionary Computation, San Antonio, TX, USA, pp. 84–88 (2000)
69. Abido, M.A.: Optimal design of power-system stabilizers using particle swarm optimization. IEEE Trans. Energy Convers. **17**(3), 406–413 (2002). September
70. Gaing, Z.L.: Particle swarm optimization to solving the economic dispatch considering the generator constraints. IEEE Trans. Power Syst. **18**(3), 11871–195 (2003). August
71. Hirotaka, Y., Kawata, K., Fukuyama, Y.: A particle swarm optimization for reactive power and voltage control considering voltage security assessment. IEEE Trans. Power Syst. **15**(4), 1232–1239 (2000). November
72. Miranda, V., Fonseca, N.: EPSO-evolutionary particle swarm optimization, a new algorithm with applications in power systems. In: IEEE Trans. Power Syst. pp. 745–750 (2000)

References

73. Shi, Y., Eberhart, R.C.: A modified particle swarm optimizer. Proceedings of IEEE International Conference on Evolutionary Computation, pp. 69–73. Anchorage (1998)
74. Zhenya, H., et al.: Extracting rules from fuzzy neural network by particle swarm optimization. Proceedings of IEEE International Conference on Evolutionary Computation, pp. 74–77. Anchorage (1998)
75. Kennedy, J., Spears, W.: Matching algorithm to problems: an expermental test of the particle swarm optimization and some genetic algorithms on the multimodal problem generator. In: Proceedings of IEEE International Conference on Evolutionary Computation, pp. 78–83. Anchorage (1998)
76. Angeline, P.: Using selection to improve particle swarm optimization. In: Proceedings of IEEE International Conference on Evolutionary Computation, pp. 84–89. Anchorage (1998)
77. Talaq, J.H., El-Hawary, F., El-Hawary, M.E.: A summary of environmental/economic dispatch algorithms. IEEE Trans. Power Syst. **9**, 1508–1516 (1994). August
78. Soliman, S.A., Al-Kandari, M.A.: Electrical Load Forecasting; Modeling and Model Construction. Elsevier, New York (2010)
79. Mantawy, H., Abdel-Magid, Y.L., Selim, S.Z., Salah, M.A.: An improved simulated annealing algorithm for unit commitment-application to Sceco-East. In: 3rd International Conference on Intelligent Applications in Communications and Power Systems, IACPS'97, pp. 133–139. UAE (1997)

Chapter 3
Economic Operation of Electric Power Systems

Objectives It is the purpose of this chapter

- To formulate the problem of optimal short-term operation of hydrothermal-nuclear systems
- To obtain the solution by using a functional analytical optimization technique that employs the minimum norm formulation
- To propose an algorithm suitable for implementing the optimal solution
- To present and formulate the fuzzy economic dispatch of all thermal power systems and explain the algorithm suitable for solution

3.1 Introduction

At present, owing to economic considerations, most of the nuclear generating stations throughout the world are operated at base load, and the length of the reactor refueling cycles for most reactors is on the order of several years. Thus it is recognized that the load scheduling of mixed power systems with nuclear plants is essentially an optimization problem of long-term nature [1–3]. The multiyear length of the optimization interval makes it virtually impossible to consider the hourly or even daily load variations of the power demand.

With nuclear plants becoming an increasing part of power generating systems it is expected that in the near future the operating role of nuclear plants will change. It is likely that in the absence of suitable energy storage methods nuclear plants will be required to operate in a load-following mode in order to achieve the flexibility needed to match the power demand [4].

The scope of the first part of this chapter, Sect. 3.2, is limited to cover power systems with thermal nuclear reactors [5] that are provided with continuous-on power refueling capabilities [6]. In this case the long-term effect of reactor fueling cycles is not present [7] and short-term scheduling becomes a meaningful problem.

The functional analytic optimization technique that is used here has been applied successfully to the analysis of hydrothermal systems [8]. Essentially this technique transforms the optimization problem into a minimum norm problem subject to a linear transformation in the context of functional analysis [9]. One of the advantages of this method is that it eliminates the multipliers associated with linear constraints. The second part of this chapter, Sect. 3.3, presents a simple technique to solve the short-term economic dispatch problem of an all-thermal electric power system, when value of the load on the system is considered to be fuzzy. The hard constraints, using this technique, are transformed to soft constraints. The membership function of the load is assumed to be triangular. A simulated example of a system consisting of two units is presented in this section to explain the main features of the proposed technique.

3.2 A Hydrothermal–Nuclear Power System

3.2.1 Problem Formulation [10–15]

The power system considered contains N_N nuclear plants, N_T fossil-fueled plants, and N_H hydro-generating stations located on the same stream. The problem is to determine the load schedule such that the total operating costs of the system in a time interval T are minimized under the following conditions.

1. All operating costs are attributed to the nuclear and fossil fuels. The total operating cost is assumed of the form

$$C = \int_0^T \left[\sum_{i=1}^{N_N} \left(\alpha_{Ni} + \beta_{Ni} P_{Ni} + \gamma_{Ni} P_{Ni}^2 \right) + \sum_{i=1}^{N_T} \left(\alpha_{Ti} + \beta_{Ti} P_{Ti} + \gamma_{Ti} P_{Ti}^2 \right) \right] dt \quad (3.1)$$

where P_{Ni} and P_{Ti} are the electric power generations at the ith nuclear and thermal plants, respectively. The coefficients α_i, β_i, and γ_i are known constants.

2. The total active power generation in the system matches the predicted power demand P_d plus the transmission losses:

$$P_d = \sum_{i=1}^{N_N} P_{Ni} + \sum_{i=1}^{N_T} P_{Ti} + \sum_{i=1}^{N_H} P_{Hi} - P_L \quad (3.2)$$

where P_{Hi} denotes the ith hydropower generation; the transmission losses P_L are expressed by the loss formula

3.2 A Hydrothermal–Nuclear Power System

$$P_L = \sum_{i=1}^{N_N}\sum_{j=1}^{N_N} P_{Ni}P_{Nj}b_{1ij} + 2\sum_{i=1}^{N_N}\sum_{j=1}^{N_T} P_{Ni}P_{Tj}b_{2ij} + 2\sum_{i=1}^{N_N}\sum_{j=1}^{N_H} P_{Ni}P_{Hj}b_{3ij}$$

$$+ \sum_{i=1}^{N_N}\sum_{j=1}^{N_T} P_{Ti}P_{Tj}b_{4ij} + 2\sum_{i=1}^{N_T}\sum_{j=1}^{N_H} P_{Ti}P_{Hj}b_{5ij} + \sum_{i=1}^{N_N}\sum_{j=1}^{N_H} P_{Hi}P_{Hj}b_{6ij} \quad (3.3)$$

with the parameters b_{ij} having known values.

3. With the water transport delay between the $(i-1)$th and ith hydro plants assumed a given constant τ_{i-1} the reservoir's dynamics are described by

$$\overset{o}{S}_1(t) = i_1(t) - q_1(t) \quad (3.4a)$$

$$\overset{o}{S}_i(t) = i_i(t) + q_{i-1}(t - \tau_{i-1}) - q_i(t) \quad i = 2, 3, \cdots, N_H \quad (3.4b)$$

where i_1 is the rate of natural water inflow to the ith reservoir, and q_i is the ith water discharge rate.

4. The water volume at the ith reservoir is given by

$$S_i = d_{1i} + d_{2i}H_i + d_{3i}H_i^2 \quad (3.5)$$

This corresponds to trapezoidal reservoirs. H_i denotes the fore bay elevation at the ith reservoir. The known parameters d_i are determined by the reservoir's dimensions [10].

5. The ith hydro plant's power generation is given by the double quadratic formula

$$P_{Hi} = \left[\alpha_{1i} + \alpha_{2i}H_i + \alpha_{3i}H_i^2\right]\left[b_{1i} + b_{2i}q_i + b_{3i}q_i^2\right] \quad (3.6)$$

This formula includes the effect of variable efficiency and variable head. The parameters α_i and b_i can be obtained by fitting Eq. 3.6 to the plant capability curves.

6. The volume of water discharge at any hydro plant is a specified constant C_i:

$$\int_0^T q_i(t)dt = C_i \quad (3.7)$$

7. Physical limitations impose the magnitude constraints

$$q_i^m \leq q_i \leq q_i^M \quad (3.8)$$

$$H_i^m \leq H_i \leq H_i^M \tag{3.9}$$

m and M are the minimum and maximum values of a variable

8. The transient changes of the average xenon 135 concentration in the ith core, which are induced by power level changes, are accounted for by the kinetic equations (3.10) and (3.11).

$$\overset{o}{I}_i = -\lambda_I I_i + \frac{\gamma_I}{e_i G} P_{Ni} \tag{3.10}$$

$$\overset{o}{X}_i = -\lambda_X X_i + \lambda_I I_i - \frac{\Gamma_X}{e_i \sum_{fi} G} X_i P_{Ni} + \frac{\gamma_X}{G e_i} P_{Ni} \tag{3.11}$$

where I_i and X_i denote the iodine 135 and xenon 135 average concentrations, respectively. The iodine and xenon radioactive decay constants are represented by λI and λ_X. $\sum_{fi}$ denotes the xenon's thermal neutron microscopic absorption cross-section. G is the energy released per fission and e_i is the plant's overall efficiency. The fission yields γ_I and γ_X denote the iodine and xenon, respectively.

9. The effects of power level and xenon concentration on the reactivity ρ_i of the ith core are described by

$$\rho_i = \rho_{0i} - \rho_{Ci} - \alpha_{di} P_{Ni} - \alpha_{Xi} X_i \tag{3.12}$$

where ρ_{0i} is a constant reference reactivity. The known parameters α_{di} and α_{Xi} denote the power and xenon feedback coefficients [5]. ρ_{Ci} is the external reactivity provided by the control devices whose compensating action against variations of power level and xenon concentration maintains the reactor criticality; that is, $\rho_I = 0$.

10. Only limited xenon override capability is available in most reactors. This is accounted for by upper and lower magnitude constraints on the external control reactivity ρ_{Ci}. Without loss of generality we assume

$$0 \leq \rho_{Ci} \leq \rho_{Ci}^M \tag{3.13}$$

11. Physical limitations impose the magnitude constraints

$$P_{Ni}^m \leq P_{Ni} \leq P_{Ni}^M \tag{3.14}$$

$$P_{Ti}^m \leq P_{Ti} \leq P_{Ti}^M \tag{3.15}$$

3.2.2 The Optimization Procedure

Here the optimization problem is reformulated as a minimum norm problem in a Hilbert space with quadratic norm. This formulation requires each of the system's variables to appear in the augmented cost functional raised to the second power [12]. With this in mind it is convenient (1) to eliminate from the system's equations the variables I_i, X_i, and S_i, which only appear linearly, (2) to introduce a change of variable in the case of ρ_{Ci}, which appears linearly and cannot be eliminated from the system's equations, and (3) to introduce pseudo control variables in order to eliminate those terms that are cubic or higher.

By differentiating Eq. 3.5 and substituting in (3.4a) and (3.4b), the variable S_i is eliminated. The resulting expression relates the fore bay elevation and the water discharge as follows.

$$0 = q_i(t) - q_{i-1}(t - \tau_{i-1}) + d_{2i}\overset{o}{H}_i(t) + 2d_{3i}H_i(t)\overset{o}{H}_i(t) - i_i(t) \tag{3.16a}$$

$$0 = q_1(t) - i_1(t) + d_{21}\overset{o}{H}_1(t) + 2d_{31}\overset{o}{H}(t)H_1(t) \tag{3.16b}$$

From Eqs. 3.10, 3.11 and 3.12, X_i and I_i can be eliminated. Making the change of variable

$$\xi_i^2 = \rho_{Ci} \tag{3.17}$$

and introducing the pseudo control variables ψ_i and ϕ_i, where

$$\Psi_i^2 = P_{Ni} \tag{3.18}$$

$$\phi_i = \Psi_i \xi_i \tag{3.19}$$

The following expression relating the power generation and the external reactivity is obtained.

$$\overset{o}{P}_{Ni}(t) = A_{1i}P_{Ni}(t) + A_{2i}P_{Ni}^2(t) + A_{3i}\phi_i^2(t) + g_i(t) + A_{4i}\xi_i^2(t)$$

$$+ A_{5i}\xi_i\overset{o}{\xi} + A_{6i}\int_0^t e^{-\lambda_l(t-\tau)}P_{Ni}(\tau)d\tau \tag{3.20}$$

where the coefficients A_{ji} and the functions $g_i(t)$ are known. The details of this derivation appear in Appendix A.1.

With the help of the pseudo control variable y_i,

$$y_i(t) = q_i(t)H_i(t) \tag{3.21}$$

the cubic terms in the hydrogenation formula (3.6) are replaced with quadratic terms. After expanding (3.6) and substituting (3.21) the hydrogenation formula is rewritten as follows.

$$\begin{aligned} P_{Hi} = {} & C_{1i} + C_{2i}H_i + C_{3i}H_i^2 + C_{4i}q_i + C_{5i}q_iH_i \\ & + C_{6i}y_iH_i + C_{7i}q_i^2 + C_{8i}q_iy_i + C_{9i}y_i^2 \end{aligned} \qquad (3.22)$$

where the coefficients C_{ji} are known.

The nonlinear equations (3.2, 3.16a, 3.16b, 3.18, 3.19, 3.20, 3.21, 3.22) and the magnitude constraints (3.8, 3.9, 3.13, 3.14, 3.15) are included in the augmented cost functional by using unknown functions as follows.

(a) The power balance equation (3.2) corresponds to the unknown function θ.
(b) Equation 3.16 corresponds to the unknown functions M_i.
(c) The pseudo constraints (3.18) and (3.19) correspond to the functions Λ_i and Ω_i, respectively.
(d) The nuclear generation equation (3.20) corresponds to R_i.
(e) The pseudo constraint (3.21) and the hydrogenation formula (3.22) correspond to the unknown functions Z_i and T_i, respectively.
(f) The right-hand sides of inequalities (3.8), (3.9), and (3.13), (3.14), and (3.15) correspond to the Kuhn–Tucker multiplier functions $\eta_{qi}^u, \eta_{Hi}^u, \eta_{pi}^u, \eta_{Ni}^u,$ and η_{Ti}^u, respectively.
(g) The left-hand sides of inequalities (3.8), (3.9), and (3.13), (3.14) and (3.15) correspond to the Kuhn–Tucker multiplier functions $\eta_{qi}^L, \eta_{Hi}^L, \eta_{pi}^L, \eta_{Ni}^L,$ and η_{Ti}^L, respectively.

The augmented objective functional is now written in a compact quadratic form. This is done by first defining the control vector

$$U = col[\phi, \Psi, \xi, P_N, P_T, P_H, H, y, q] \qquad (3.23)$$

(where the notation X represents a column vector with elements x_i) and then eliminating all those terms that do not explicitly depend on U:

$$J(U) = \int_0^T \left[U^T F U + L^T U \right] dt \qquad (3.24)$$

here F is a symmetric matrix, required by this formulation to be positive definite, and L is a column vector. The relations presented in Appendix A.2 are used through the derivation of (3.24).

The matrix F is partitioned in the form

$$F = \text{diag} \, [F_1, F_2, F_3, F_4] \qquad (3.25)$$

3.2 A Hydrothermal–Nuclear Power System

where F is a diagonal matrix of N_Nth order with elements

$$F_{1ii} = R_i A_{3i} \tag{3.26}$$

F_2 is a symmetric matrix of $2N_N$th order

$$F_2 = \begin{bmatrix} F_{21} & F_{22} \\ F_{22} & F_{23} \end{bmatrix} \tag{3.27}$$

where

$$F_{21} = \text{diag}\{\Lambda_i\}, i = 1, 2, N_N \tag{3.28}$$

$$F_{22} = \{-\Omega/2\} \tag{3.29}$$

and

$$F_{24} = \text{diag}\left\{A_{4i}R_i - A_{5i}\overset{o}{R}_i\right\} \tag{3.30}$$

F_3 is a symmetric matrix of $(N_N + N_T + N_H)$th order partitioned as follows.

$$F_3 = \begin{bmatrix} F_{31} & F_{32} & F_{33} \\ F_{32}^T & F_{34} & F_{35} \\ F_{33}^T & F_{35}^T & F_{36}^T \end{bmatrix} \tag{3.31}$$

where F_{31} is an N_Nth-order symmetric matrix with elements

$$F_{31ij} = \begin{cases} R_i A_{2i} + \gamma_{Ni} - \theta b_{ij}, & i = j \\ -\theta b_{1ij} & i \neq j \end{cases} \tag{3.32}$$

F_{32} is an $N_N \times N_T$ rectangular matrix with elements

$$F_{32ij} = b_{2ij}\theta \tag{3.33}$$

F_{33} is an $N_N \times N_H$ rectangular matrix with elements

$$F_{33ij} = -b_{3ij}\theta \tag{3.34}$$

F_{34} is a symmetric matrix of order N_T, where

$$F_{34ij} = \begin{cases} \gamma_{Ti} - \theta b_{4ij} \\ -b_{4ij} \end{cases} \tag{3.35}$$

F_{35} is an $N_T \times N_H$ rectangular matrix with elements

$$F_{35ij} = -\theta b_{5ij} \tag{3.36}$$

F_{36} is an N_Hth-order symmetric matrix:

$$F_{36ij} = -\theta b_{6ij} \tag{3.37}$$

Finally, F_4 is a symmetric matrix of three N_Hth-order partitioned in the form

$$F_4 \begin{bmatrix} F_{41} & F_{42} & F_{43} \\ F_{42} & F_{44} & F_{45} \\ F_{43} & F_{45} & F_{46} \end{bmatrix} \tag{3.38}$$

where every partition forms an N_Hth-order diagonal matrix:

$$F_{41} = diag\{T_1 C_{3i} - d_{3i}\overset{o}{M}_i\} \tag{3.39}$$

$$F_{42} = diag\{\tfrac{1}{2} T_i C_{6i}\} \tag{3.40}$$

$$F_{43} = diag\{\tfrac{1}{2} C(T_i C_{5i} + Z_i)\} \tag{3.41}$$

$$F_{44} = diag\{T_i C_{9i}\} \tag{3.42}$$

$$F_{45} = diag\{\tfrac{1}{2} T_i C_{8i}\} \tag{3.43}$$

$$F_{46} = diag\{T_{7i} C_i\} \tag{3.44}$$

The column vector L is partitioned in the form

$$L = col.[L_1, L_2, \cdots, L_9] \tag{3.45}$$

where

$$L_{1i} = \Omega_i, \quad i = 1, 2, \cdots, N_N \tag{3.46}$$

$$L_{2i} = 0, \quad i = 1, 2, \cdots, N_N \tag{3.47}$$

$$L_{3i} = \eta^u_{\rho i} - \eta^L_{\rho i} \quad i = 1, 2, \cdots, N_N \tag{3.48}$$

$$L_{4i} = \beta_{Ni} + \overset{o}{R}_i + A_{1i} R_i + A_{6i} \int_t^T e^{\lambda_1 (t-\tau)} R_i(\tau) d\tau$$

$$+ \theta + \eta^u_{Ni} - \eta^L_{Ni}, \quad i = 1, 2, \cdots, N_N \tag{3.49}$$

3.2 A Hydrothermal–Nuclear Power System

$$L_{5i} = \beta_{Ti} + \theta + \eta_{Ti}^u - \eta_{Ti}^L, \qquad i = 1, 2, \cdots, N_T \tag{3.50}$$

$$L_{6i} = \theta - T_i, \qquad i = 1, 2, \cdots, N_H \tag{3.51}$$

$$L_{7i} = T_i C_{2i} - d_{2i} \overset{o}{M}_i + \eta_{Hi}^u - \eta_{Hi}^L \qquad i = 1, 2, \cdots, N_H \tag{3.52}$$

$$L_{8i} = -Z_i, \qquad i = 1, 2, \cdots, N_H \tag{3.53}$$

$$L_{9i} = T_i C_{4i} + M - \tilde{\phi}(t, T, \tau_i) M_{i+1}(t + \tau_i) + \eta_{qi}^u - \eta_{qi}^L$$
$$i = 1, 2, \cdots, N_{H-1} \tag{3.54}$$

$$L_{9i} = T_i C_{4i} + M_i + \eta_{qi}^u - \eta_{qi}^L, \quad i = N_H$$

Here $\tilde{\phi}(t, T, \tau)$ denotes the function

$$\tilde{\phi}(t, T, \tau) = \begin{cases} 0, & t > T - \tau \\ 1, & t \leq T - \tau \end{cases} \tag{3.55}$$

The unknown functions R_i and M_i satisfy the homogeneous boundary conditions $R_i(T) = 0$ and $M_i(T) = 0$.

The problem is reformulated now as minimizing the objective functional (3.24) subject to the linear constraints (3.7). For convenience, Eq. 3.7 is expressed in terms of U:

$$C = \int_0^T EU dt \tag{3.56}$$

where E is an $N_H \times (4 N_N + 4 N_H + N_T)$ rectangular matrix partitioned as

$$E = [0|I] \tag{3.57}$$

with I being the N_Hth-order identity matrix.

3.2.3 The Optimal Solution Using Minimum Norm Technique

It has been shown [8] that the problem of minimizing an objective functional in the form of (3.24) subject to a linear constraint can be reduced to the following minimum norm problem in the framework of functional analysis [9].

Let B and D be Banach spaces and T a bounded linear transformation defined on B with values in D. For each α in the range of T and for $\hat{\mu} \in B$, the unique element $\mu_\alpha \in B$ that satisfies

$$\alpha = T_\mu \tag{3.58}$$

and minimizing the norm

$$J(\mu) = \|\mu - \hat{\mu}\| \tag{3.59}$$

is given by

$$\mu_\alpha = T^+[\alpha - T\hat{\mu}] + \hat{\mu} \tag{3.60}$$

where T^+ denotes the pseudo inverse of T.

Here the vector U defined in (3.23) is associated with the element μ of the minimum norm problem, the objective functional (3.56) corresponds to (3.58), and the vector $-\frac{1}{2}F^{-1}L$ is associated with $\hat{\mu}$.

The solution to our problem is therefore given by (3.60) in the form

$$U_{opt} = F^{-1}E^T \left[\int_0^T EF^{-1}E^T d_1 \right] \left[C + \frac{1}{2} \int_0^T EF^{-1}L dt \right] - \frac{1}{2}F^{-1}L \tag{3.61}$$

Substituting F, E, and L in (3.61) the conditions for optimality are obtained:

(a) $$\Omega_i = -2R_i \phi_I A_{3i} \tag{3.62}$$

(b) $$\Omega_i \xi_i - 2\Lambda_i \Psi_i = 0 \tag{3.63}$$

(c) $$\Omega_i \Psi_i - 2A_{4i} R_i \xi_i + 2A_{5i} \xi_i \overset{o}{R}_i - \eta_{pi}^u + \eta_{pi}^L = 0 \tag{3.64}$$

(d) $$\overset{o}{R}_i = -(A_{1i} + 2A_{2i} P_{Ni}) R_i - A_{6i} \int_t^T e^{\lambda_l(t-\tau)}$$

$$\times R_i(\tau) d\tau - 2\gamma_{Ni} P_{Ni} + 2\theta \sum_{i=1}^{N_N} b_{1ij} P_{Nj}$$

$$- \theta - \eta_{Ni}^u + \eta_{Ni}^L + 2\theta \sum_{j=1}^{N_T} b_{2ij} P_{Tj} + 2\theta \sum_{j=1}^{N_H} b_{3ij} P_{Hj} \tag{3.65}$$

3.2 A Hydrothermal–Nuclear Power System

with $R_i(T) = 0$

(e)
$$2\theta \left[\sum_{j=1}^{N_N} b_{2ji} P_{Nj} + \sum_{j=1}^{N_T} b_{4ij} P_{Tj} + \sum_{j=1}^{N_H} b_{5ij} P_{Hj} \right]$$
$$+ \eta_{Ti}^u - \eta_{Ti}^L - \partial \gamma_{Ti} P_{Ti} - \beta_{Ti} - \theta = 0 \tag{3.66}$$

(f)
$$T_i = \left[1 - 2 \left(\sum_{j=1}^{N_T} b_{3ji} P_{Nj} + \sum_{j=1}^{N_T} b_{5ji} P_{Tj} + \sum_{j=1}^{N_H} b_{6ij} P_{Hj} \right) \right] \theta \tag{3.67}$$

(g)
$$Z_i = T_i C_{6i} H_i + 2 T_i (q_i y_i) + T_i C_{8i} q_i \tag{3.68}$$

(h) $\overset{o}{M}_i (2 H_i d_{2i} + d_{2i}) = T_i (C_{2i} + 2 C_{3i} H_i + C_{6i} y_i + C_{5i} q_i) + \eta_H^u - \eta_{Hi}^L$
$$- Z_i q_i \text{ with } M_i(t) = 0 \tag{3.69}$$

(i) $q_i = G_{6i} \left[\int_0^T G_{6i} dt \right]^{-1} \left[C_i + \frac{1}{2} \int_0^T (G_{3i} L_{7i} + G_{5i} L_{8i} + G_{6i} L_{qi}) \right]$
$$- \frac{1}{2} \left[G_{3i} L_{7i} + G_{5i} L_{8i} + G_{6i} L_{qi} \right] \tag{3.70}$$

where

$$G_{3i} = \frac{T_i^2 \left(\frac{1}{4} C_{6i} C_{8i} - C_{5i} C_{9i} \frac{1}{2} \right) - \frac{1}{2} T_i Z_i C_{9i}}{\Delta_i} \tag{3.71}$$

$$\Delta_i = T_i^3 \left(C_{3i} C_{9i} C_{7i} + \frac{1}{4} C_{6i} C_{5i} C_{8i} - \frac{1}{4} C_{5i}^2 C_{9i} - \frac{1}{4} C_{8i}^2 C_{3i} - \frac{1}{4} C_{6i}^2 C_{7i} \right)$$
$$+ T_i^2 \overset{o}{M} \left(\frac{1}{4} C_{8i}^2 d_{3i} - C_{9i} C_{7i} d_{3i} \right) + T_i^2 Z_i \left(\frac{1}{4} C_{6i} C_{8i} - \frac{1}{2} C_{5i} C_{9i} \right) - Z_i^2 T_i \left(\frac{1}{4} C_{9i} \right)$$
$$\tag{3.72}$$

$$G_{5i} = \left[T_i^2 \left(\frac{1}{4} C_{6i} C_{5i} - \frac{1}{2} T_i^2 C_{8i} C_{3i} \right) + \frac{1}{4} T_i Z_i C_{6i} \right.$$
$$\left. + \frac{1}{2} d_{3i} C_{8i} \overset{o}{M}_i T_i \right] / \Delta_i \tag{3.73}$$

$$G_{6i} = \left[T_i^2 (C_{3i} C_{9i} - \frac{1}{4} C_{6i}^2) - T_i C_{9i} C_{9i} d_{3i} \overset{o}{M}_i \right] \Delta_i \tag{3.74}$$

The Kuhn–Tucker multiplier functions η_i^u and η_i^L are obtained such that the exclusion equations associated with the inequality constraints are satisfied.

The conditions for optimality can be further simplified by solving for $\Omega_i, \Lambda_i, \Psi_i,$ and ϕ_i from Eqs. 3.18, 3.19, 3.62, and 3.63 and then substituting in (3.64) in order to solve for ξ_i. By doing this, the following expression is obtained,

$$\xi_i = \frac{\left(\eta_p^u - \eta_{\rho i}^L\right)}{2\left[A_{5i}\overset{o}{R}_i - (A_{3i}P_{Ni} + A_{4i})R_i\right]} \tag{3.75}$$

Using (3.75) in conjunction with the exclusion equations

$$\eta_{\rho i}^u\left[\xi_i - \left(\rho_i^M\right)^{1/2}\right] = 0, \qquad \eta_{\rho i}^u \geq 0 \tag{3.76}$$

and

$$\eta_{\rho i}^L[-\xi_i] = 0, \qquad \eta_{\rho i}^L \geq 0 \tag{3.77}$$

η_ρ^L and $\eta_{\rho i}^u$ are eliminated yielding the final result:

$$\xi_i = \begin{cases} \left(\rho_i^M\right)^{1/2} & \text{if } A_{5i}\overset{o}{R}_i - (A_{3i}P_{Ni} - A_{4i})R_i > 0 \\ 0 & \text{if } A_{5i}\overset{o}{R}_i - (A_{3i}P_{Ni} - A_{4i})R_i > 0 \end{cases} \tag{3.78}$$

3.2.4 A Feasible Multilevel Approach

A two-level iterative approach is proposed for implementing the optimal solution [13]. We ensure that at each iteration step between levels the system constraints are satisfied. Therefore, every iteration constitutes a feasible suboptimal solution.

The iterative scheme is composed of two levels with each level divided into three sublevels as shown in Fig. 3.1. The variables attached to arriving arrows are treated at each level as known functions and only those variables appearing inside the blocks are treated as unknown. Here an iteration cycle is described.

Given an initial guess $U_0(t)$, θ is computed at sublevel (2b) from the Kth equation (3.66):

$$\theta(t) = \frac{\left(2\gamma_{TK}P_{TK} + \beta_{TK} + \eta_{TK}^L - \eta_{TK}^u\right)}{2\left(\sum_{j=1}^{N_N} b_{2jk}P_{nj} + \sum_{j=1}^{N_T} b_{4kj}P_{Tj} + \sum_{j=1}^{N_H} b_{5kj}P_{Hj} - \frac{1}{2}\right)} \tag{3.79}$$

3.2 A Hydrothermal–Nuclear Power System

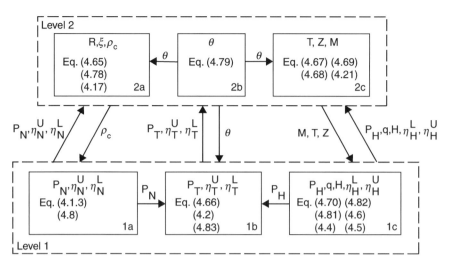

Fig. 3.1 A two-level iteration scheme

This result is then fed into sublevels (2a) and (2c). At sublevel (2a) $R_i(t)$ is computed by solving Eq. 3.66 backward in time.

ξ_i and ρ_{Ci} are obtained from Eqs. 3.78 and 3.17, respectively. At sublevel (2c) the variables T_i, Z_i, and M_i are obtained from Eqs. 3.67, 3.68 and 3.21. M_i is computed by solving (3.69) backward in time.

The information obtained at level 2 is then fed back into level 1 as shown in the figure. At sublevel (1a) the nuclear power generations are obtained from Eq. A.1.3 in Appendix A.1 and the exclusion equations

$$\eta_{Ni}^u [P_{Ni} - P_{Ni}^M] = 0, \qquad \eta_{Ni}^u \geq 0$$
$$\eta_{Ni}^L [P_{Ni}^m - P_{Ni}] = 0, \qquad \eta_{Ni}^L \geq 0 \qquad (3.80)$$

At sublevel (1c) the variables q_i are obtained from Eq. 3.70 and the exclusion equations

$$\eta_{qi}^u [q_i - q_i^M] = 0, \qquad \eta_{qi}^u \geq 0$$
$$\eta_{qi}^L [q_i^m - q_i] = 0, \qquad \eta_{qi}^L \geq 0 \qquad (3.81)$$

The fore bay elevations H_i are obtained by solving Eqs. 3.4 and 3.5 and the exclusion equations

$$\eta_{Hi}^u [H_i - H_i^M] = 0, \qquad \eta_{Hi}^u \geq 0$$
$$\eta_{Hi}^L [P_i^m - P_i] = 0, \qquad \eta_{Hi}^L \geq 0 \qquad (3.82)$$

The hydropower generations P_{Hi} are given by Eq. 3.6.

Finally the fossil plant generations are computed at level (1b) by solving simultaneously the remaining $N_T - 1$ equation (3.66), the power balance equation (3.2), and the exclusion equations

$$\eta_{Ti}^u [P_{Ti} - P_{Ti}^M] = 0, \qquad \eta_{Ti}^u \geq 0$$
$$\eta_{Ti}^L [P_{Ti}^m - P_{Ti}] = 0, \qquad \eta_{Ti}^L \geq 0 \qquad (3.83)$$

3.2.5 Conclusions and Comments

The optimal solution to the problem of short-term operation of hydrothermal–nuclear systems is obtained by using an optimization technique that employs the minimum norm formulation. A two-level iterative scheme of the so-called feasible type is proposed for implementing the optimal solution.

The optimality conditions presented here are applicable to power systems with thermal nuclear reactors provided with continuous fueling capabilities and hydro plants located on the same stream with trapezoidal reservoirs.

The nuclear model considered is only a crude approximation to the very complex processes that take place in a nuclear reactor core. It would be impractical, if not impossible, to consider in an application of this nature a very detailed distributed model describing the reactor core processes. It is thought that the present methodology can be applied to more sophisticated (but still relatively simple) reactor models where the spatial effects in the core are considered.

3.3 All-Thermal Power Systems

This section presents a technique to solve the short-term economic dispatch problem of an all-thermal electric power system, when the load on the system is considered to be fuzzy. The hard constraints, using this technique, are transformed to soft constraints. The membership function of the load is assumed to be triangular. A simulated example of a system consisting of two units is presented in this section to explain the main features of the proposed technique.

3.3.1 Conventional All-Thermal Power Systems; Problem Formulation [17, 19, 20, 22, 27, 37–38]

Given a power system that consists of m thermal units, it is required to supply the load on the system, this value of load is assumed to be fuzzy. The objective is to minimize total fuel cost that supplies the load demand. For a quadratic fuel cost function, the problem can be mathematically stated as [8]

3.3 All-Thermal Power Systems

Minimize

$$C_{total} = \sum_{i=1}^{NG} C_i = \sum_{i=1}^{NG} \alpha_i + \beta_i P_{Gi} + \gamma_i P_{Gi}^2 \qquad (3.84)$$

Subject to satisfying the following constraints on the system.

- Active power balance equation (APBE)

$$\sum_{i=1}^{NG} P_{Gi} = P_D + P_L \qquad (3.85)$$

where
$P_D =$ is the total system demand of the network.
$P_L =$ is the system transmission losses, which is a function of the generation of each unit and system parameters related to the network model.

- The power output of any generator should not exceed its rating nor should it be lower than the minimum value necessary for stable boiler operation. Thus, the generations are restricted to lie within given minimum and maximum limits expressed as

$$P_i(\min) \leq P_i \leq P_i(\max) \quad i = 1, \ldots \ldots NG$$

The problem formulated above is a classical economic dispatch problem. It is well known and many techniques have been developed to solve it. Reference [16] gives a comprehensive survey of the techniques used in solving the economic dispatch problem and recent developments to improve the solution. The unit commitment problem and optimal power flow may be included in the problem formulation to overcome the difficulty of including the system losses in the formulation. Some techniques use the B-coefficients to express system losses.

3.3.2 Fuzzy All-Thermal Power Systems; Problem Formulation [39]

It may be possible that fuzzy formulation for the economic dispatch problem would overcome difficulties involved in solving the problem [18, 20, 21, 23–26, 28–32, 40–46]. In the next section, we offer this formulation deriving the necessary equations based on the principle of equal incremental cost-neglecting losses, where the solution of this type of problem can be found using closed-form expressions. However, when losses are considered, the resulting equations as seen in the next section are nonlinear and must be solved iteratively. In this section these losses are neglected. The fuzzy economic dispatch problem can be stated as

1. The power load on the system is Fuzzy $\tilde{P}_D$.
2. The power generated from each unit will be fuzzy $\tilde{P}_{Gi}$.

Then the optimization problem in this case is given as
Minimize

$$\tilde{C}_{total} = \sum_{i=1}^{NG} \tilde{C}_i = \sum_{i=1}^{NG} \alpha_i + \beta_i \tilde{P}_{Gi} + \gamma_i \tilde{P}_{Gi}^2 \qquad (3.86)$$

Subject to satisfying the following constraints.

$$\sum_{i=1}^{NG} \tilde{P}_{Gi} - \tilde{P}_D \geq 0 \qquad (3.87)$$

$$\tilde{P}_{Gi}(\min) \leq \tilde{P}_{Gi} \leq \tilde{P}_{Gi}(\max) \quad i = 1, \ldots \ldots NG \qquad (3.88)$$

From Eq. 3.87 **IF** a fuzzy system load $\tilde{P}_D = (\bar{P}_D, \alpha_D, \beta_D)$ is assumed to be a triangular membership function as shown in Fig. 3.2a, where its middle value is represented by $\bar{P}_D$ and the left and right spread are α_D, β_D, respectively, **THEN** the fuzzy generation $\tilde{P}_{Gi} = (\bar{P}_{Gi}, a_i, b_i)$ will be of the same triangular membership function representation as shown in Fig. 3.2b. The middle crisp value is represented by $\bar{P}_{Gi}$ and the left, right spread of a_i, b_i respectively, where $i = 1, \ldots NG$ represents the number of generation sources committed to the system network. The left, right side of the triangular membership function for the load demand can be calculated as $L_{\tilde{P}_D} = (\bar{P}_D - \alpha_D), R_{\tilde{P}_D} = (\bar{P}_D + \beta_D)$, respectively, as shown in Fig. 3.2a. the left and right sides of the power generation are $L_{\tilde{P}_{Gi}} = (\bar{P}_{Gi} - a_{Gi})$, $R_{\tilde{P}_{Gi}} = (\bar{P}_{Gi} + b_{Gi})$ as shown in Fig. 3.2b.

In this formulation, we have translated the fuzzy load into a triangular membership function by assigning a degree of membership to each possible α-cut value of the load, which means mapping the fuzzy variable on the [0, 1] interval. The solution of Eq. 3.86 will provide the generation possibility distributions corresponding to fuzzy loads for the minimum cost of operations. In Eq. 3.87 the hard constraints mentioned are transferred to soft constraints by using the Lagrange multiplier. Using such equality constraints implicitly includes the demand. The approach used in this section is to assume the fuzzy demand and fuzzy generation with different representations of their α-cut are expressed by (0, 0.5, 0.75, and 1) where the α-cut is used to create a family of crisp sets in order to be used in fuzzy mathematical operations.

3.3 All-Thermal Power Systems

Fig. 3.2 (a) Membership function for power demand. (b) Membership function for power generation

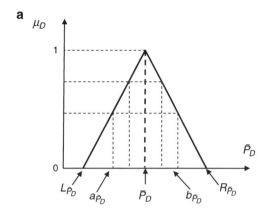

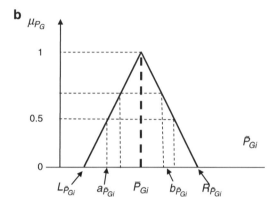

The membership formula for the load demand is expressed as

$$\mu(\tilde{P}_D) = \begin{cases} 0 & \tilde{P}_D < L_{\tilde{P}_D} \\ \dfrac{\tilde{P}_D - L_{\tilde{P}_D}}{a_{\tilde{P}_D}} & L_{\tilde{P}_D} \leq \tilde{P}_D \leq \bar{P}_D \\ \dfrac{R_{\tilde{P}_D} - \tilde{P}_D}{a_{\tilde{P}_D}} & \bar{P}_D \leq \tilde{P}_D \leq R_{\tilde{P}_D} \\ 0 & \tilde{P}_D > R_{\tilde{P}_D} \end{cases} \qquad (3.89)$$

The membership formula for the generator becomes:

$$\mu(\tilde{P}_{Gi}) = \begin{cases} 0 & \tilde{P}_{Gi} \leq L_{\tilde{P}_{Gi}} \\ \dfrac{\tilde{P}_{Gi} - L_{\tilde{P}_{Gi}}}{a_{\tilde{P}_{Gi}}} & L_{\tilde{P}_{Gi}} \leq \tilde{P}_{Gi} \leq \bar{P}_{Gi} \\ \dfrac{R_{\tilde{P}_{Gi}} - \tilde{P}_{Gi}}{b_{\tilde{P}_{Gi}}} & \bar{P}_{Gi} \leq \tilde{P}_{Gi} \leq R_{\tilde{P}_{Gi}} \\ 0 & \tilde{P}_{Gi} \geq R_{\tilde{P}_{Gi}} \end{cases} \qquad (3.90)$$

Below is a review of the crisp case to obtain the optimal solution using the Lagrange multiplier formula to relax "system wide constraints" into an unconstrained form that matches the original objective function at feasible points.

$$L = \sum_{i=1}^{NG} (\alpha_i + \beta_i P_{Gi} + \gamma_i P_{Gi}^2) + \lambda(P_D - \sum_{i=1}^{NG} P_{Gi})$$
$$+ \mu_i(P_{Gi}^m - P_{Gi}) + \Psi_i(P_{Gi} - P_{Gi}^M) \tag{3.91}$$

where λ, μ_i, and Ψ_i are fuzzy Kuhn–Tucker multipliers.

Optimizing the formula by setting partial derivative to zero,

$$\frac{\partial L}{\partial P_{Gi}} = \beta_i + 2\gamma_i P_{Gi} + \lambda(-1) - \mu_i + \Psi_i = 0 \tag{3.92}$$

$$\beta_i + 2\gamma_i P_{Gi} - \lambda - \mu_i + \Psi_i = 0$$

$$\frac{\partial L}{\partial \lambda} = \lambda(\sum_{i=1}^{NG} P_{Gi} - P_D) = 0 \tag{3.93}$$

$$\frac{\partial L}{\partial \mu_i} = \mu_i(P_{Gi}^m - P_{Gi}) = 0 \tag{3.94}$$

$$\frac{\partial L}{\partial \Psi_i} = \Psi_i(P_{Gi} - P_{Gi}^M) = 0 \tag{3.95}$$

Assuming that unit i is operating within the specified limits, μ_i and Ψ_i will be equal to zero.

Then from Eq. 3.92 we can obtain the incremental fuel cost λ as

$$\lambda = \beta_i + 2\gamma_i P_{Gi} \tag{3.96}$$

Thus, fuzzifying the optimal solution obtained from the crisp optimization problem, the incremental cost $\tilde{\lambda}$ can then be written as

$$\tilde{\lambda} = \beta_i + 2\gamma_i \tilde{P}_{Gi} \tag{3.97}$$

Solving for the power generation we get:

$$\tilde{P}_{Gi} = \frac{\tilde{\lambda} - \beta_i}{2\gamma_i} \quad i = 1, .., NG \tag{3.98}$$

3.3 All-Thermal Power Systems

Then replacing Eq. 3.87 with (3.98) we get:

$$\sum_{i=1}^{NG} \frac{\tilde{\lambda} - \beta_i}{2\gamma_i} = \tilde{P}_D \tag{3.99}$$

Solving for $\tilde{\lambda}$ we get:

$$\tilde{\lambda} = \frac{2\tilde{P}_D + \sum_{i=1}^{NG} \frac{\beta_i}{\gamma_i}}{\sum_{i=1}^{NG} \frac{1}{\gamma_i}} \tag{3.100}$$

Substituting the middle, left, and right spread representation into Eq. 3.100:

$$\tilde{\lambda}(\bar{\lambda}, a_{\tilde{\lambda}}, b_{\tilde{\lambda}}) = \frac{2(\bar{P}_D, a_{\tilde{P}_D}, b_{\tilde{P}_D}) + \sum_{i=1}^{NG} \frac{\bar{\beta}_i}{\bar{\gamma}_i}}{\sum_{i=1}^{NG} \frac{1}{\bar{\gamma}_i}} \tag{3.101}$$

In the above equation, we assume that the unit coefficients β_i and γ_i are crisp values. Using Tables 2.1 and 2.2 from Chap. 2 to implement the operation of fuzzy numbers such as addition, subtraction, division, multiplication, and inversion by their α-cut operation, the crisp values in Eq. 3.101 are then obtained by collecting all the crisp middle values of the fuel incremental cost which can be written as

$$\bar{\lambda} = \frac{2\bar{P}_D + \sum_{i=1}^{NG} \left(\frac{\bar{\beta}_i}{\bar{\gamma}_i}\right)}{\sum_{i=1}^{NG} \frac{1}{\bar{\gamma}_i}} \tag{3.102}$$

and the left spread can be calculated from (3.100) to become:

$$a_{\tilde{\lambda}} = \frac{2(a_{\tilde{P}_D})}{\sum_{i=1}^{NG} \left(\frac{1}{\bar{\gamma}_i}\right)} \tag{3.103}$$

and the right spread can be calculated as

$$b_{\tilde{\lambda}} = \frac{2(b_{\tilde{P}_D})}{\sum_{i=1}^{NG} \left(\frac{1}{\bar{\gamma}_i}\right)} \tag{3.104}$$

Equations 3.103 and 3.104 describe the fuzzy incremental fuel cost. Substituting the middle, left, and right spreads into the fuzzy generation of each unit from Eq. 3.98 we get

$$\tilde{P}_{Gi} = (\bar{P}_{Gi}, a_{\tilde{P}_{Gi}}, b_{\tilde{P}_{Gi}}) = \frac{(\bar{\lambda}, a_{\tilde{\lambda}}, b_{\tilde{\lambda}}) - \bar{\beta}_i}{2\bar{\gamma}_i} i = 1,..,NG \quad (3.105)$$

The middle of the generation is calculated as

$$\bar{P}_{Gi} = \frac{\bar{\lambda} - \bar{\beta}_i}{2\bar{\gamma}_i} = \bar{P}_{Gi} \; i = 1,..,NG \quad (3.106)$$

whereas the left side of the generation can be calculated as

$$a_{\tilde{P}_{Gi}} = \frac{a_{\tilde{\lambda}}}{2\bar{\gamma}_i} \cong \tilde{P}_{Gi}(\min) \quad (3.107)$$

and the right side of the generator can be calculated as

$$b_{\tilde{P}_{Gi}} = \frac{b_{\tilde{\lambda}}}{2\bar{\gamma}_i} \cong \tilde{P}_{Gi}(\max) \quad (3.108)$$

The left and right sides of the generation given by Eqs. 3.107 and 3.108 may equal the maximum and minimum limits of each thermal generator unit, or they may be included within the membership. This setting should not lead to any violation of the limit restricted on the generation as shown in Eq. 3.88. This means the load will be distributed evenly between the two units and satisfy the quality constraints given in Eq. 3.87. Using such a simplification reduces the cost calculation in the iterative method that considers the transmission line losses, even if there are some approximations. Furthermore, there is no crisp load in real-time; the value of the load changes from minute to minute.

Equations 3.87 and 3.88 can be rewritten to be

$$\sum_{i=1}^{NG} (\bar{P}_{Gi}, a_{\tilde{P}_{Gi}}, b_{\tilde{P}_{Gi}}) - (\bar{P}_D, a_{\tilde{P}_D}, b_{\tilde{P}_D}) \geq 0 \quad (3.109)$$

$$a_{\tilde{P}_{Gi}} \cong \frac{a_{\tilde{\lambda}}}{2\bar{\gamma}_i} \leq \bar{P}_{Gi} \leq b_{\tilde{P}_{Gi}} \cong \frac{b_{\tilde{\lambda}}}{2\bar{\gamma}_i} i = 1,........NG \quad (3.110)$$

The total fuzzy optimal cost function can be calculated using Eq. 3.86 after substituting the right, left spread of the power generation of each unit and calculating the cost of each unit individually. Then the total cost of all units is added to obtain the complete cost. This can be verified in the following equations.

3.3 All-Thermal Power Systems

$$\tilde{C}_1 = (\bar{C}_1, L_{\tilde{C}_1}, R_{\tilde{C}_1})$$
$$= \alpha + \beta(\bar{P}_{G1}, L_{\tilde{P}_{G1}}, R_{\tilde{P}_{G1}}) + \gamma(\bar{P}_{G1}, L_{\tilde{P}_{G1}}, R_{\tilde{P}_{G1}})(\bar{P}_{G1}, L_{\tilde{P}_{G1}}, R_{\tilde{P}_{G1}}) \quad (3.111)$$

$$\tilde{C}_2 = (\bar{C}_2, L_{\tilde{C}_2}, R_{\tilde{C}_2})$$
$$= \alpha + \beta(\bar{P}_{G2}, L_{\tilde{P}_{G2}}, R_{\tilde{P}_{G2}}) + \gamma(\bar{P}_{G2}, L_{\tilde{P}_{G2}}, R_{\tilde{P}_{G2}})(\bar{P}_{G2}, L_{\tilde{P}_{G2}}, R_{\tilde{P}_{G2}}) \quad (3.112a)$$

$$\tilde{C}_t = (\bar{C}_t, L_{\tilde{C}_t}, R_{\tilde{C}_t}) = (\bar{C}_1, L_{\tilde{C}_1}, R_{\tilde{C}_1}) + (\bar{C}_2, L_{\tilde{C}_2}, R_{\tilde{C}_2}) \quad (3.112b)$$

Using Table 2.2 from Chap. 2 the middle, left, and right side of the total cost become:

$$\bar{C}_t = \sum_{i=1}^{n} [\alpha_i + \beta_i(\bar{P}_{Gi}) + \gamma_i(\bar{P}_{Gi}^2)]$$

$$L_{\tilde{C}_t} = \sum_{i=1}^{n} [\alpha_i + \beta_i(L_{\tilde{P}_{Gi}}) + \gamma_i((L_{\tilde{P}_{Gi}})(L_{\tilde{P}_{Gi}}))]$$

$$R_{\tilde{C}_t} = \sum_{i=1}^{n} [\alpha_i + \beta_i(R_{\tilde{P}_{Gi}}) + \gamma_i((R_{\tilde{P}_{Gi}})(R_{\tilde{P}_{Gi}}))] \quad (3.113)$$

And if we use Tables 2.1 and 2.2, the middle, left, and right spread become:

$$\bar{C}_t = \sum_{i=1}^{n} [\alpha_i + \beta_i(\bar{P}_{Gi}) + \gamma_i(\bar{P}_{Gi}^2)]$$

$$a_{\tilde{C}_t} = \sum_{i=1}^{n} [\beta_i(a_{\tilde{P}_{Gi}}) + \gamma_i((a_{\tilde{P}_{Gi}}\bar{P}_{Gi}) + (a_{\tilde{P}_{Gi}}\bar{P}_{Gi}))]$$

$$b_{\tilde{C}_t} = \sum_{i=1}^{n} [\beta_i(b_{\tilde{P}_{Gi}}) + \gamma_i((b_{\tilde{P}_{Gi}}\bar{P}_{Gi}) + (b_{\tilde{P}_{Gi}}\bar{P}_{Gi}))] \quad (3.114)$$

It is worthwhile to state here that trapezoidal membership functions can be used for the load demand, the fuel incremental cost, and power generation of each unit instead of a triangular membership function representation as shown in Fig. 3.3a–c. In Fig. 3.3a the peak value of the power load will occur between P_{-D} and $\bar{P}_D$. The minimum estimated load is a -5% deviation which will be P_{-D}, and the maximum estimated load is a 10% deviation which will be $1.1\bar{P}_D$. The fuel incremental cost membership function shown in Fig. 3.3b shows the calculated incremental fuel cost λ_L corresponding to the value of $0.95 P_{-D}$, whereas $\underline{\lambda}$ is calculated from P_{-D}, $\bar{\lambda}$ from $\bar{P}_D$, and λ_R from the maximum allowable 10% load deviation represented by $1.1\bar{P}_D$. Finally, the calculated trapezoidal membership function for the power generation describing the obtained fuzzy generation from a fuzzy load is shown in Fig. 3.3c.

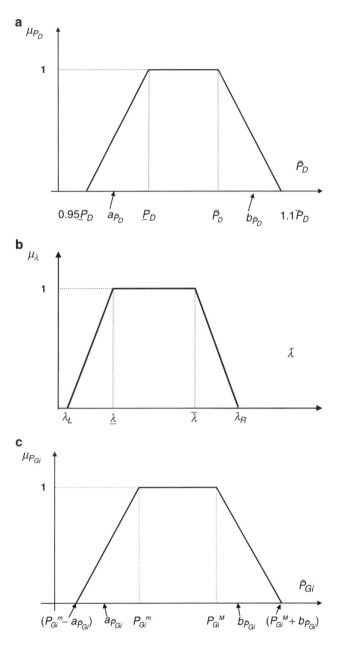

Fig. 3.3 (a) A trapezoidal membership function for power load. (b) A trapezoidal incremental cost membership function. (c) A trapezoidal membership function for power generation

3.3.3 Solution Algorithm

The load demand data were obtained from an estimated short-term load forecasting model, developed on the basis of fuzzy multiple linear regressions, to minimize the spread of the fuzzy coefficients that exist in the fuzzy winter model for weekdays and a weekend with a 20% deviation in the load demand in a 24-h period. The load demand will have upper, middle, and lower limits. Table 3.1 and Fig. 3.4 show the actual load demand on the system when α-cuts are equal to zero and the other left and right sides for different α-cuts values are calculated from the membership formula for the load demand in Eq. 3.89. In addition, the number of m thermal units feeding the load and the crisp characteristic coefficients of each unit $\alpha_i, \beta_i,$ and γ_i are known. Then a solution to the ED problem can be obtained using the following steps.

1. Apply the principle of equal incremental cost to determine the optimal fuzzy dispatch and the total fuzzy cost. Then calculate the fuzzy fuel incremental cost, middle, and spread, using Eqs. 3.102, 3.103, and 3.104 for different α-cut values represented by (0, 0.5, 0.75, and 1) for the hour in question.
2. For each fuzzy incremental fuel cost, determine the fuzzy generation of each unit, middle, and spread, using Eqs. 3.106, 3.107, and 3.108.
3. Calculate the fuel cost of each unit that corresponds to its generation and hence the total fuel cost using Eq. 3.114.

3.3.4 Examples

The above steps are applied to a simulated example, consisting of two generating units. The input/out fuel cost functions for each unit are given as

$$F(P_{G1}) = 200 + 7P_{G1} + 0.008P_{G1}^2 \text{ kJ/h}$$
$$F(P_{G2}) = 180 + 6.3P_{G2} + 0.009P_{G2}^2 \text{ kJ/h}$$

The generation limit for each unit is:

$$P_{G1\min} \leq P_{G1} \leq P_{G1\max} \text{ MW}$$
$$P_{G2\min} \leq P_{G2} \leq P_{G2\max} \text{ MW}$$

The fuzzy load demand on the system at different α-cut ($\alpha = 0, 0.5, 0.75$) is given in Figs. 3.4, 3.5, and 3.6, whereas Fig. 3.7 gives the membership functions of these fuzzy loads at all α-cut levels.

Replacing the minimum and maximum limits with the left and right sides of power generation, it can then be written as

Table 3.1 Membership function of load demand for (0, 0.5, 0.75, 1) α-cut representation for model "A" weekdays with 20% deviation

Membership function	$\mu_{P_{Load}} = 0$			$\mu_{P_{Load}} = 0.5$			$\mu_{P_{Load}} = 0.75$			$\mu_{P_{Load}} = 1$		
Daily hours	Left load (MW)	Mid load (MW)	Right load (MW)	Left load (MW)	Mid load (MW)	Right load (MW)	Left load (MW)	Mid load (MW)	Right load (MW)	Left load (MW)	Mid load (MW)	Right load (MW)
1	257	735.9	1,316	496.4	735.9	1,026	616.2	735.9	880.9	735.9	735.9	735.9
2	277.6	650.6	1,336	464.1	650.6	993.5	557.3	650.6	822	650.6	650.6	650.6
3	269.8	613.1	1,329	441.4	613.1	970.8	527.3	613.1	792	613.1	613.1	613.1
4	274.9	599.6	1,334	437.3	599.6	966.7	518.4	599.6	783.1	599.6	599.6	599.6
5	279.6	604.8	1,338	442.2	604.8	971.6	523.5	604.8	788.2	604.8	604.8	604.8
6	290.4	617.1	1,349	453.8	617.1	983.2	535.4	617.1	800.1	617.1	617.1	617.1
7	301.5	635.1	1,360	468.3	635.1	997.7	551.7	635.1	816.4	635.1	635.1	635.1
8	295.4	731.5	1,354	513.5	731.5	1,043	622.5	731.5	887.2	731.5	731.5	731.5
9	299.6	915.8	1,358	607.7	915.8	1,137	761.7	915.8	1,026	915.8	915.8	915.8
10	317.9	1,002	1,377	659.9	1,002	1,189	830.8	1,002	1,096	1,002	1,002	1001.8
11	320.2	1,013	1,379	666.6	1,013	1,196	839.8	1,013	1,105	1,013	1,013	1,013
12	322	1,015	1,381	668.3	1,015	1,198	841.4	1,015	1,106	1,015	1,015	1014.6
13	338.4	1,021	1,397	679.7	1,021	1,209	850.3	1,021	1,115	1,021	1,021	1020.9
14	348.1	995.1	1,407	671.6	995.1	1,201	833.4	995.1	1,098	995.1	995.1	995.1
15	377.4	979.7	1,436	678.6	979.7	1,208	829.1	979.7	1,094	979.7	979.7	979.7
16	396.1	965.5	1,455	680.8	965.5	1,210	823.2	965.5	1,088	965.5	965.5	965.5
17	393.7	975.1	1,453	684.4	975.1	1,214	829.8	975.1	1,094	975.1	975.1	975.1
18	384.4	1,030	1,443	707.1	1,030	1,236	868.4	1,030	1,133	1,030	1,030	1029.7
19	394.2	1,025	1,453	709.5	1,025	1,239	867.2	1,025	1,132	1,025	1,025	1024.8
20	380	968.3	1,439	674.2	968.3	1,204	821.2	968.3	1,086	968.3	968.3	968.3
21	393.9	955.2	1,453	674.5	955.2	1,204	814.9	955.2	1,080	955.2	955.2	955.2
22	432.2	960	1,491	696.1	960	1,226	828.1	960	1,093	960	960	960
23	453.1	950.7	1,512	701.9	950.7	1,231	826.3	950.7	1,091	950.7	950.7	950.7
24	507.8	858.3	1,567	683	858.3	1,212	770.7	858.3	1,035	858.3	858.3	858.3

3.3 All-Thermal Power Systems

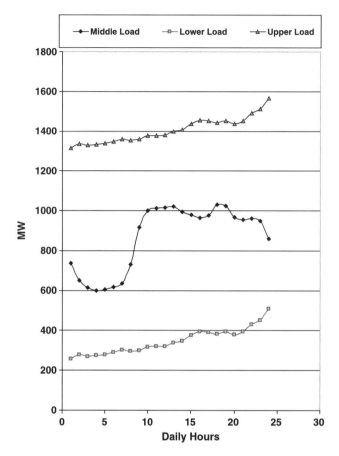

Fig. 3.4 Fuzzy load for (0-α-cut) representation

$$L_{P_{G1\min}} \leq P_{G1} \leq R_{P_{G1\max}} \quad \text{MW}$$
$$L_{P_{G2\min}} \leq P_{G2} \leq R_{P_{G2\max}} \quad \text{MW}$$

Following the solution of the algorithm step by step in a simulation program for different α-cut values, a number of tables are obtained and figures are plotted to show the outcome that influences the generation and cost function when the load varies hour by hour. As an example, the load demand at the tenth hour is a triangular membership function with middle, left, and right spread. Those values can be calculated using the membership formula (3.89) for each α-cut representation; then for each α-cut the incremental fuel cost is calculated for that particular hour.

The power generation middle, left, and right spread of each unit is obtained from Eqs. 3.106, 3.107, and 3.108, respectively; then the total generation is added and tested with the load demand middle, left, and right values. If they are equal then no

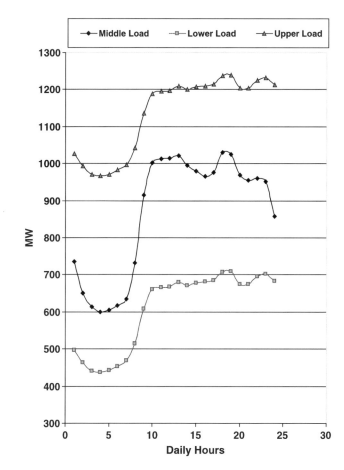

Fig. 3.5 Fuzzy load for (0.5-α-cut) representation

violation has occurred. This solution algorithm is known as the analytical method. Another technique to obtain the solution is gradient methods where an iterative search solution for the fuel incremental cost is given as a guess initially and then the search continues until the total generations are equal to the load demand. In our example, the total sum of the generations of units 1 and 2 at the tenth hour is a triangular membership function with middle, left, and right spreads equal to the load demand at the tenth hour, which proves that there was no violation to the generation limit because the upper and lower values of the generation are within the 10% deviation of the load. Comparing the triangular membership representation of the load demand and the total power generation (Figs. 3.15 and 3.16, respectively), we can see that they are identical and satisfy Eq. 3.87.

The middle, left, and right spread value of the cost function for the tenth hour is calculated for each unit from Eq. 3.114; the total sum of the two units is then obtained. This range of cost value is important because the variation of load happens suddenly

3.3 All-Thermal Power Systems

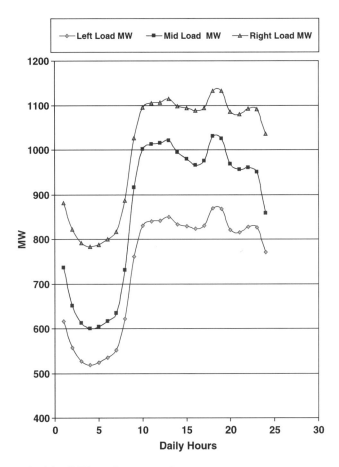

Fig. 3.6 Fuzzy load for (0.75-α-cut) representation

and the calculations using the standard method take a lot of computation time if a large interconnected network is involved. In fuzzy methods, variations are included in the analysis and the range of cost value is calculated hour by hour as the load changes.

Examining these figures reveals the following remarks.

- Figures 3.8, 3.9, and 3.10 show the fuzzy generation of each unit at different α-cut values for model A with 20% deviation on weekdays and their fuzzy triangular representation of the fuzzy load shown in Fig. 3.7. It is clear that the load changes hour by hour and the left, right spread is getting closer as the α-cut increases between [0,1].
- Tables 3.2 and 3.3 show the result of the two-unit generation committed to the system for different α-cut values. Figs. 3.8, 3.9, and 3.10 show the generation values change according to the changes in load demand to satisfy the equality constraint given by Eq. 3.87.

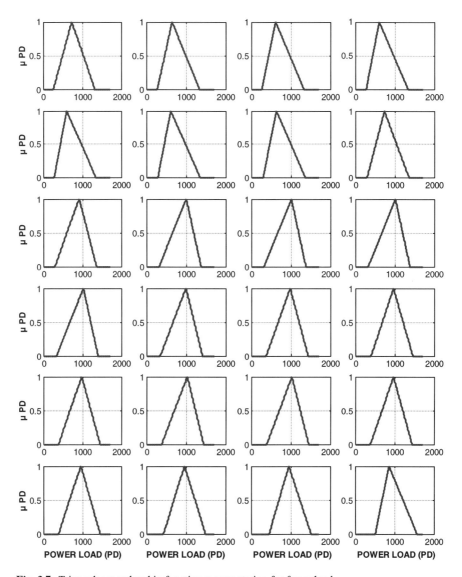

Fig. 3.7 Triangular membership function representation for fuzzy load

- A triangular membership function representation is shown in Figs. 3.11 and 3.12 for each generation unit committed to the system. The left and right spread cover the limit violation restricted on the generation. If a violation occurs then it can be overcome by committing additional units to the system.
- Table 3.4 shows the total generation for different α-cut values. In addition Figs. 3.13 and 3.14 show the satisfaction of the equality constraint in Eq. 3.87 where the total generation committed is equal to the power load demand.
- A triangular membership function representation of the total generation is shown in Fig. 3.15.

3.3 All-Thermal Power Systems

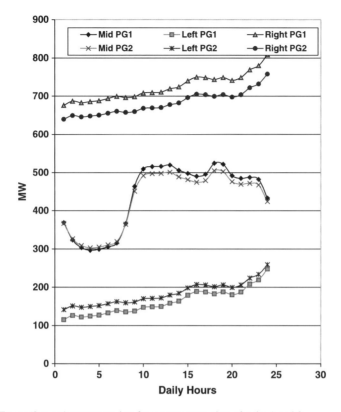

Fig. 3.8 Fuzzy (0-α-cut) representation for power generation of units 1 and 2

- The total fuel costs of different α-cut values calculated from Eq. 3.114 are shown in Figs. 3.16, 3.17, and 3.18. At the hour considered, there is a range of fuel costs for each unit as well as the total cost. Clearly the maximum and minimum values are valuable information to the operator supplying the load to know the cost of the power generated hour by hour Table 3.5.
- A triangular membership function representation of the total cost is shown in Fig. 3.19.
- Note the different values of each α-cut representation. At an α-cut equal to 1, the middle, left, and right spreads have the same value, which defines the crisp value case. The extreme case is defined at α-cut equal to 0 and the average case is represented at α-cut equal to 0.5.

3.3.5 Conclusion

This section presents the economic dispatch problem using fuzzy sets. The load on the system is fuzzy and thus the fuel incremental cost, the cost of generation of each unit, as well as the total cost, is all fuzzy. The simulated example shows the

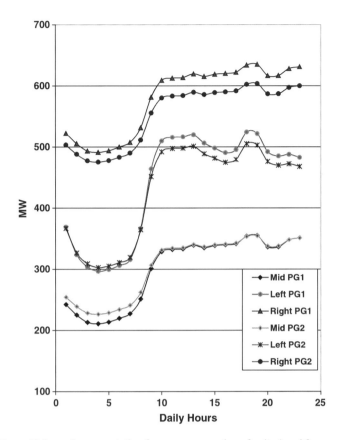

Fig. 3.9 Fuzzy (0.5-α-cut) representation for power generation of units 1 and 2

effectiveness of the algorithm in dealing with the system constraints. Using such an algorithm in the deregulation environment leads to reliable system operation, because the hard constraints on the system are transferred to soft constraints.

3.4 All-Thermal Power Systems with Fuzzy Load and Cost Function Parameters

In this section the parameters $\tilde{\alpha}_i$, $\tilde{\beta}_i$, and $\tilde{\gamma}_i$ of the polynomial nonlinear cost function are considered to be fuzzy. The section starts with a simple application where the power losses in the transmission system that affect minimization of the cost function and the system overall performance are ignored. In the next section the effects of the transmission losses are taken into account in the equality constraint to achieve a more realistic economic dispatch problem.

3.4 All-Thermal Power Systems with Fuzzy Load and Cost Function Parameters

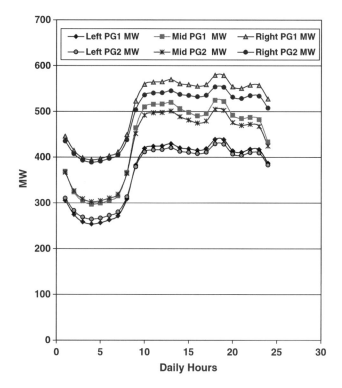

Fig. 3.10 Fuzzy (0.75-α-cut) representation for power generation of units 1 and 2

3.4.1 Problem Formulation

The objective is to find the minimum value of the total operating cost at the same time the equality and inequality operation constraints on the system are satisfied. This can be expressed mathematically as follows.

Minimize

$$\tilde{C}_{total} = \sum_{i=1}^{NG} \tilde{C}_i = \sum_{i=1}^{NG} \tilde{\alpha}_i + \tilde{\beta}_i \tilde{P}_{Gi} + \tilde{\gamma}_i \tilde{P}_{Gi}^2 \quad (3.115)$$

Subject to satisfying the fuzzy constraints given by

$$\sum_{i=1}^{NG} \tilde{P}_{Gi} \geq \tilde{P}_{Demand} \quad (3.116)$$

$$\tilde{P}_{G_i}(\min) \leq \tilde{P}_{G_i} \leq \tilde{P}_{G_i}(\max) \quad i = 1, \ldots \ldots NG \quad (3.117)$$

Table 3.2 Membership function of generator #1 for (0, 0.5, 0.75, 1) α-cut representation model "A" weekdays with 20% deviation

Membership function	$\mu_{PG1}=0$			$\mu_{PG1}=0.5$			$\mu_{PG1}=0.75$			$\mu_{PG1}=1$		
Daily hours	Left PG1 (MW)	Mid PG1 (MW)	Right PG1 (MW)	Left PG1 (MW)	Mid PG1 (MW)	Right PG1 (MW)	Left PG1 (MW)	Mid PG1 (MW)	Right PG1 (MW)	Left PG1 (MW)	Mid PG1 (MW)	Right PG1 (MW)
1	115.5	369	676	242.2	369	522.5	305.6	369	445.8	369	369	369
2	126.4	323.8	686.9	225.1	323.8	505.4	274.5	323.8	414.6	323.8	323.8	323.8
3	122.2	304	682.8	213.1	304	493.4	258.6	304	398.7	304	304	304
4	125	296.8	685.5	210.9	296.8	491.2	253.9	296.8	394	296.8	296.8	296.8
5	127.4	299.6	688	213.5	299.6	493.8	256.6	299.6	396.7	299.6	299.6	299.6
6	133.2	306.1	693.7	219.6	306.1	499.9	262.9	306.1	403	306.1	306.1	306.1
7	139	315.6	699.6	227.3	315.6	507.6	271.5	315.6	411.6	315.6	315.6	315.6
8	135.8	366.7	696.4	251.2	366.7	531.5	309	366.7	449.1	366.7	366.7	366.7
9	138	464.2	698.6	301.1	464.2	581.4	382.7	464.2	522.8	464.2	464.2	464.2
10	147.7	509.8	708.3	328.8	509.8	609	419.3	509.8	559.4	509.8	509.8	509.8
11	148.9	515.7	709.5	332.3	515.7	612.6	424	515.7	564.1	515.7	515.7	515.7
12	149.9	516.6	710.4	333.2	516.6	613.5	424.9	516.6	565	516.6	516.6	516.6
13	158.6	519.9	719.1	339.2	519.9	619.5	429.6	519.9	569.7	519.9	519.9	519.9
14	163.7	506.2	724.2	335	506.2	615.2	420.6	506.2	560.7	506.2	506.2	506.2
15	179.2	498.1	739.8	338.6	498.1	618.9	418.4	498.1	558.5	498.1	498.1	498.1
16	189.1	490.6	749.7	339.8	490.6	620.1	415.2	490.6	555.3	490.6	490.6	490.6
17	187.9	495.6	748.4	341.8	495.6	622	418.7	495.6	558.8	495.6	495.6	495.6
18	182.9	524.5	743.5	353.7	524.5	634	439.1	524.5	579.3	524.5	524.5	524.5
19	188.1	522	748.7	355	522	635.3	438.5	522	578.6	522	522	522
20	180.6	492	741.2	336.3	492	616.6	414.2	492	554.3	492	492	492
21	187.9	485.1	748.5	336.5	485.1	616.8	410.8	485.1	550.9	485.1	485.1	485.1
22	208.2	487.6	768.8	347.9	487.6	628.2	417.8	487.6	557.9	487.6	487.6	487.6
23	219.3	482.7	779.8	351	482.7	631.3	416.9	482.7	557	482.7	482.7	482.7
24	248.2	433.8	808.8	341	433.8	621.3	387.4	433.8	527.5	433.8	433.8	433.8

3.4 All-Thermal Power Systems with Fuzzy Load and Cost Function Parameters

Table 3.3 Membership function of generator #2 for (0, 0.5, 0.75, 1) α-cut representation model "A" weekdays with 20% deviation

Membership function	$\mu_{P_{G2}} = 0$			$\mu_{P_{G2}} = 0.5$			$\mu_{P_{G2}} = 0.75$			$\mu_{P_{G2}} = 1$		
Daily hours	Left PG2 (MW)	Mid PG2 (MW)	Right PG2 (MW)	Left PG2 (MW)	Mid PG2 (MW)	Right PG2 (MW)	Left PG2 (MW)	Mid PG2 (MW)	Right PG2 (MW)	Left PG2 (MW)	Mid PG2 (MW)	Right PG2 (MW)
1	141.5	366.9	639.8	254.2	366.9	503.3	310.5	366.9	435.1	366.9	366.9	366.9
2	151.2	326.8	649.5	239	326.8	488.1	282.9	326.8	407.4	326.8	326.8	326.8
3	147.5	309.1	645.8	228.3	309.1	477.5	268.7	309.1	393.3	309.1	309.1	309.1
4	150	302.8	648.2	226.4	302.8	475.5	264.6	302.8	389.1	302.8	302.8	302.8
5	152.2	305.2	650.4	228.7	305.2	477.8	266.9	305.2	391.5	305.2	305.2	305.2
6	157.2	311	655.5	234.1	311	483.2	272.6	311	397.1	311	311	311
7	162.5	319.5	660.7	241	319.5	490.1	280.2	319.5	404.8	319.5	319.5	319.5
8	159.6	364.8	657.9	262.2	364.8	511.3	313.5	364.8	438.1	364.8	364.8	364.8
9	161.6	451.6	659.8	306.6	451.6	555.7	379.1	451.6	503.6	451.6	451.6	451.6
10	170.2	492	668.5	331.1	492	580.2	411.6	492	536.1	492	492	492
11	171.3	497.3	669.5	334.3	497.3	583.4	415.8	497.3	540.4	497.3	497.3	497.3
12	172.1	498	670.4	335.1	498	584.2	416.6	498	541.1	498	498	498
13	179.8	501	678.1	340.4	501	589.6	420.7	501	545.3	501	501	501
14	184.4	488.9	682.7	336.6	488.9	585.8	412.8	488.9	537.3	488.9	488.9	488.9
15	198.2	481.6	696.4	339.9	481.6	589	410.8	481.6	535.3	481.6	481.6	481.6
16	207	474.9	705.3	341	474.9	590.1	408	474.9	532.5	474.9	474.9	474.9
17	205.9	479.5	704.1	342.7	479.5	591.8	411.1	479.5	535.6	479.5	479.5	479.5
18	201.5	505.2	699.8	353.3	505.2	602.5	429.2	505.2	553.8	505.2	505.2	505.2
19	206.1	502.8	704.4	354.5	502.8	603.6	428.7	502.8	553.2	502.8	502.8	502.8
20	199.4	476.3	697.7	337.8	476.3	587	407.1	476.3	531.6	476.3	476.3	476.3
21	205.9	470.1	704.2	338	470.1	587.1	404.1	470.1	528.6	470.1	470.1	470.1
22	224	472.4	722.2	348.2	472.4	597.3	410.3	472.4	534.8	472.4	472.4	472.4
23	233.8	468	732.1	350.9	468	600	409.4	468	534	468	468	468
24	259.5	424.5	757.8	342	424.5	591.1	383.3	424.5	507.8	424.5	424.5	424.5

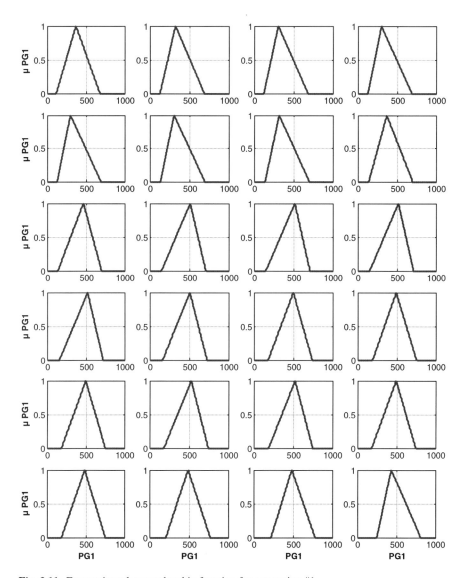

Fig. 3.11 Fuzzy triangular membership function for generation #1

The cost function coefficients, load, and power generators are assumed to be fuzzy and are given as follows.

1. Fuzzy $\tilde{\alpha}_i = (\bar{\alpha}_i, L_{\tilde{\alpha}_i}, R_{\tilde{\alpha}_i})$.
2. Fuzzy $\tilde{\beta}_i = (\bar{\beta}_i, L_{\tilde{\beta}_i}, R_{\tilde{\beta}_i})$.
3. Fuzzy $\tilde{\gamma}_i = (\bar{\gamma}_i, L_{\tilde{\gamma}_i}, R_{\tilde{\gamma}_i})$.
4. Fuzzy load demand $(\bar{P}_D, L_{\tilde{P}_D}, R_{\tilde{P}_D})$.

3.4 All-Thermal Power Systems with Fuzzy Load and Cost Function Parameters

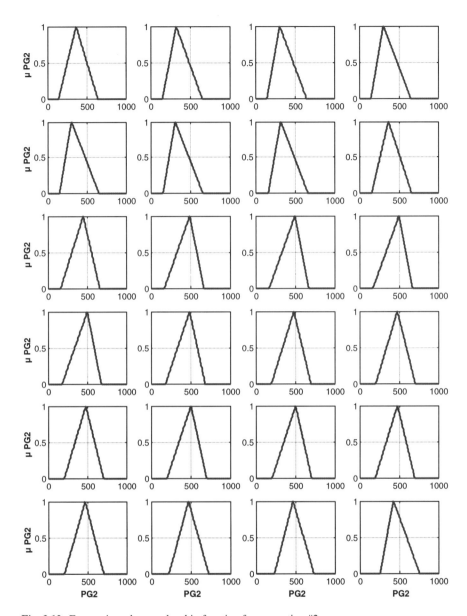

Fig. 3.12 Fuzzy triangular membership function for generation #2

Table 3.4 Membership function of total generator for (0, 0.5, 0.75, 1) α-cut representation model "A" weekdays with 20% deviation

Membership function	$\mu_{tP_G}=0$			$\mu_{tP_G}=0.5$			$\mu_{tP_G}=0.75$			$\mu_{tP_G}=1$		
Daily hours	Left tPG (MW)	Mid tPG (MW)	Right tPG (MW)	Left tPG (MW)	Mid tPG (MW)	Right tPG (MW)	Left tPG (MW)	Mid tPG (MW)	Right tPG (MW)	Left tPG (MW)	Mid tPG (MW)	Right tPG (MW)
1	257	735.9	1,316	496.4	735.9	1,026	616.2	735.9	880.9	735.9	735.9	735.9
2	277.6	650.6	1,336	464.1	650.6	993.5	557.3	650.6	822	650.6	650.6	650.6
3	269.8	613.1	1,329	441.4	613.1	970.8	527.3	613.1	792	613.1	613.1	613.1
4	274.9	599.6	1,334	437.3	599.6	966.7	518.4	599.6	783.1	599.6	599.6	599.6
5	279.6	604.8	1,338	442.2	604.8	971.6	523.5	604.8	788.2	604.8	604.8	604.8
6	290.4	617.1	1,349	453.8	617.1	983.2	535.4	617.1	800.1	617.1	617.1	617.1
7	301.5	635.1	1,360	468.3	635.1	997.7	551.7	635.1	816.4	635.1	635.1	635.1
8	295.4	731.5	1,354	513.5	731.5	1,043	622.5	731.5	887.2	731.5	731.5	731.5
9	299.6	915.8	1,358	607.7	915.8	1,137	761.7	915.8	1,026	915.8	915.8	915.8
10	317.9	1,002	1,377	659.9	1,002	1,189	830.8	1,002	1,096	1,002	1,002	1,002
11	320.2	1,013	1,379	666.6	1,013	1,196	839.8	1,013	1,105	1,013	1,013	1,013
12	322	1,015	1,381	668.3	1,015	1,198	841.4	1,015	1,106	1,015	1,015	1,015
13	338.4	1,021	1,397	679.7	1,021	1,209	850.3	1,021	1,115	1,021	1,021	1,021
14	348.1	995.1	1,407	671.6	995.1	1,201	833.4	995.1	1,098	995.1	995.1	995.1
15	377.4	979.7	1,436	678.6	979.7	1,208	829.1	979.7	1,094	979.7	979.7	979.7
16	396.1	965.5	1,455	680.8	965.5	1,210	823.2	965.5	1,088	965.5	965.5	965.5
17	393.7	975.1	1,453	684.4	975.1	1,214	829.8	975.1	1,094	975.1	975.1	975.1
18	384.4	1,030	1,443	707.1	1,030	1,236	868.4	1,030	1,133	1,030	1,030	1,030
19	394.2	1,025	1,453	709.5	1,025	1,239	867.2	1,025	1,132	1,025	1,025	1,025
20	380	968.3	1,439	674.2	968.3	1,204	821.2	968.3	1,086	968.3	968.3	968.3
21	393.9	955.2	1,453	674.5	955.2	1,204	814.9	955.2	1,080	955.2	955.2	955.2
22	432.2	960	1,491	696.1	960	1,226	828.1	960	1,093	960	960	960
23	453.1	950.7	1,512	701.9	950.7	1,231	826.3	950.7	1,091	950.7	950.7	950.7
24	507.8	858.3	1,567	683	858.3	1,212	770.7	858.3	1,035	858.3	858.3	858.3

3.4 All-Thermal Power Systems with Fuzzy Load and Cost Function Parameters 119

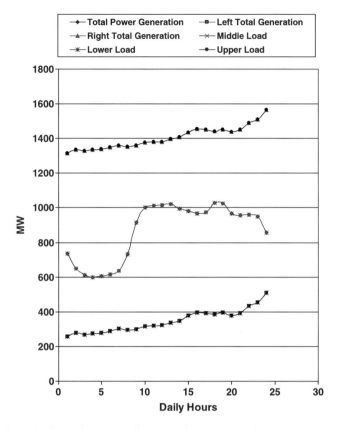

Fig. 3.13 Fuzzy (0.75-α-cut) representation for total power generation

5. Fuzzy power generator $(\bar{P}_{Gi}, L_{\tilde{P}_{Gi}}, R_{\tilde{P}_{Gi}})$.

Substituting the middle, left, and right side values into the cost function equation we get:

$$\tilde{C}_{total} = \sum_{i=1}^{NG} (\bar{C}_{ti}, L_{\tilde{C}_{ti}}, R_{\tilde{C}_{ti}}) = \sum_{i=1}^{NG} (\bar{\alpha}_i, L_{\tilde{\alpha}_i}, R_{\tilde{\alpha}_i}) + (\bar{\beta}_i, L_{\tilde{\beta}_i}, R_{\tilde{\beta}_i})(\bar{P}_{Gi}, L_{\tilde{P}_{Gi}}, R_{\tilde{P}_{Gi}})$$
$$+ (\bar{\gamma}_i, L_{\tilde{\gamma}_i}, R_{\tilde{\gamma}_i})(\bar{P}_{Gi}, L_{\tilde{P}_{Gi}}, R_{\tilde{P}_{Gi}})(\bar{P}_{Gi}, L_{\tilde{P}_{Gi}}, R_{\tilde{P}_{Gi}})$$

(3.118)

Subject to satisfying

$$\sum_{i=1}^{NG} (\bar{P}_{Gi}, L_{\tilde{P}_{Gi}}, R_{\tilde{P}_{Gi}}) \geq (\bar{P}_D, L_{\tilde{P}_D}, R_{\tilde{P}_D})$$ (3.119)

$$L_{\tilde{P}_{Gi}} \leq \bar{P}_{Gi} \leq R_{\tilde{P}_{Gi}} \quad i = 1, \ldots\ldots NG$$ (3.120)

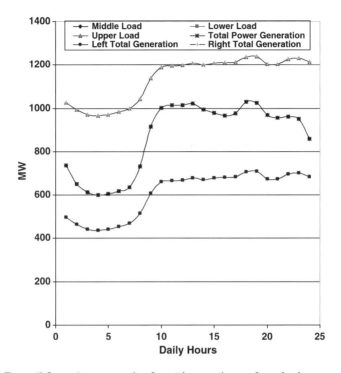

Fig. 3.14 Fuzzy (0.5-α-cut) representation for total generation vs. fuzzy load

The same method as in the previous section is applied where the lower and upper limits of each unit generator are substituted by the left, right spreads or the left and right side of the membership function for the fuzzy generation as shown in Eq. 3.20. Fuzzy triangular membership functions for all the fuzzy parameter values stated earlier are chosen. The load demand and the power generation triangular membership function were shown in Fig. 3.3a, b. The percentage of deviation among the three parameters is taken into consideration to explore the outcome of the minimum cost of the network.

The mathematical formula of $\tilde{\alpha}_i$ membership function is:

$$\mu(\tilde{\alpha}_i) = \begin{cases} 0 & \tilde{\alpha}_i < L_{\tilde{\alpha}_i} \\ \dfrac{\tilde{\alpha}_i - L_{\tilde{\alpha}_i}}{a_{\alpha_i}} & L_{\tilde{\alpha}_i} \leq \tilde{\alpha}_i \leq \bar{\alpha}_i \\ \dfrac{R_{\tilde{\alpha}_i} - \tilde{\alpha}_i}{b_{\alpha_i}} & \bar{\alpha}_i \leq \tilde{\alpha}_i \leq R_{\tilde{\alpha}_i} \\ 0 & \tilde{\alpha}_i > R_{\tilde{\alpha}_i} \end{cases} \tag{3.121}$$

The same mathematical formula representation applies to $\tilde{\beta}_i$ and $\tilde{\gamma}_i$. Using the Lagrange multiplier formula, to relax systemwide constraints into unconstrained form as in the crisp economic dispatch problem, a crisp optimization of the minimum cost function is obtained. The procedure in this section is to transform

3.4 All-Thermal Power Systems with Fuzzy Load and Cost Function Parameters

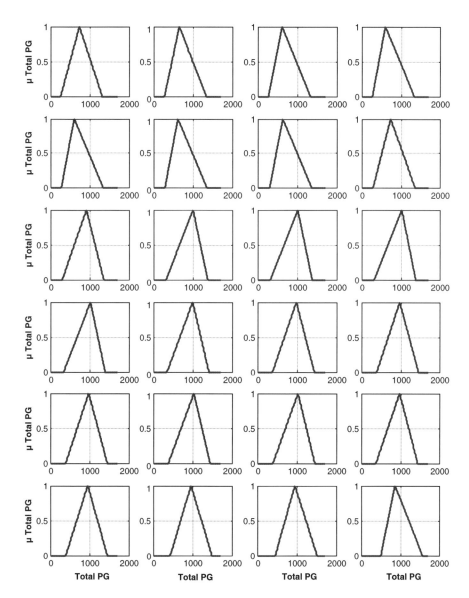

Fig. 3.15 Fuzzy triangular membership function for total generation

the fuzzy load and fuzzy cost function coefficients $\tilde{\alpha}_i$, $\tilde{\beta}_i$, and $\tilde{\gamma}_i$ into a triangular membership function by assigning a degree of membership to each possible α-cut value of the load and the cost function coefficients, which means mapping the fuzzy variable on the [0, 1] interval and then performing the fuzzy arithmetic operation to obtain the minimum total cost value.

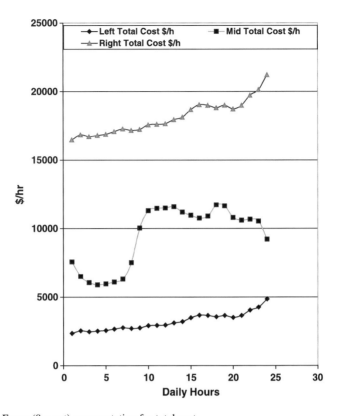

Fig. 3.16 Fuzzy (0-α-cut) representation for total cost

Applying the fuzzy parameters to the crisp Eq. 3.84 we get:

$$\tilde{\lambda} = \tilde{\beta}_i + 2\tilde{\gamma}_i \tilde{P}_{G_i} \qquad (3.122)$$

Setting the value of $\tilde{P}_{G_i}$ equal to $\tilde{P}_D$ as in Eq. 3.116 we get:

$$\sum_{i=1}^{NG} \frac{\tilde{\lambda} - \tilde{\beta}_i}{2\tilde{\gamma}_i} = \tilde{P}_D \qquad (3.123)$$

Solving for $\tilde{\lambda}$,

$$\tilde{\lambda} = \frac{2\tilde{P}_D + \sum_{i=1}^{NG} \frac{\tilde{\beta}_i}{\tilde{\gamma}_i}}{\sum_{i=1}^{NG} \frac{1}{\tilde{\gamma}_i}} \qquad (3.124)$$

3.4 All-Thermal Power Systems with Fuzzy Load and Cost Function Parameters 123

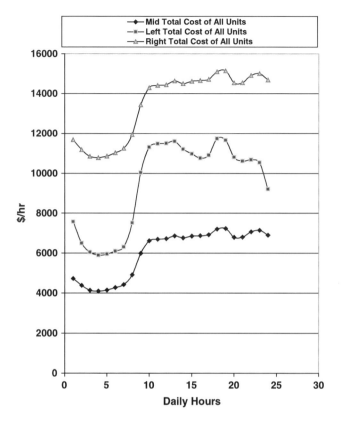

Fig. 3.17 Fuzzy (0.5-α-cut) representation for total cost

Evaluating the middle, left, and right sides of the incremental cost in the equation

$$(\bar{\lambda}, L_{\tilde{\lambda}}, R_{\tilde{\lambda}}) = \frac{2(\bar{P}_D, L_{\tilde{P}_D}, R_{\tilde{P}_D}) + \sum_{i=1}^{NG} \frac{(\bar{\beta}_i, L_{\tilde{\beta}_i}, R_{\tilde{\beta}_i})}{(\bar{\gamma}_i, L_{\tilde{\gamma}_i}, R_{\tilde{\gamma}_i})}}{\sum_{i=1}^{NG} \frac{1}{(\bar{\gamma}_i, L_{\tilde{\gamma}_i}, R_{\tilde{\gamma}_i})}} \quad (3.125)$$

3.4.2 Fuzzy Interval Arithmetic Representation on Triangular Fuzzy Numbers

Equation 3.125 has a number of arithmetic operations such as addition, subtraction, multiplication, division, and inverse function which are all in fuzzy form. Applying fuzzy interval arithmetic operations implemented by their α-cut operation on triangular fuzzy numbers is shown in Table 2.5, taking into consideration that *X* and *Y*

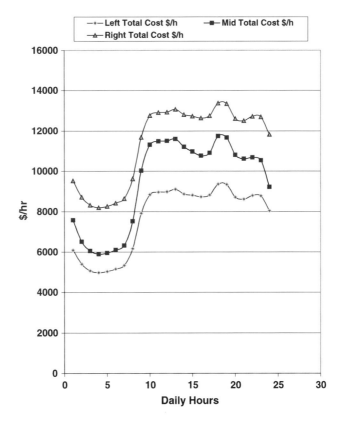

Fig. 3.18 Fuzzy (0.75-α-cut) representation for total cost

have greater than zero parameter formulation because all the input data values are greater than zero in the economical dispatch formulation.

Then the middle incremental cost becomes:

$$\bar{\lambda} = \frac{2\bar{P}_D + \sum_{i=1}^{NG} \dfrac{\bar{\beta}_i}{\bar{\gamma}_i}}{\sum_{i=1}^{NG} \dfrac{1}{\bar{\gamma}_i}} \tag{3.126}$$

The left equation for the incremental cost is:

$$L_{\tilde{\lambda}} = \frac{2L_{\tilde{P}_D} + \sum_{i=1}^{NG} \dfrac{L_{\tilde{\beta}_i}}{R_{\tilde{\gamma}_i}}}{\sum_{i=1}^{NG} \dfrac{1}{R_{\tilde{\gamma}_i}}} \tag{3.127}$$

3.4 All-Thermal Power Systems with Fuzzy Load and Cost Function Parameters

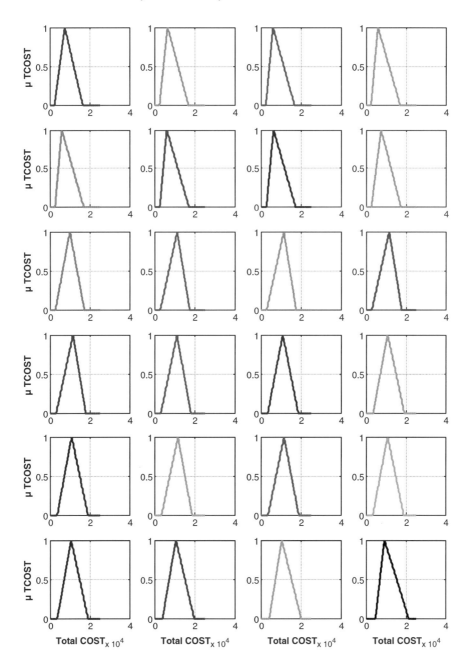

Fig. 3.19 Fuzzy triangular membership function for total cost

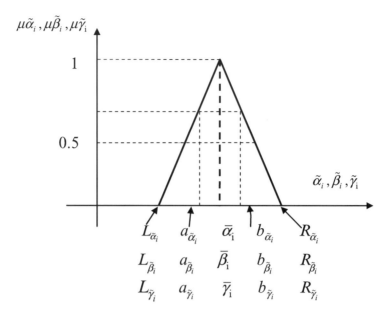

Fig. 3.20 A triangular membership function for fuzzy cost function coefficient

The right-side equation for the incremental cost is:

$$R_{\tilde{\lambda}} = \frac{2r_{\tilde{P}_D} + \sum_{i=1}^{NG} \frac{R_{\tilde{\beta}_i}}{L_{\tilde{\gamma}_i}}}{\sum_{i=1}^{NG} \frac{1}{L_{\tilde{\gamma}_i}}} \qquad (3.128)$$

The fuzzy generation of each unit can be calculated as

$$\tilde{P}_{Gi} = \frac{\tilde{\lambda} - \tilde{\beta}_i}{2\tilde{\gamma}_i} \quad i = 1,..,NG \qquad (3.129)$$

Evaluating the middle, left, and right sides of the generation into the equation

$$\tilde{P}_{Gi} = (\bar{P}_{Gi}, L_{\tilde{P}_{Gi}}, R_{\tilde{P}_{Gi}}) = \frac{(\bar{\lambda}, L_{\tilde{\lambda}}, R_{\tilde{\lambda}}) - (\bar{\beta}_i, L_{\tilde{\beta}_i}, R_{\tilde{\beta}_i})}{2(\bar{\gamma}_i, L_{\tilde{\gamma}_i}, R_{\tilde{\gamma}_i})} \quad i = 1,..,NG \qquad (3.130)$$

The middle, left, and right sides of all the generator equations are:

$$\bar{P}_i = \frac{\bar{\lambda}_i - \bar{\beta}_i}{2(\bar{\gamma}_i)} \quad i = 1,..,NG \qquad (3.131)$$

3.4 All-Thermal Power Systems with Fuzzy Load and Cost Function Parameters 127

Table 3.5 Membership function of total cost for (0, 0.5, 0.75, 1) α-cut representation model "A" weekdays with 20% deviation

Membership function	$\mu_C = 0$			$\mu_C = 0.5$			$\mu_C = 0.75$			$\mu_C = 1$		
Daily hours	Left cost ($/h)	Mid cost ($/h)	Right cost ($/h)	Left cost ($/h)	Mid cost ($/h)	Right cost ($/h)	Left cost ($/h)	Mid cost ($/h)	Right cost ($/h)	Left cost ($/h)	Mid cost ($/h)	Right cost ($/h)
1	2,366.5	7,575	16,482	4,728	7,575	11,673	6,091	7,575	9,535	7,575	7,575	7,575
2	2,550.7	6,505	16,851	4,381	6,505	11,180	5,406	6,505	8,718	6,505	6,505	6,505
3	2,480.7	6,055	16,711	4,143	6,055	10,841	5,067	6,055	8,312	6,055	6,055	6,055
4	2,527	5,895	16,804	4,099	5,895	10,779	4,969	5,895	8,194	5,895	5,895	5,895
5	2,568.9	5,956	16,887	4,151	5,956	10,852	5,026	5,956	8,262	5,956	5,956	5,956
6	2,667.1	6,102	17,083	4,272	6,102	11,025	5,159	6,102	8,422	6,102	6,102	6,102
7	2,768.7	6,318	17,283	4,425	6,318	11,244	5,342	6,318	8,641	6,318	6,318	6,318
8	2,713.2	7,519	17,174	4,915	7,519	11,936	6,166	7,519	9,624	7,519	7,519	7,519
9	2,751.3	10,034	17,249	5,990	10,034	13,434	7,912	10,034	11,682	10,034	10,034	10,034
10	2,921.7	11,306	17,584	6,619	11,306	14,296	8,839	11,306	12,764	11,306	11,306	11,306
11	2,942.9	11,476	17,626	6,701	11,476	14,409	8,962	11,476	12,907	11,476	11,476	11,476
12	2,959.5	11,501	17,658	6,722	11,501	14,437	8,984	11,501	12,933	11,501	11,501	11,501
13	3,115.4	11,597	17,961	6,863	11,597	14,629	9,107	11,597	13,076	11,597	11,597	11,597
14	3,208.1	11,205	18,141	6,763	11,205	14,493	8,873	11,205	12,804	11,205	11,205	11,205
15	3,493.5	10,973	18,689	6,849	10,973	14,610	8,815	10,973	12,737	10,973	10,973	10,973
16	3,679.8	10,761	19,043	6,877	10,761	14,649	8,734	10,761	12,642	10,761	10,761	10,761
17	3,655.9	10,904	18,998	6,922	10,904	14,710	8,824	10,904	12,747	10,904	10,904	10,904
18	3,563.2	11,732	18,822	7,207	11,732	15,096	9,359	11,732	13,369	11,732	11,732	11,732
19	3,660.9	11,657	19,008	7,238	11,657	15,138	9,342	11,657	13,349	11,657	11,657	11,657
20	3,519.6	10,803	18,739	6,795	10,803	14,536	8,707	10,803	12,611	10,803	10,803	10,803
21	3,657.2	10,609	19,001	6,799	10,609	14,543	8,621	10,609	12,510	10,609	10,609	10,609
22	4,047.1	10,680	19,734	7,069	10,680	14,909	8,800	10,680	12,720	10,680	10,680	10,680
23	4,264.6	10,543	20,139	7,141	10,543	15,007	8,776	10,543	12,692	10,543	10,543	10,543
24	4,851.9	9,218	21,217	6,905	9,218	14,686	8,029	9,218	11,819	9,218	9,218	9,218

$$L_{\tilde{P}_{Gi}} = \frac{L_{\tilde{\lambda}} - R_{\tilde{\beta}_i}}{2(R_{\tilde{\gamma}_i})} \quad i = 1, .., NG \tag{3.132}$$

$$R_{\tilde{P}_{Gi}} = \frac{R_{\tilde{\lambda}} - L_{\tilde{\beta}_i}}{2(L_{\tilde{\gamma}_i})} \quad i = 1, .., NG \tag{3.133}$$

Substituting the power generation into cost function equation we get:

$$\sum_{i=1}^{NG} \tilde{C}_i = \sum_{i=1}^{NG} \bar{\alpha}_i + \bar{\beta}_i \bar{P}_{Gi} + \bar{\gamma}_i \bar{P}_{Gi}^2$$

$$\sum_{i=1}^{NG} L_{\tilde{C}_i} = \sum_{i=1}^{NG} L_{\tilde{\alpha}_i} + L_{\tilde{\beta}_i} L_{\tilde{P}_{Gi}} + L_{\tilde{\gamma}_i} L_{\tilde{P}_{Gi}} L_{\tilde{P}_{Gi}}$$

$$\sum_{i=1}^{NG} R_{\tilde{C}_i} = \sum_{i=1}^{NG} R_{\tilde{\alpha}_i} + R_{\tilde{\beta}_i} R_{\tilde{P}_{Gi}} + R_{\tilde{\gamma}_i} R_{\tilde{P}_{Gi}} R_{\tilde{P}_{Gi}} \tag{3.134}$$

3.4.3 Fuzzy Arithmetic on Triangular L–R Representation of Fuzzy Numbers

Using Table 2.4 in Chap. 2 to perform the fuzzy arithmetic calculation on triangular L–R representation of fuzzy numbers in Eqs. 3.124 and 3.129 we get the following. The middle or crisp value of the incremental cost function is:

$$\bar{\lambda} = \frac{2\bar{P}_D + \sum_{i=1}^{NG} \frac{\bar{\beta}_i}{\bar{\gamma}_i}}{\sum_{i=1}^{NG} \frac{1}{\bar{\gamma}_i}} \tag{3.135}$$

The left spread of the incremental cost becomes:

$$a_{\lambda i} = \frac{(2\bar{P}_D + \sum_{i=1}^{NG} \frac{\bar{\beta}_i}{\bar{\gamma}_i})(\sum_{i=1}^{NG} a_{\gamma i} \bar{\gamma}_i^2) + (\sum_{i=1}^{NG} \frac{1}{\bar{\gamma}_i})(2a_{Di} + (\sum_{i=1}^{NG} (\bar{\beta}_i b_{\gamma i} + a_{\beta i} \bar{\gamma}_i)/\bar{\gamma}_i^2))}{(\sum_{i=1}^{NG} \frac{1}{\bar{\gamma}_i})^2}$$

$$\tag{3.136}$$

The right spread of the incremental cost is:

$$b_{\lambda i} = \frac{(2\bar{P}_D + \sum_{i=1}^{NG} \frac{\bar{\beta}_i}{\bar{\gamma}_i})(\sum_{i=1}^{NG} b_{\gamma i} \bar{\gamma}_i^2) + (\sum_{i=1}^{NG} \frac{1}{\bar{\gamma}_i})(2b_{Di} + (\sum_{i=1}^{NG} (\bar{\beta}_i a_{\gamma i} + b_{\beta i} \bar{\gamma}_i)/\bar{\gamma}_i^2))}{(\sum_{i=1}^{NG} \frac{1}{\bar{\gamma}_i})^2} \tag{3.137}$$

After applying the *L–R* representation method, the middle power generation in Eq. 3.129 becomes:

$$\bar{P}_i = \frac{\bar{\lambda}_i - \bar{\beta}_i}{2(\bar{\gamma}_i)} \quad (3.138)$$

The left spread of the power generation is:

$$a_{pi} = \frac{(\bar{\lambda}_i - \bar{\beta}_i)2b_{\gamma i} + 2\bar{\gamma}_i(a_{\lambda i} + b_{\beta i})}{(2\bar{\gamma}_i)^2} \quad (3.139)$$

The right spread of the power generation is:

$$b_{pi} = \frac{(\bar{\lambda}_i - \bar{\beta}_i)2a_{\gamma i} + 2\bar{\gamma}_i(b_{\lambda i} + a_{\beta i})}{(2\bar{\gamma}_i)^2} \quad (3.140)$$

Substituting the power generation into the middle, left, and right spread cost equation we get:

$$\sum_{i=1}^{NG} \bar{C}_i = \sum_{i=1}^{NG} \bar{\alpha}_i + \bar{\beta}_i \bar{P}_i + \bar{\gamma}_i \bar{P}_i^2$$

$$\sum_{i=1}^{NG} a_{c_i} = \sum_{i=1}^{NG} a_{\alpha_i} + \bar{\beta}_i a_{P_i} + \bar{P}_i a_{\beta_i} + 2\bar{\gamma}_i \bar{P}_i a_{P_i} + \bar{P}_i^2 a_{\gamma_i}$$

$$\sum_{i=1}^{NG} b_{c_i} = \sum_{i=1}^{NG} b_{\alpha_i} + \bar{\beta}_i b_{P_i} + \bar{P}_i b_{\beta_i} + 2\bar{\gamma}_i \bar{P}_i b_{P_i} + \bar{P}_i^2 b_{\gamma_i} \quad (3.141)$$

3.4.4 Example

In this section a simulated example is presented to perform the economical dispatch (ED) operation of power systems when the load demand on the system is fuzzy for 24 h and the cost function coefficients are fuzzy while ignoring the transmission line losses that affect the minimization of the cost function and the system's overall performance. The load demand is chosen to be a triangular membership with a 10% deviation as tabulated in Table 3.6 and plotted in Fig. 3.21. In addition the fuzzy cost function coefficients are a triangular membership function with different percentages of deviation. The selected synthetic system example contains three thermal units and the input/out fuel cost functions, for each unit, are given as follows.

Table 3.6 Membership function of load demand for (0, 0.5, 0.7, 1) α-cut representation for model "A" weekdays with 10% deviation for (P_D) and 3% for (α, β, γ)

Membership function	$\mu_{P_{Load}} = 0$			$\mu_{P_{Load}} = 0.5$			$\mu_{P_{Load}} = 0.75$			$\mu_{P_{Load}} = 1$		
Daily hours	Left load (MW)	Mid load (MW)	Right load (MW)	Left load (MW)	Mid load (MW)	Right load (MW)	Left load (MW)	Mid load (MW)	Right load (MW)	Left load (MW)	Mid load (MW)	Right load (MW)
1	1,005.8	1,117.6	1,229.4	1,061.7	1,117.6	1,173.5	1,089.7	1,117.6	1,145.5	1,117.6	1,117.6	1,117.6
2	905.76	1,006.4	1,107	956.08	1,006.4	1,056.7	981.24	1,006.4	1,031.6	1,006.4	1,006.4	1,006.4
3	849.24	943.6	1,038	896.42	943.6	990.78	920.01	943.6	967.19	943.6	943.6	943.6
4	784.05	871.17	958.29	827.61	871.17	914.73	849.39	871.17	892.95	871.17	871.17	871.17
5	731.7	813	894.3	772.35	813	853.65	792.68	813	833.33	813	813	813
6	782.73	869.7	956.67	826.22	869.7	913.19	847.96	869.7	891.44	869.7	869.7	869.7
7	823.32	914.8	1,006.3	869.06	914.8	960.54	891.93	914.8	937.67	914.8	914.8	914.8
8	880.83	978.7	1,076.6	929.77	978.7	1,027.6	954.23	978.7	1,003.2	978.7	978.7	978.7
9	1,041.6	1,157.3	1,273	1,099.4	1,157.3	1,215.2	1,128.4	1,157.3	1,186.2	1,157.3	1,157.3	1,157.3
10	1,101.4	1,223.8	1,346.2	1,162.6	1,223.8	1,285	1,193.2	1,223.8	1,254.4	1,223.8	1,223.8	1,223.8
11	1,095.1	1,216.8	1,338.5	1,156	1,216.8	1,277.6	1,186.4	1,216.8	1,247.2	1,216.8	1,216.8	1,216.8
12	1,155.9	1,284.3	1,412.7	1,220.1	1,284.3	1,348.5	1,252.2	1,284.3	1,316.4	1,284.3	1,284.3	1284.3
13	1,132.7	1,258.6	1,384.5	1,195.7	1,258.6	1,321.5	1,227.1	1,258.6	1,290.1	1,258.6	1,258.6	1,258.6
14	1,087	1,207.8	1,328.6	1,147.4	1,207.8	1,268.2	1,177.6	1,207.8	1,238	1,207.8	1,207.8	1,207.8
15	1,039.9	1,155.4	1,270.9	1,097.6	1,155.4	1,213.2	1,126.5	1,155.4	1,184.3	1,155.4	1,155.4	1,155.4
16	999.54	1,110.6	1,221.7	1,055.1	1,110.6	1,166.1	1,082.8	1,110.6	1,138.4	1,110.6	1,110.6	1,110.6
17	984.69	1,094.1	1,203.5	1,039.4	1,094.1	1,148.8	1,066.7	1,094.1	1,121.5	1,094.1	1,094.1	1,094.1
18	1,002	1,113.3	1,224.6	1,057.6	1,113.3	1,169	1,085.5	1,113.3	1,141.1	1,113.3	1,113.3	1,113.3
19	1,067.8	1,186.4	1,305	1,127.1	1,186.4	1,245.7	1,156.7	1,186.4	1,216.1	1,186.4	1,186.4	1,186.4
20	1,025.1	1,139	1,252.9	1,082.1	1,139	1,196	1,110.5	1,139	1,167.5	1,139	1,139	1,139
21	1,037.1	1,152.3	1,267.5	1,094.7	1,152.3	1,209.9	1,123.5	1,152.3	1,181.1	1,152.3	1,152.3	1,152.3
22	1,103.9	1,226.5	1,349.2	1,165.2	1,226.5	1,287.8	1,195.8	1,226.5	1,257.2	1,226.5	1,226.5	1,226.5
23	1,078.6	1,198.4	1,318.2	1,138.5	1,198.4	1,258.3	1,168.4	1,198.4	1,228.4	1,198.4	1,198.4	1,198.4
24	1,000.6	1,111.8	1,223	1,056.2	1,111.8	1,167.4	1,084	1,111.8	1,139.6	1,111.8	1,111.8	1,111.8

3.4 All-Thermal Power Systems with Fuzzy Load and Cost Function Parameters 131

Table 3.7 Membership function of incremental cost for (0, 0.5, 0.7, 1) α-cut representation for model "A" weekdays with 10% deviation for (P_D) and 3% for (α, β, γ)

Membership function	$\mu_\lambda = 0$			$\mu_\lambda = 0.5$			$\mu_\lambda = 0.75$			$\mu_\lambda = 1$		
Daily hours	Left λ ($MW/h)	Mid λ ($MW/h)	Right λ ($MW/h)	Left λ ($MW/h)	Mid λ ($MW/h)	Right λ ($MW/h)	Left λ ($MW/h)	Mid λ ($MW/h)	Right λ ($MW/h)	Left λ ($MW/h)	Mid λ ($MW/h)	Right λ ($MW/h)
1	11.985	12.617	13.214	12.306	12.617	12.92	12.463	12.617	12.77	12.617	12.617	12.617
2	11.441	12.031	12.588	11.753	12.031	12.305	11.895	12.031	12.166	12.031	12.031	12.031
3	11.134	11.699	12.234	11.443	11.699	11.954	11.574	11.699	11.824	11.699	11.699	11.699
4	10.78	11.317	11.827	11.083	11.317	11.55	11.202	11.317	11.432	11.317	11.317	11.317
5	10.495	11.01	11.499	10.793	11.01	11.227	10.903	11.01	11.117	11.01	11.01	11.01
6	10.773	11.309	11.818	11.079	11.309	11.54	11.194	11.309	11.424	11.309	11.309	11.309
7	10.993	11.547	12.072	11.305	11.547	11.789	11.427	11.547	11.668	11.547	11.547	11.547
8	11.306	11.884	12.432	11.626	11.884	12.143	11.755	11.884	12.014	11.884	11.884	11.884
9	12.18	12.827	13.438	12.522	12.827	13.132	12.674	12.827	12.98	12.827	12.827	12.827
10	12.505	13.178	13.812	12.855	13.178	13.501	13.016	13.178	13.339	13.178	13.178	13.178
11	12.471	13.141	13.773	12.82	13.141	13.462	12.98	13.141	13.302	13.141	13.141	13.141
12	12.801	13.497	14.153	13.158	13.497	13.836	13.328	13.497	13.667	13.497	13.497	13.497
13	12.675	13.362	14.008	13.03	13.362	13.694	13.196	13.362	13.528	13.362	13.362	13.362
14	12.427	13.094	13.722	12.775	13.094	13.412	12.934	13.094	13.253	13.094	13.094	13.094
15	12.17	12.817	13.427	12.512	12.817	13.122	12.665	12.817	12.969	12.817	12.817	12.817
16	11.951	12.581	13.175	12.287	12.581	12.874	12.434	12.581	12.727	12.581	12.581	12.581
17	11.87	12.493	13.082	12.205	12.493	12.782	12.349	12.493	12.638	12.493	12.493	12.493
18	11.964	12.595	13.19	12.301	12.595	12.889	12.448	12.595	12.742	12.595	12.595	12.595
19	12.322	12.981	13.602	12.668	12.981	13.294	12.824	12.981	13.137	12.981	12.981	12.981
20	12.09	12.73	13.335	12.43	12.73	13.031	12.58	12.73	12.881	12.73	12.73	12.73
21	12.155	12.801	13.41	12.497	12.801	13.105	12.649	12.801	12.953	12.801	12.801	12.801
22	12.518	13.192	13.827	12.869	13.192	13.516	13.03	13.192	13.354	13.192	13.192	13.192
23	12.381	13.044	13.669	12.728	13.044	13.36	12.886	13.044	13.202	13.044	13.044	13.044
24	11.957	12.587	13.182	12.294	12.587	12.88	12.44	12.587	12.734	12.587	12.587	12.587

Table 3.8 Membership function of generation #1 for (0, 0.5, 0.7, 1) α-cut representation for model "A" weekdays with 10% deviation for (P_D) and 3% for (α, β, γ)

Membership function	$\mu_{PG1} = 0$			$\mu_{PG1} = 0.5$			$\mu_{PG1} = 0.75$			$\mu_{PG1} = 1$		
Daily hours	Left PG1 (MW)	Mid PG1 (MW)	Right PG1 (MW)	Left PG1 (MW)	Mid PG1 (MW)	Right PG1 (MW)	Left PG1 (MW)	Mid PG1 (MW)	Right PG1 (MW)	Left PG1 (MW)	Mid PG1 (MW)	Right PG1 (MW)
1	289.77	351.09	413.93	320.25	351.09	382.32	335.63	351.09	366.66	351.09	351.09	351.09
2	256.76	314.42	373.59	291.56	314.42	337.36	304.54	314.42	324.29	314.42	314.42	314.42
3	238.11	293.7	350.8	275	293.7	312.42	285.53	293.7	301.88	293.7	293.7	293.7
4	216.61	269.81	324.52	253.87	269.81	285.76	262.53	269.81	277.09	269.81	269.81	269.81
5	199.34	250.62	303.42	236.43	250.62	264.82	243.89	250.62	257.35	250.62	250.62	250.62
6	216.18	269.33	323.99	254.59	269.33	284.06	262.15	269.33	276.5	269.33	269.33	269.33
7	229.56	284.2	340.35	268.92	284.2	299.49	276.66	284.2	291.75	284.2	284.2	284.2
8	248.53	305.28	363.54	289.04	305.28	321.52	297.21	305.28	313.35	305.28	305.28	305.28
9	301.55	364.19	428.34	345.05	364.19	383.32	354.65	364.19	373.73	364.19	364.19	364.19
10	321.29	386.12	452.47	365.92	386.12	406.33	376.03	386.12	396.21	386.12	386.12	386.12
11	319.22	383.81	449.93	363.73	383.81	403.89	373.78	383.81	393.85	383.81	383.81	383.81
12	339.25	406.08	474.42	384.89	406.08	427.27	395.49	406.08	416.67	406.08	406.08	406.08
13	331.62	397.6	465.09	376.84	397.6	418.36	387.22	397.6	407.98	397.6	397.6	397.6
14	316.54	380.85	446.66	360.92	380.85	400.77	370.89	380.85	390.81	380.85	380.85	380.85
15	300.99	363.56	427.65	344.51	363.56	382.62	354.03	363.56	373.09	363.56	363.56	363.56
16	287.69	348.78	411.39	330.47	348.78	367.1	339.63	348.78	357.94	348.78	348.78	348.78
17	282.79	343.34	405.41	325.3	343.34	361.39	334.32	343.34	352.36	343.34	343.34	343.34
18	288.49	349.68	412.37	331.31	349.68	368.04	340.5	349.68	358.86	349.68	349.68	349.68
19	310.19	373.79	438.9	354.22	373.79	393.35	364	373.79	383.57	373.79	373.79	373.79
20	296.12	358.15	421.7	339.37	358.15	376.94	348.76	358.15	367.54	358.15	358.15	358.15
21	300.07	362.54	426.52	343.54	362.54	381.54	353.04	362.54	372.04	362.54	362.54	362.54
22	322.09	387.01	453.45	366.79	387.01	407.24	376.9	387.01	397.13	387.01	387.01	387.01
23	313.75	377.75	443.25	357.98	377.75	397.51	367.86	377.75	387.63	377.75	377.75	377.75
24	288.04	349.18	411.83	330.84	349.18	367.52	340.01	349.18	358.35	349.18	349.18	349.18

3.4 All-Thermal Power Systems with Fuzzy Load and Cost Function Parameters

Table 3.9 Membership function of generation #2 for (0, 0.5, 0.7, 1) α-cut representation for model "A" weekdays with 10% deviation for (P_D) and 3% for (α, β, γ)

Membership function	$\mu_{P_{G2}} = 0$			$\mu_{P_{G2}} = 0.5$			$\mu_{P_{G2}} = 0.75$			$\mu_{P_{G2}} = 1$		
Daily hours	Left PG2 (MW)	Mid PG2 (MW)	Right PG2 (MW)	Left PG2 (MW)	Mid PG2 (MW)	Right PG2 (MW)	Left PG2 (MW)	Mid PG2 (MW)	Right PG2 (MW)	Left PG2 (MW)	Mid PG2 (MW)	Right PG2 (MW)
1	296.46	350.97	406.83	323.56	350.97	378.73	337.22	350.97	364.81	350.97	350.97	350.97
2	267.12	318.37	370.97	298.06	318.37	338.76	309.6	318.37	327.15	318.37	318.37	318.37
3	250.55	299.96	350.71	283.33	299.96	316.6	292.69	299.96	307.22	299.96	299.96	299.96
4	231.43	278.72	327.35	264.55	278.72	292.89	272.25	278.72	285.19	278.72	278.72	278.72
5	216.08	261.67	308.59	249.05	261.67	274.28	255.68	261.67	267.65	261.67	261.67	261.67
6	231.04	278.29	326.88	265.19	278.29	291.39	271.91	278.29	284.67	278.29	278.29	278.29
7	242.95	291.51	341.42	277.93	291.51	305.1	284.81	291.51	298.22	291.51	291.51	291.51
8	259.81	310.25	362.03	295.81	310.25	324.68	303.07	310.25	317.42	310.25	310.25	310.25
9	306.94	362.61	419.63	345.6	362.61	379.62	354.13	362.61	371.09	362.61	362.61	362.61
10	324.48	382.11	441.08	364.15	382.11	400.07	373.14	382.11	391.08	382.11	382.11	382.11
11	322.64	380.06	438.82	362.21	380.06	397.91	371.14	380.06	388.98	380.06	380.06	380.06
12	340.45	399.85	460.59	381.01	399.85	418.68	390.43	399.85	409.26	399.85	399.85	399.85
13	333.67	392.31	452.3	373.86	392.31	410.77	383.09	392.31	401.54	392.31	392.31	392.31
14	320.26	377.42	435.92	359.71	377.42	395.13	368.57	377.42	386.27	377.42	377.42	377.42
15	306.43	362.05	419.02	345.12	362.05	378.99	353.59	362.05	370.52	362.05	362.05	362.05
16	294.61	348.92	404.57	332.64	348.92	365.2	340.78	348.92	357.06	348.92	348.92	348.92
17	290.26	344.08	399.25	328.04	344.08	360.12	336.06	344.08	352.1	344.08	344.08	344.08
18	295.32	349.71	405.44	333.39	349.71	366.03	341.55	349.71	357.87	349.71	349.71	349.71
19	314.61	371.14	429.02	353.75	371.14	388.54	362.45	371.14	379.84	371.14	371.14	371.14
20	302.11	357.25	413.73	340.55	357.25	373.94	348.9	357.25	365.6	357.25	357.25	357.25
21	305.62	361.15	418.02	344.25	361.15	378.04	352.7	361.15	369.59	361.15	361.15	361.15
22	325.2	382.9	441.95	364.92	382.9	400.88	373.91	382.9	391.89	382.9	382.9	382.9
23	317.78	374.66	432.89	357.09	374.66	392.23	365.88	374.66	383.45	374.66	374.66	374.66
24	294.93	349.27	404.96	332.97	349.27	365.57	341.12	349.27	357.42	349.27	349.27	349.27

Table 3.10 Membership function of generation #3 for (0, 0.5, 0.7, 1) α-cut representation for model "A" weekdays with 10% deviation for (P_D) and 3% for (α, β, γ)

Membership function	$\mu_{P_{G3}} = 0$			$\mu_{P_{G3}} = 0.5$			$\mu_{P_{G3}} = 0.75$			$\mu_{P_{G3}} = 1$		
Daily hour	Left PG2 (MW)	Mid PG2 (MW)	Right PG2 (MW)	Left PG2 (MW)	Mid PG2 (MW)	Right PG2 (MW)	Left PG2 (MW)	Mid PG2 (MW)	Right PG2 (MW)	Left PG2 (MW)	Mid PG2 (MW)	Right PG2 (MW)
1	345.45	415.54	487.35	380.29	415.54	451.22	397.86	415.54	433.32	415.54	415.54	415.54
2	307.72	373.62	441.24	347.5	373.62	399.84	362.34	373.62	384.9	373.62	373.62	373.62
3	286.42	349.94	415.2	328.57	349.94	371.34	340.6	349.94	359.29	349.94	349.94	349.94
4	261.84	322.64	385.17	304.42	322.64	340.86	314.32	322.64	330.96	322.64	322.64	322.64
5	242.11	300.71	361.05	284.49	300.71	316.94	293.02	300.71	308.4	300.71	300.71	300.71
6	261.34	322.09	384.56	305.24	322.09	338.93	313.88	322.09	330.29	322.09	322.09	322.09
7	276.64	339.09	403.26	321.62	339.09	356.55	330.46	339.09	347.71	339.09	339.09	339.09
8	298.32	363.18	429.76	344.62	363.18	381.73	353.95	363.18	372.4	363.18	363.18	363.18
9	358.92	430.5	503.81	408.63	430.5	452.37	419.59	430.5	441.41	430.5	430.5	430.5
10	381.48	455.57	531.39	432.47	455.57	478.66	444.04	455.57	467.1	455.57	455.57	455.57
11	379.1	452.93	528.49	429.98	452.93	475.88	441.46	452.93	464.4	452.93	452.93	452.93
12	402	478.38	556.48	454.16	478.38	502.59	466.27	478.38	490.48	478.38	478.38	478.38
13	393.28	468.69	545.82	444.96	468.69	492.41	456.83	468.69	480.55	468.69	468.69	468.69
14	376.05	449.54	524.76	426.77	449.54	472.3	438.16	449.54	460.92	449.54	449.54	449.54
15	358.27	429.78	503.03	408.01	429.78	451.56	418.9	429.78	440.67	429.78	429.78	429.78
16	343.07	412.9	484.45	391.96	412.9	433.83	402.43	412.9	423.36	412.9	412.9	412.9
17	337.47	406.68	477.61	386.05	406.68	427.3	396.37	406.68	416.99	406.68	406.68	406.68
18	343.99	413.91	485.57	392.93	413.91	434.9	403.42	413.91	424.41	413.91	413.91	413.91
19	368.79	441.47	515.88	419.11	441.47	463.83	430.29	441.47	452.65	441.47	441.47	441.47
20	352.71	423.6	496.23	402.13	423.6	445.07	412.87	423.6	434.34	423.6	423.6	423.6
21	357.22	428.62	501.74	406.9	428.62	450.33	417.76	428.62	439.48	428.62	428.62	428.62
22	382.39	456.59	532.51	433.47	456.59	479.7	445.03	456.59	468.15	456.59	456.59	456.59
23	372.86	445.99	520.86	423.41	445.99	468.58	434.7	445.99	457.29	445.99	445.99	445.99
24	343.48	413.35	484.95	392.39	413.35	434.3	402.87	413.35	423.83	413.35	413.35	413.35

3.4 All-Thermal Power Systems with Fuzzy Load and Cost Function Parameters 135

Table 3.11 Membership function of total generator for (0, 0.5, 0.7, 1) α-cut representation for model "A" weekdays with 10% deviation for (P_D) and 3% for (α, β, γ)

Membership function	$\mu_{tP_G} = 0$			$\mu_{tP_G} = 0.5$			$\mu_{tP_G} = 0.75$			$\mu_{tP_G} = 1$		
Daily hours	Left tPG (MW)	Mid tPG (MW)	Right tPG (MW)	Left tPG (MW)	Mid tPG (MW)	Right tPG (MW)	Left tPG (MW)	Mid tPG (MW)	Right tPG (MW)	Left tPG (MW)	Mid tPG (MW)	Right tPG (MW)
1	931.67	1,117.6	1,308.1	1,024.1	1,117.6	1,212.3	1,070.7	1,117.6	1,164.8	1,117.6	1,117.6	1,117.6
2	831.59	1,006.4	1,185.8	937.12	1,006.4	1,076	976.47	1,006.4	1,036.3	1,006.4	1,006.4	1,006.4
3	775.07	943.6	1,116.7	886.91	943.6	1,000.4	918.82	943.6	968.38	943.6	943.6	943.6
4	709.89	871.17	1,037	822.85	871.17	919.51	849.09	871.17	893.25	871.17	871.17	871.17
5	657.53	813	973.06	769.97	813	856.04	792.6	813	833.4	813	813	813
6	708.56	869.7	1,035.4	825.02	869.7	914.38	847.94	869.7	891.46	869.7	869.7	869.7
7	749.15	914.8	1,085	868.46	914.8	961.14	891.93	914.8	937.68	914.8	914.8	914.8
8	806.66	978.7	1,155.3	929.47	978.7	1,027.9	954.23	978.7	1,003.2	978.7	978.7	978.7
9	967.4	1,157.3	1,351.8	1,099.3	1,157.3	1,215.3	1,128.4	1,157.3	1,186.2	1,157.3	1,157.3	1,157.3
10	1,027.3	1,223.8	1,424.9	1,162.5	1,223.8	1,285.1	1,193.2	1,223.8	1,254.4	1,223.8	1,223.8	1,223.8
11	1,021	1,216.8	1,417.2	1,155.9	1,216.8	1,277.7	1,186.4	1,216.8	1,247.2	1,216.8	1,216.8	1,216.8
12	1,081.7	1,284.3	1,491.5	1,220.1	1,284.3	1,348.5	1,252.2	1,284.3	1,316.4	1,284.3	1,284.3	1,284.3
13	1,058.6	1,258.6	1,463.2	1,195.7	1,258.6	1,321.5	1,227.1	1,258.6	1,290.1	1,258.6	1,258.6	1,258.6
14	1,012.9	1,207.8	1,407.3	1,147.4	1,207.8	1,268.2	1,177.6	1,207.8	1,238	1,207.8	1,207.8	1,207.8
15	965.69	1,155.4	1,349.7	1,097.6	1,155.4	1,213.2	1,126.5	1,155.4	1,184.3	1,155.4	1,155.4	1,155.4
16	925.37	1,110.6	1,300.4	1,055.1	1,110.6	1,166.1	1,082.8	1,110.6	1,138.4	1,110.6	1,110.6	1,110.6
17	910.52	1,094.1	1,282.3	1,039.4	1,094.1	1,148.8	1,066.7	1,094.1	1,121.5	1,094.1	1,094.1	1,094.1
18	927.8	1,113.3	1,303.4	1,057.6	1,113.3	1,169	1,085.5	1,113.3	1,141.1	1,113.3	1,113.3	1,113.3
19	993.59	1,186.4	1,383.8	1,127.1	1,186.4	1,245.7	1,156.7	1,186.4	1,216.1	1,186.4	1,186.4	1,186.4
20	950.93	1,139	1,331.7	1,082.1	1,139	1,196	1,110.5	1,139	1,167.5	1,139	1,139	1,139
21	962.9	1,152.3	1,346.3	1,094.7	1,152.3	1,209.9	1,123.5	1,152.3	1,181.1	1,152.3	1,152.3	1,152.3
22	1,029.7	1,226.5	1,427.9	1,165.2	1,226.5	1,287.8	1,195.8	1,226.5	1,257.2	1,226.5	1,226.5	1,226.5
23	1,004.4	1,198.4	1,397	1,138.5	1,198.4	1,258.3	1,168.4	1,198.4	1,228.4	1,198.4	1,198.4	1,198.4
24	926.45	1,111.8	1,301.7	1,056.2	1,111.8	1,167.4	1,084	1,111.8	1,139.6	1,111.8	1,111.8	1,111.8

Table 3.12 Membership function of total cost for (0, 0.5, 0.7, 1) α-cut representation for model "A" weekdays with 10% deviation for (P_D) and 3% for (α, β, γ)

Membership function	$\mu_C = 0$			$\mu_C = 0.5$			$\mu_C = 0.75$			$\mu_C = 1$		
Daily hours	Left cost ($/h)	Mid cost ($/h)	Right cost ($/h)	Left cost ($/h)	Mid cost ($/h)	Right cost ($/h)	Left cost ($/h)	Mid cost ($/h)	Right cost ($/h)	Left cost ($/h)	Mid cost ($/h)	Right cost ($/h)
1	8,754.8	11,318	14,268	9,990.4	11,318	12,742	10,642	11,318	12,018	11,318	11,318	11,318
2	7,650.9	9,947.4	12,593	9,049.1	9,947.4	10,887	9,569.5	9,947.4	10,332	9,947.4	9,947.4	9,947.4
3	7,050	9,202.3	11,682	8,510.9	9,202.3	9,916.5	8,909.2	9,202.3	9,498.9	9,202.3	9,202.3	9,202.3
4	6,377.4	8,368.8	10,665	7,811.1	8,368.8	8,941	8,119.1	8,368.8	8,621.1	8,368.8	8,368.8	8,368.8
5	5,852.9	7,719.4	9,872.1	7,242.5	7,719.4	8,207	7,495.6	7,719.4	7,945.4	7,719.4	7,719.4	7,719.4
6	6,364	8,352.1	10,644	7,847.9	8,352.1	8,867.4	8,107.2	8,352.1	8,599.6	8,352.1	8,352.1	8,352.1
7	6,780	8,867.6	11,273	8,335.9	8,867.6	9,410.8	8,604.8	8,867.6	9,133.1	8,867.6	8,867.6	8,867.6
8	7,383.8	9,616.2	12,188	9,036.3	9,616.2	10,209	9,327	9,616.2	9,908.6	9,616.2	9,616.2	9,616.2
9	9,161.4	11,823	14,886	11,087	11,823	12,577	11,454	11,823	12,196	11,823	11,823	11,823
10	9,857.1	12,688	15,945	11,890	12,688	13,505	12,287	12,688	13,093	12,688	12,688	12,688
11	9,783	12,595	15,832	11,805	12,595	13,405	12,198	12,595	12,998	12,595	12,595	12,595
12	10,506	13,495	16,933	12,638	13,495	14,373	13,064	13,495	13,931	13,495	13,495	13,495
13	10,228	13,149	16,510	12,319	13,149	14,001	12,732	13,149	13,572	13,149	13,149	13,149
14	9,688	12,477	15,687	11,696	12,477	13,278	12,084	12,477	12,875	12,477	12,477	12,477
15	9,141.8	11,799	14,857	11,067	11,799	12,548	11,431	11,799	12,171	11,799	11,799	11,799
16	8,683.8	11,230	14,161	10,539	11,230	11,936	10,882	11,230	11,581	11,230	11,230	11,230
17	8,517.3	11,023	13,907	10,347	11,023	11,714	10,683	11,023	11,367	11,023	11,023	11,023
18	8,711.2	11,264	14,202	10,571	11,264	11,973	10,915	11,264	11,616	11,264	11,264	11,264
19	9,463.6	12,198	15,346	11,438	12,198	12,978	11,816	12,198	12,586	12,198	12,198	12,198
20	8,973.2	11,589	14,600	10,873	11,589	12,323	11,229	11,589	11,954	11,589	11,589	11,589
21	9,109.8	11,759	14,808	11,030	11,759	12,505	11,392	11,759	12,130	11,759	11,759	11,759
22	9,885.7	12,723	15,988	11,924	12,723	13,542	12,321	12,723	13,130	12,723	12,723	12,723
23	9,589.2	12,355	15,537	11,582	12,355	13,146	11,966	12,355	12,748	12,355	12,355	12,355
24	8,696	11,245	14,179	10,553	11,245	11,953	10,897	11,245	11,597	11,245	11,245	11,245

3.4 All-Thermal Power Systems with Fuzzy Load and Cost Function Parameters 137

Table 3.13 Membership function of load demand for (0, 0.5, 0.75, 1) α-cut representation for model "A" weekdays with 10% deviation for (P_D) and 3% for (α, β, γ)

Membership function	$\mu_{P_{Load}} = 0$			$\mu_{P_{Load}} = 0.5$			$\mu_{P_{Load}} = 0.75$			$\mu_{P_{Load}} = 1$		
Daily hours	Left load (MW)	Mid load (MW)	Right load (MW)	Left load (MW)	Mid load (MW)	Right load (MW)	Left load (MW)	Mid load (MW)	Right load (MW)	Left load (MW)	Mid load (MW)	Right load (MW)
1	1,006	1,118	1,229	1,062	1,118	1,173	1,090	1,118	1,146	1,118	1,118	1,118
2	905.8	1,006	1,107	956.1	1,006	1,057	981.2	1,006	1,032	1,006	1,006	1,006
3	849.2	943.6	1,038	896.4	943.6	990.8	920	943.6	967.2	943.6	943.6	943.6
4	784.1	871.2	958.3	827.6	871.2	914.7	849.4	871.2	892.9	871.2	871.2	871.2
5	731.7	813	894.3	772.4	813	853.7	792.7	813	833.3	813	813	813
6	782.7	869.7	956.7	826.2	869.7	913.2	848	869.7	891.4	869.7	869.7	869.7
7	823.3	914.8	1,006	869.1	914.8	960.5	891.9	914.8	937.7	914.8	914.8	914.8
8	880.8	978.7	1,077	929.8	978.7	1,028	954.2	978.7	1,003	978.7	978.7	978.7
9	1,042	1,157	1,273	1,099	1,157	1,215	1,128	1,157	1,186	1,157	1,157	1,157
10	1,101	1,224	1,346	1,163	1,224	1,285	1,193	1,224	1,254	1,224	1,224	1,224
11	1,095	1,217	1,338	1,156	1,217	1,278	1,186	1,217	1,247	1,217	1,217	1,217
12	1,156	1,284	1,413	1,220	1,284	1,349	1,252	1,284	1,316	1,284	1,284	1,284
13	1,133	1,259	1,384	1,196	1,259	1,322	1,227	1,259	1,290	1,259	1,259	1,259
14	1,087	1,208	1,329	1,147	1,208	1,268	1,178	1,208	1,238	1,208	1,208	1,208
15	1,040	1,155	1,271	1,098	1,155	1,213	1,127	1,155	1,184	1,155	1,155	1,155
16	999.5	1,111	1,222	1,055	1,111	1,166	1,083	1,111	1,138	1,111	1,111	1,111
17	984.7	1,094	1,204	1,039	1,094	1,149	1,067	1,094	1,121	1,094	1,094	1,094
18	1,002	1,113	1,225	1,058	1,113	1,169	1,085	1,113	1,141	1,113	1,113	1,113
19	1,068	1,186	1,305	1,127	1,186	1,246	1,157	1,186	1,216	1,186	1,186	1,186
20	1,025	1,139	1,253	1,082	1,139	1,196	1,111	1,139	1,167	1,139	1,139	1,139
21	1,037	1,152	1,268	1,095	1,152	1,210	1,123	1,152	1,181	1,152	1,152	1,152
22	1,104	1,227	1,349	1,165	1,227	1,288	1,196	1,227	1,257	1,227	1,227	1,227
23	1,079	1,198	1,318	1,138	1,198	1,258	1,168	1,198	1,228	1,198	1,198	1,198
24	1,001	1,112	1,223	1,056	1,112	1,167	1,084	1,112	1,140	1,112	1,112	1,112

Table 3.14 Membership function of incremental cost for (0, 0.5, 0.75, 1) α-cut representation for model "A" weekdays with 10% deviation for (P_D) and 3% for ($α, β, γ$)

Membership function	$μ_λ = 0$			$μ_λ = 0.5$			$μ_λ = 0.75$			$μ_λ = 1$	
Daily hours	Left λ $MW/h for ($μ=0$)	Mid λ $MW/h for ($μ=0$)	Right λ $MW/h for ($μ=0$)	Left λ $MW/h for ($μ=0.5$)	Mid λ $MW/h for ($μ=0.5$)	Right λ $MW/h for ($μ=0.5$)	Left λ $MW/h for ($μ=0.75$)	Mid λ $MW/h for ($μ=0.75$)	Right λ $MW/h for ($μ=0.75$)	Left λ $MW/h for ($μ=1$)	Mid λ $MW/h for ($μ=1$)	Right λ $MW/h for ($μ=1$)
1	11.99	12.62	13.214	12.31	12.62	12.92	12.463	12.617	12.77	12.62	12.62	12.617
2	11.44	12.03	12.588	11.75	12.03	12.305	11.895	12.031	12.166	12.03	12.03	12.031
3	11.13	11.7	12.234	11.44	11.7	11.954	11.574	11.699	11.824	11.7	11.7	11.699
4	10.78	11.32	11.827	11.08	11.32	11.55	11.202	11.317	11.432	11.32	11.32	11.317
5	10.5	11.01	11.499	10.79	11.01	11.227	10.903	11.01	11.117	11.01	11.01	11.01
6	10.77	11.31	11.818	11.08	11.31	11.54	11.194	11.309	11.424	11.31	11.31	11.309
7	10.99	11.55	12.072	11.31	11.55	11.789	11.427	11.547	11.668	11.55	11.55	11.547
8	11.31	11.88	12.432	11.63	11.88	12.143	11.755	11.884	12.014	11.88	11.88	11.884
9	12.18	12.83	13.438	12.52	12.83	13.132	12.674	12.827	12.98	12.83	12.83	12.827
10	12.51	13.18	13.812	12.86	13.18	13.501	13.016	13.178	13.339	13.18	13.18	13.178
11	12.47	13.14	13.773	12.82	13.14	13.462	12.98	13.141	13.302	13.14	13.14	13.141
12	12.8	13.5	14.153	13.16	13.5	13.836	13.328	13.497	13.667	13.5	13.5	13.497
13	12.68	13.36	14.008	13.03	13.36	13.694	13.196	13.362	13.528	13.36	13.36	13.362
14	12.43	13.09	13.722	12.78	13.09	13.412	12.934	13.094	13.253	13.09	13.09	13.094
15	12.17	12.82	13.427	12.51	12.82	13.122	12.665	12.817	12.969	12.82	12.82	12.817
16	11.95	12.58	13.175	12.29	12.58	12.874	12.434	12.581	12.727	12.58	12.58	12.581
17	11.87	12.49	13.082	12.21	12.49	12.782	12.349	12.493	12.638	12.49	12.49	12.493
18	11.96	12.6	13.19	12.3	12.6	12.889	12.448	12.595	12.742	12.6	12.6	12.595
19	12.32	12.98	13.602	12.67	12.98	13.294	12.824	12.981	13.137	12.98	12.98	12.981
20	12.09	12.73	13.335	12.43	12.73	13.031	12.58	12.73	12.881	12.73	12.73	12.73
21	12.16	12.8	13.41	12.5	12.8	13.105	12.649	12.801	12.953	12.8	12.8	12.801
22	12.52	13.19	13.827	12.87	13.19	13.516	13.03	13.192	13.354	13.19	13.19	13.192
23	12.38	13.04	13.669	12.73	13.04	13.36	12.886	13.044	13.202	13.04	13.04	13.044
24	11.96	12.59	13.182	12.29	12.59	12.88	12.44	12.587	12.734	12.59	12.59	12.587

3.4 All-Thermal Power Systems with Fuzzy Load and Cost Function Parameters

Table 3.15 Membership function of generator #1 for (0, 0.5, 0.75, 1) α-cut representation for model "A" weekdays with 10% deviation for (P_D) and 3% for (α, β, γ)

Membership function	$\mu_{PG1} = 0$			$\mu_{PG1} = 0.5$			$\mu_{PG1} = 0.75$			$\mu_{PG1} = 1$		
Daily hours	Left PG1 MW	Mid PG1 MW	Right PG1 MW	Left PG1 MW	Mid PG1 MW	Right PG1 MW	Left PG1 MW	Mid PG1 MW	Right PG1 MW	Left PG1 MW	Mid PG1 MW	Right PG1 MW
1	341.4	384.6	428.8	362.9	384.6	406.6	373.7	384.6	395.6	384.6	384.6	384.6
2	303.4	341.6	380.5	322.5	341.6	361	332	341.6	351.3	341.6	341.6	341.6
3	282.2	317.7	353.7	299.9	317.7	335.6	308.8	317.7	326.6	317.7	317.7	317.7
4	258.1	290.4	323.2	274.2	290.4	306.7	282.3	290.4	298.6	290.4	290.4	290.4
5	238.8	268.8	299	253.8	268.8	283.8	261.3	268.8	276.3	268.8	268.8	268.8
6	257.6	289.9	322.6	273.7	289.9	306.2	281.8	289.9	298	289.9	289.9	289.9
7	272.6	306.8	341.5	289.7	306.8	324.1	298.2	306.8	315.4	306.8	306.8	306.8
8	294.1	331	368.7	312.5	331	349.8	321.7	331	340.4	331	331	331
9	355.1	400.2	446.4	377.5	400.2	423.2	388.8	400.2	411.7	400.2	400.2	400.2
10	378.3	426.6	476.1	402.3	426.6	451.2	414.4	426.6	438.8	426.6	426.6	426.6
11	375.8	423.8	472.9	399.7	423.8	448.2	411.7	423.8	435.9	423.8	423.8	423.8
12	399.6	450.9	503.5	425.1	450.9	477	437.9	450.9	463.9	450.9	450.9	450.9
13	390.5	440.5	491.8	415.3	440.5	466	427.9	440.5	453.2	440.5	440.5	440.5
14	372.7	420.2	468.9	396.3	420.2	444.4	408.2	420.2	432.2	420.2	420.2	420.2
15	354.4	399.5	445.5	376.8	399.5	422.4	388.1	399.5	410.9	399.5	399.5	399.5
16	339	381.9	425.8	360.3	381.9	403.7	371.1	381.9	392.8	381.9	381.9	381.9
17	333.3	375.5	418.5	354.3	375.5	396.9	364.9	375.5	386.2	375.5	375.5	375.5
18	339.9	383	426.9	361.3	383	404.8	372.1	383	393.9	383	383	383
19	365.2	411.7	459.3	388.3	411.7	435.4	400	411.7	423.5	411.7	411.7	411.7
20	348.8	393	438.3	370.8	393	415.5	381.9	393	404.2	393	393	393
21	353.4	398.2	444.1	375.7	398.2	421.1	386.9	398.2	409.6	398.2	398.2	398.2
22	379.2	427.6	477.3	403.3	427.6	452.3	415.4	427.6	439.9	427.6	427.6	427.6
23	369.4	416.5	464.7	392.8	416.5	440.4	404.6	416.5	428.4	416.5	416.5	416.5
24	339.4	382.4	426.3	360.8	382.4	404.2	371.6	382.4	393.3	382.4	382.4	382.4

Table 3.16 Membership function of generator #2 for (0, 0.5, 0.75, 1) α-cut representation for model "A" weekdays with 10% deviation for (P_D) and 3% for (α, β, γ)

Membership function	$\mu_{P_{G2}} = 0$			$\mu_{P_{G2}} = 0.5$			$\mu_{P_{G2}} = 0.75$			$\mu_{P_{G2}} = 1$	
Daily hours	Left PG2 MW	Mid PG2 MW	Right PG2 MW	Left PG2 MW	Mid PG2 MW	Right PG2 MW	Left PG2 MW	Mid PG2 MW	Right PG2 MW	Left PG2 MW	Mid PG2 MW	Right PG2 MW
1	337.8	377.1	417.2	357.3	377.1	397	367.2	377.1	387	377.1	377.1	377.1
2	303.6	338.2	373.5	320.8	338.2	355.8	329.5	338.2	347	338.2	338.2	338.2
3	284.5	316.7	349.3	300.5	316.7	333	308.6	316.7	324.8	316.7	316.7	316.7
4	262.9	292.1	321.8	277.4	292.1	307	284.7	292.1	299.6	292.1	292.1	292.1
5	245.6	272.7	300.1	259	272.7	286.4	265.9	272.7	279.5	272.7	272.7	272.7
6	262.4	291.6	321.3	276.9	291.6	306.5	284.3	291.6	299.1	291.6	291.6	291.6
7	275.9	306.9	338.3	291.3	306.9	322.6	299	306.9	314.7	306.9	306.9	306.9
8	295.2	328.7	362.8	311.8	328.7	345.8	320.2	328.7	337.2	328.7	328.7	328.7
9	350.2	391.1	433.1	370.5	391.1	412.1	380.8	391.1	401.6	391.1	391.1	391.1
10	371.1	415	460	392.8	415	437.4	403.9	415	426.2	415	415	415
11	368.9	412.5	457.1	390.5	412.5	434.7	401.4	412.5	423.6	412.5	412.5	412.5
12	390.4	437	484.9	413.5	437	460.8	425.2	437	448.9	437	437	437
13	382.1	427.6	474.3	404.7	427.6	450.8	416.1	427.6	439.2	427.6	427.6	427.6
14	366	409.2	453.5	387.4	409.2	431.3	398.3	409.2	420.2	409.2	409.2	409.2
15	349.6	390.5	432.3	369.8	390.5	411.3	380.1	390.5	400.9	390.5	390.5	390.5
16	335.6	374.6	414.4	354.9	374.6	394.5	364.7	374.6	384.5	374.6	374.6	374.6
17	330.5	368.8	407.9	349.5	368.8	388.3	359.1	368.8	378.5	368.8	368.8	368.8
18	336.5	375.6	415.5	355.8	375.6	395.5	365.7	375.6	385.5	375.6	375.6	375.6
19	359.3	401.5	444.8	380.2	401.5	423.1	390.8	401.5	412.3	401.5	401.5	401.5
20	344.4	384.6	425.7	364.4	384.6	405.1	374.5	384.6	394.9	384.6	384.6	384.6
21	348.6	389.4	431.1	368.8	389.4	410.2	379.1	389.4	399.7	389.4	389.4	389.4
22	371.9	416	461.1	393.7	416	438.4	404.8	416	427.2	416	416	416
23	363.1	405.8	449.7	384.3	405.8	427.7	395	405.8	416.7	405.8	405.8	405.8
24	336	375	414.9	355.3	375	394.9	365.2	375	384.9	375	375	375

3.4 All-Thermal Power Systems with Fuzzy Load and Cost Function Parameters

Table 3.17 Membership function of generator #3 for (0, 0.5, 0.75, 1) α-cut representation for model "A" weekdays with 10% deviation for (P_D) and 3% for (α, β, γ)

Membership function	$\mu_{P_{G3}} = 0$			$\mu_{P_{G3}} = 0.5$			$\mu_{P_{G3}} = 0.75$			$\mu_{P_{G3}} = 1$		
Daily hours	Left PG3 MW	Mid PG3 MW	Right PG3 MW	Left PG3 MW	Mid PG3 MW	Right PG3 MW	Left PG3 MW	Mid PG3 MW	Right PG3 MW	Left PG3 MW	Mid PG3 MW	Right PG3 MW
1	407.7	458.1	509.5	432.8	458.1	483.7	445.4	458.1	470.8	458.1	458.1	458.1
2	363.4	407.8	453.1	385.5	407.8	430.3	396.6	407.8	419	407.8	407.8	407.8
3	338.6	379.9	421.7	359.2	379.9	400.7	369.5	379.9	390.2	379.9	379.9	379.9
4	310.5	348.1	386.2	329.3	348.1	367	338.7	348.1	357.5	348.1	348.1	348.1
5	288.1	322.9	358.1	305.5	322.9	340.4	314.2	322.9	331.6	322.9	322.9	322.9
6	309.9	347.4	385.5	328.7	347.4	366.3	338.1	347.4	356.9	347.4	347.4	347.4
7	327.4	367.2	407.5	347.3	367.2	387.2	357.2	367.2	377.2	367.2	367.2	367.2
8	352.5	395.5	439.2	373.9	395.5	417.2	384.7	395.5	406.3	395.5	395.5	395.5
9	423.8	476.3	530	450	476.3	502.9	463.1	476.3	489.6	476.3	476.3	476.3
10	450.9	507.1	564.8	478.9	507.1	535.7	493	507.1	521.4	507.1	507.1	507.1
11	448	503.9	561.1	475.9	503.9	532.2	489.8	503.9	518	503.9	503.9	503.9
12	475.9	535.6	597	505.7	535.6	566	520.6	535.6	550.8	535.6	535.6	535.6
13	465.2	523.5	583.2	494.2	523.5	553.1	508.8	523.5	538.2	523.5	523.5	523.5
14	444.3	499.7	556.4	471.9	499.7	527.8	485.8	499.7	513.7	499.7	499.7	499.7
15	423	475.4	529	449.1	475.4	502	462.2	475.4	488.7	475.4	475.4	475.4
16	404.9	454.9	505.9	429.8	454.9	480.2	442.3	454.9	467.5	454.9	454.9	454.9
17	398.3	447.4	497.5	422.8	447.4	472.2	435	447.4	459.7	447.4	447.4	447.4
18	406	456.1	507.3	431	456.1	481.5	443.5	456.1	468.8	456.1	456.1	456.1
19	435.6	489.7	545.2	462.6	489.7	517.2	476.1	489.7	503.4	489.7	489.7	489.7
20	416.3	467.9	520.5	442.1	467.9	494	454.9	467.9	480.9	467.9	467.9	467.9
21	421.7	474	527.4	447.8	474	500.5	460.9	474	487.2	474	474	474
22	452	508.4	566.2	480.1	508.4	537	494.2	508.4	522.7	508.4	508.4	508.4
23	440.5	495.3	551.5	467.8	495.3	523.1	481.5	495.3	509.2	495.3	495.3	495.3
24	405.4	455.4	506.5	430.4	455.4	480.8	442.9	455.4	468.1	455.4	455.4	455.4

Table 3.18 Membership function of total P_G for (0, 0.5, 0.75, 1) α-cut representation for model "A" weekdays with 10% deviation for (P_D) and 3% for (α, β, γ)

Membership function	$\mu_{tP_G} = 0$			$\mu_{tP_G} = 0.5$			$\mu_{tP_G} = 0.75$			$\mu_{tP_G} = 1$		
Daily hours	Left tPG (MW)	Mid tPG (MW)	Right tPG (MW)	Left tPG (MW)	Mid tPG (MW)	Right tPG (MW)	Left tPG (MW)	Mid tPG (MW)	Right tPG (MW)	Left tPG (MW)	Mid tPG (MW)	Right tPG (MW)
1	1,087	1,220	1,356	1,153	1,220	1,287	1,186	1,220	1,253	1,220	1,220	1,220
2	970.4	1,088	1,207	1,029	1,088	1,147	1,058	1,088	1,117	1,088	1,088	1,088
3	905.3	1,014	1,125	959.6	1,014	1,069	986.8	1,014	1,042	1,014	1,014	1,014
4	831.4	930.6	1,031	880.9	930.6	980.8	905.7	930.6	955.7	930.6	930.6	930.6
5	772.6	864.3	957.2	818.4	864.3	910.6	841.3	864.3	887.4	864.3	864.3	864.3
6	829.9	929	1,029	879.3	929	979	904.1	929	953.9	929	929	929
7	875.8	980.8	1,087	928.2	980.8	1,034	954.5	980.8	1,007	980.8	980.8	980.8
8	941.7	1,055	1,171	998.2	1,055	1,113	1,027	1,055	1,084	1,055	1,055	1,055
9	1,129	1,268	1,409	1,198	1,268	1,338	1,233	1,268	1,303	1,268	1,268	1,268
10	1,200	1,349	1,501	1,274	1,349	1,424	1,311	1,349	1,386	1,349	1,349	1,349
11	1,193	1,340	1,491	1,266	1,340	1,415	1,303	1,340	1,378	1,340	1,340	1,340
12	1,266	1,423	1,585	1,344	1,423	1,504	1,384	1,423	1,464	1,423	1,423	1,423
13	1,238	1,392	1,549	1,314	1,392	1,470	1,353	1,392	1,431	1,392	1,392	1,392
14	1,1,83	1,329	1,479	1,256	1,329	1,403	1,292	1,329	1,366	1,329	1,329	1,329
15	1,127	1,265	1,407	1,196	1,265	1,336	1,230	1,265	1,300	1,265	1,265	1,265
16	1,080	1,211	1,346	1,145	1,211	1,278	1,178	1,211	1,245	1,211	1,211	1,211
17	1,062	1,192	1,324	1,127	1,192	1,257	1,159	1,192	1,224	1,192	1,192	1,192
18	1,082	1,215	1,350	1,148	1,215	1,282	1,181	1,215	1,248	1,215	1,215	1,215
19	1,160	1,303	1,449	1,231	1,303	1,376	1,267	1,303	1,339	1,303	1,303	1,303
20	1,110	1,246	1,385	1,177	1,246	1,315	1,211	1,246	1,280	1,246	1,246	1,246
21	1,124	1,262	1,403	1,192	1,262	1,332	1,227	1,262	1,297	1,262	1,262	1,262
22	1,203	1,352	1,505	1,277	1,352	1,428	1,314	1,352	1,390	1,352	1,352	1,352
23	1,173	1,318	1,466	1,245	1,318	1,391	1,281	1,318	1,354	1,318	1,318	1,318
24	1,081	1,213	1,348	1,146	1,213	1,280	1,180	1,213	1,246	1,213	1,213	1,213

3.4 All-Thermal Power Systems with Fuzzy Load and Cost Function Parameters 143

Table 3.19 Membership function of the power losses for (0, 0.5, 0.75, 1) α-cut representation for model "A" weekdays with 10% deviation for (P_D) and 3% for (α, β, γ)

Membership function	$\mu_{P_L} = 0$			$\mu_{P_L} = 0.5$			$\mu_{P_L} = 0.75$			$\mu_{P_L} = 1$		
Daily hours	Left PL	Mid PL	Right PL	Left PL	Mid PL	Right PL	Left PL	Mid PL	Right PL	Left PL	Mid PL	Right PL
1	81.179	102.2	126.25	91.35	102.2	113.854	96.7	102.2	107.95	102.2	102.2	102.23
2	64.723	81.3	100.13	72.74	81.3	90.42	76.95	81.3	85.786	81.3	81.3	81.296
3	56.343	70.69	86.92	63.29	70.69	78.567	66.93	70.69	74.568	70.69	70.69	70.69
4	47.527	59.53	73.085	53.34	59.53	66.114	56.39	59.53	62.774	59.53	59.53	59.533
5	41.054	51.36	62.973	46.05	51.36	57.002	48.66	51.36	54.139	51.36	51.36	51.36
6	47.357	59.32	72.818	53.15	59.32	65.875	56.19	59.32	62.547	59.32	59.32	59.318
7	52.735	66.12	81.25	59.22	66.12	73.466	62.62	66.12	69.737	66.12	66.12	66.121
8	60.957	76.52	94.179	68.48	76.52	85.078	72.44	76.52	80.731	76.52	76.52	76.517
9	87.587	110.4	136.48	98.61	110.4	123.021	104.4	110.4	116.61	110.4	110.4	110.41
10	98.973	125	154.75	111.5	125	139.364	118.1	125	132.04	125	125	124.97
11	97.736	123.4	152.76	110.1	123.4	137.584	116.6	123.4	130.36	123.4	123.4	123.38
12	110.11	139.2	172.66	124.1	139.2	155.371	131.5	139.2	147.15	139.2	139.2	139.21
13	105.27	133	164.89	118.7	133	148.424	125.7	133	140.6	133	133	133.03
14	96.158	121.4	150.22	108.3	121.4	135.316	114.7	121.4	128.22	121.4	121.4	121.36
15	87.274	110	135.98	98.25	110	122.572	104	110	116.19	110	110	110.01
16	80.079	100.8	124.49	90.1	100.8	112.282	95.38	100.8	106.46	100.8	100.8	100.83
17	77.52	97.57	120.42	87.21	97.57	108.63	92.3	97.57	103.01	97.57	97.57	97.567
18	80.502	101.4	125.17	90.58	101.4	112.887	95.89	101.4	107.03	101.4	101.4	101.37
19	92.468	116.6	144.3	104.1	116.6	130.018	110.3	116.6	123.22	116.6	116.6	116.64
20	84.598	106.6	131.7	95.22	106.6	118.742	100.8	106.6	112.57	106.6	106.6	106.59
21	86.764	109.4	135.17	97.68	109.4	121.843	103.4	109.4	115.5	109.4	109.4	109.36
22	99.453	125.6	155.52	112.1	125.6	140.054	118.7	125.6	132.7	125.6	125.6	125.58
23	94.527	119.3	147.6	106.5	119.3	132.973	112.8	119.3	126.01	119.3	119.3	119.28
24	80.267	101.1	124.79	90.32	101.1	112.551	95.6	101.1	106.72	101.1	101.1	101.07

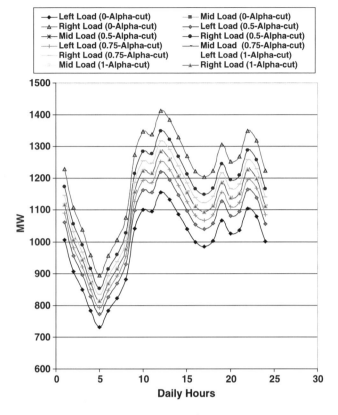

Fig. 3.21 Fuzzy load demand for all α-cut representation

$$F(P_{G_1}) = 200 + 7.0P_1 + 0.008P_1^2 \text{ kJ/h}$$
$$F(P_{G_2}) = 180 + 6.3P_2 + 0.009P_2^2 \text{ kJ/h}$$
$$F(P_{G3}) = 140 + 6.8P_3 + 0.007P_3^2 \text{ kJ/h}$$

The generation limits are given by the left and right sides of each unit:

$$L_{\tilde{P}_{G1}} \leq \tilde{P}_{G_1} \leq R_{\tilde{P}_{G1}} \quad I = 1, \ldots\ldots NG$$
$$L_{\tilde{P}_{G2}} \leq \tilde{P}_{G_2} \leq R_{\tilde{P}_{G2}} \quad I = 1, \ldots\ldots NG$$
$$L_{\tilde{P}_{G3}} \leq \tilde{P}_{G_3} \leq R_{\tilde{P}_{G3}} \quad I = 1, \ldots\ldots NG$$

This example is implemented on the generalized interval arithmetic to fuzzy numbers. Using the principle of equal incremental cost, we determine the optimal fuzzy dispatch and total fuzzy cost applying the generalized method to perform mathematical addition, subtraction, division, inversion, and multiplication on the equation as explained in Chap. 2 on fuzzy sets. Two simulation programs were created using MATLAB® software. In the first program all the equations in Sect. 3.5.2 were analyzed and debugged in a complete program set and in the second program a MATLAB toolbox was used to simulate the mathematical formula. The results of the two programs were identical, which was expected. Examining the tables calculated (Tables 3.6–3.19) and the graphs plotted (Figs. 3.20–3.44), Table 3.6 shows the fuzzy load with 10% deviation for model A representing the weekdays in a month. Figure 3.21 shows the middle, left, and right sides for all α-cut representations of the fuzzy load. The analysis was tested on a 3% deviation for $\tilde{\alpha}$, $\tilde{\beta}$, and $\tilde{\gamma}$, then tested again on a 10% deviation. All the results were tabulated and plotted, including the fuzzy incremental cost, fuzzy generation of each unit, total fuzzy generation vs. fuzzy load, and total fuzzy cost (Fig. 3.22).

The results were exactly as expected: the total fuzzy generation is approximately greater than or equal to the fuzzy load. The crisp value of the total power generation and load demand are equal and satisfy the load demand in Eq. 3.119. The fuzzy total generation represented by the left and right sides is greater than the expected load due to the increased range of the cost function coefficients. The triangular membership function of the fuzzy load, fuzzy generation, total generation, and fuzzy minimum cost are presented in Figs. 3.28, 3.29, 3.30, 3.31, 3.32, 3.33, and 3.34 to justify that the committed unit did not violate the limits given in the example. If the deviation of $\tilde{\beta}$ and $\tilde{\gamma}$ increased more than 5%, then the minimum and maximum total fuzzy generation are much larger than the minimum and maximum fuzzy load which means that it is important to keep the coefficient values in control and not to exceed their expected values. The calculated results show that the fuzzy parameters in the cost function play a great role in the performance of the network to obtain a minimum cost for the committed thermal generation. This procedure outlines all the possibilities that could be encountered hour by hour for 24 h including the minimum cost of the sudden increased load which is the main objective of the economical dispatch method.

3.5 Fuzzy Economical Dispatch Including Losses

It is important to consider the losses in transmission lines due to the large interconnected network where power is transmitted over long distances with low load density areas. In addition, determining the distribution of load between plants needs to take into consideration transmission line losses where, for a given distribution of loads, usually the plant with low incremental fuel cost rate has greater transmission losses than the other plants which will affect the overall economy of the system. It is therefore wise to lower the load at that plant to achieve a minimum fuel cost.

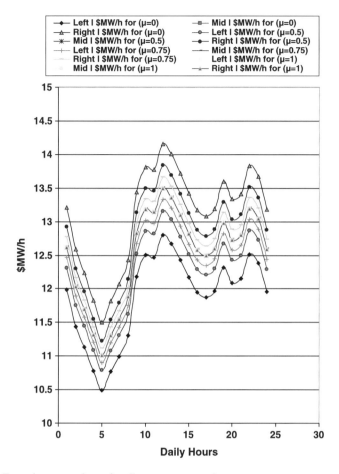

Fig. 3.22 Fuzzy incremental cost for all α-cut representation

3.5.1 Problem Formulation

The objective is to find the minimum value of the total cost function subject to the equality and inequality constraints.

Minimize

$$\tilde{C}_{total} = \sum_{i=1}^{NG} \tilde{C}_i = \sum_{i=1}^{NG} \tilde{\alpha}_i + \tilde{\beta}_i \tilde{P}_{Gi} + \tilde{\gamma}_i \tilde{P}_{Gi}^{2} \tag{3.142}$$

Subject to satisfying

$$\sum_{i=1}^{NG} \tilde{P}_{Gi} \geq \tilde{P}_D + \tilde{P}_L \tag{3.143}$$

3.5 Fuzzy Economical Dispatch Including Losses

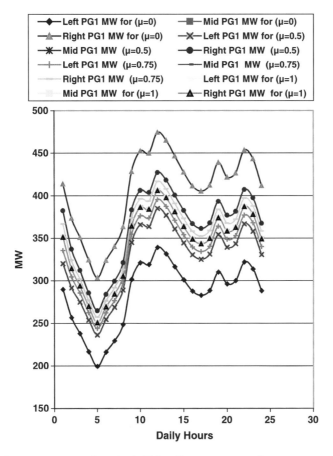

Fig. 3.23 Fuzzy power generation of unit #1 for all α-cut representation

$$\tilde{P}_{Gi}(\min) \leq \tilde{P}_{Gi} \leq \tilde{P}_{Gi}(\max) \quad i = 1, \ldots\ldots NG \tag{3.144}$$

The fuzzy variable added in this case is the power losses $\tilde{P}_L = (\bar{P}_L, L_{\tilde{L}}, R_{\tilde{L}})$ denoting the middle, left, and right sides of the power losses. The total transmission loss formula is a quadratic function of the generator power output expressed as

$$\tilde{P}_L = \sum_{i=1}^{NG} \sum_{j=1}^{NG} \tilde{P}_{Gi} B_{ij} \tilde{P}_{Gj} \tag{3.145}$$

A more general formula containing linear terms is known as Kron's loss formula:

$$\tilde{P}_L = \sum_{i=1}^{NG} \sum_{j=1}^{NG} \tilde{P}_{Gi} B_{ij} \tilde{P}_{Gj} + \sum_{i=1}^{NG} B_{0i} \tilde{P}_{Gi} + B_{00} \tag{3.146}$$

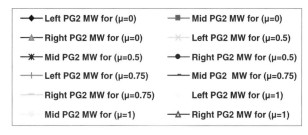

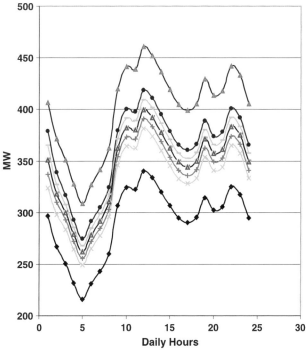

Fig. 3.24 Fuzzy power generation of unit #2 for all α-cut representation

Applying fuzzy interval arithmetic operations implemented by their α-cut operation to obtain the power loss formula that includes the middle, left, and right sides of the triangular membership function, it becomes:

$$\tilde{P}_L(\bar{P}_L, L_{\tilde{L}}, R_{\tilde{L}}) = \sum_{i=1}^{NG} \sum_{j=1}^{NG} (\bar{P}_{Gi}, L_{\tilde{P}_{Gi}}, R_{\tilde{P}_{Gi}}) B_{ij} (\bar{P}_{Gj}, L_{\tilde{P}_{Gj}}, R_{\tilde{P}_{Gj}})$$
$$+ \sum_{i=1}^{NG} B_{0i} (\bar{P}_{Gi}, L_{\tilde{P}_{Gi}}, R_{\tilde{P}_{Gi}}) + B_{00} \qquad (3.147)$$

3.5 Fuzzy Economical Dispatch Including Losses

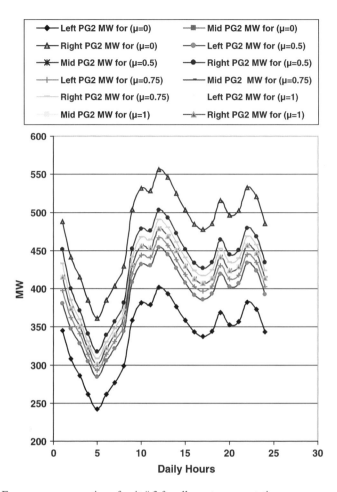

Fig. 3.25 Fuzzy power generation of unit # 3 for all α-cut representation

Using the simplest quadratic form we get:

$$\tilde{P}_L = \sum_{i=1}^{NG} B_{ii}\tilde{P}_{Gi}^{2} \qquad (3.148)$$

Substituting the middle, left, and right sides of the generation triangular membership function into the equation we get:

$$(\bar{P}_L, L_{\tilde{L}}, R_{\tilde{L}}) = \sum_{i=1}^{NG} (B_{ii}\bar{P}_i^{2}, B_{ii}L_{\tilde{P}_i}^{2}, B_{ii}R_{\tilde{P}_i}^{2}) \qquad (3.149)$$

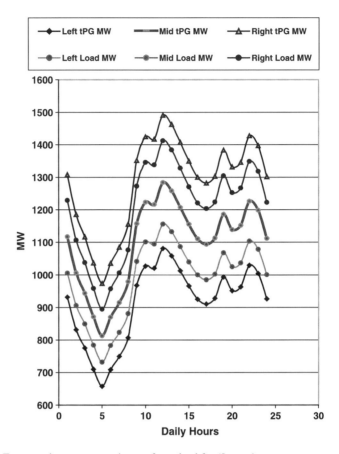

Fig. 3.26 Fuzzy total power generation vs. fuzzy load for (0-α-cut)

Then using Table 3.2 to obtain the middle, left, and right sides of the equation:

$$\bar{P}_L = \sum_{i=1}^{NG} B_{ii} \bar{P}_{Gi}^{2} \tag{3.150}$$

$$L_{\tilde{L}} = \sum_{i=1}^{NG} B_{ii} L_{\tilde{P}_{Gi}}^{2} \tag{3.151}$$

$$R_{\tilde{L}} = \sum_{i=1}^{NG} B_{ii} R_{\tilde{P}_{Gi}}^{2} \tag{3.152}$$

The power generation of each unit can be calculated from

3.5 Fuzzy Economical Dispatch Including Losses

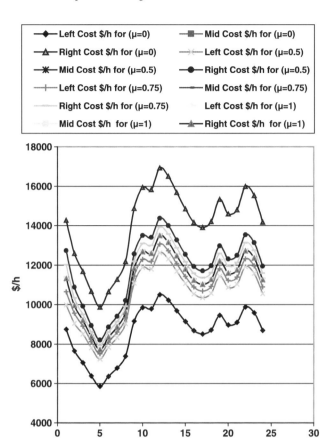

Fig. 3.27 Fuzzy minimum total cost for all α-cut representation

$$(\bar{P}_{Gi}, L_{\tilde{P}_{Gi}}, R_{\tilde{P}_{Gi}})^{[k]} = \frac{(\bar{\lambda}, L_{\tilde{\lambda}}, R_{\tilde{\lambda}})^{[k]} - (\bar{\beta}_i, L_{\tilde{\beta}_i}, R_{\tilde{\beta}_i})}{2((\bar{\gamma}_i, L_{\tilde{\gamma}_i}, R_{\tilde{\gamma}_i}) + (\bar{\lambda}, L_{\tilde{\lambda}}, R_{\tilde{\lambda}})^{[k]} B_{ii})} \quad (3.153)$$

Using Table 2.4 to perform the fuzzy set arithmetic calculation, the middle crisp, left, and right values of the equation become:

$$\bar{P}_{Gi}^{[k]} = \frac{\bar{\lambda}_i^{[k]} - \bar{\beta}_i}{2(\bar{\gamma}_i + \bar{\lambda}_i^{[k]} B_{ii})} \quad (3.154)$$

$$L_{\tilde{P}_{Gi}}^{[k]} = \frac{L_{\tilde{\lambda}}^{[k]} - R_{\tilde{\beta}_i}}{2(R_{\tilde{\gamma}_i} + r_{\tilde{\lambda}}^{[k]} B_{ii})} \quad (3.155)$$

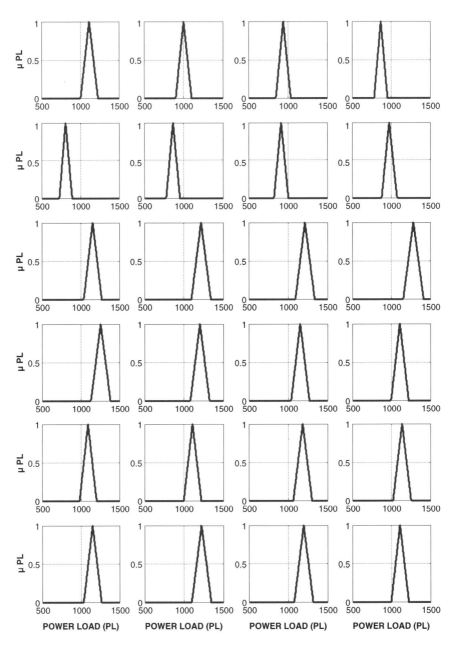

Fig. 3.28 A triangular membership function for power load

3.5 Fuzzy Economical Dispatch Including Losses

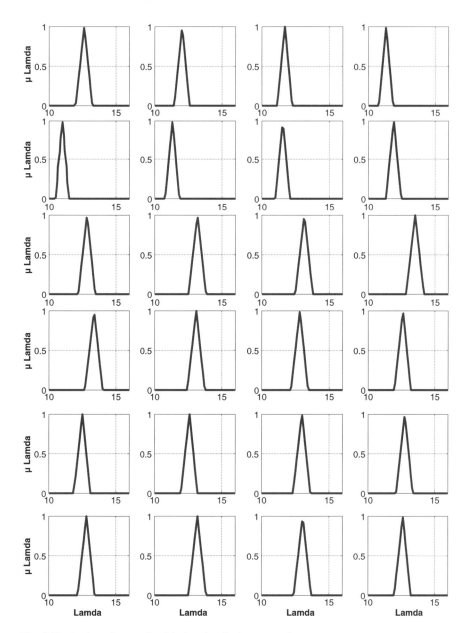

Fig. 3.29 A triangular membership function for incremental cost

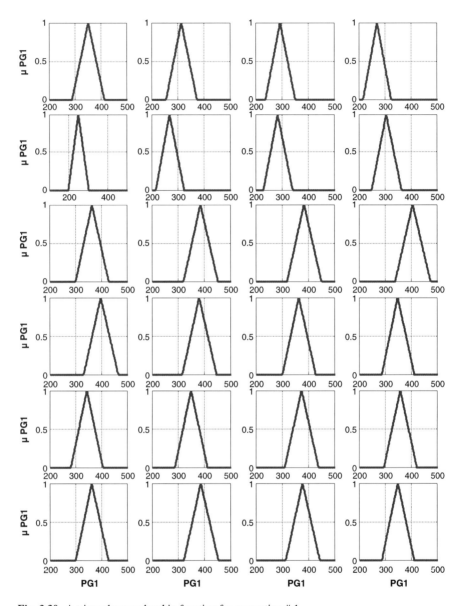

Fig. 3.30 A triangular membership function for generation # 1

$$R_{\tilde{P}_{Gi}}{}^{[k]} = \frac{R_{\tilde{\lambda}}{}^{[k]} - L_{\tilde{\beta}_i}}{2(L_{\tilde{\gamma}_i} + L_{\tilde{\lambda}}{}^{[k]} B_{ii})} \tag{3.156}$$

Substitute the generation values into the loss formula to calculate the power losses and then check the equality constraints to see if they are satisfied. If they are not satisfied we use the iterative method shown in Chart 3.1, where

3.5 Fuzzy Economical Dispatch Including Losses 155

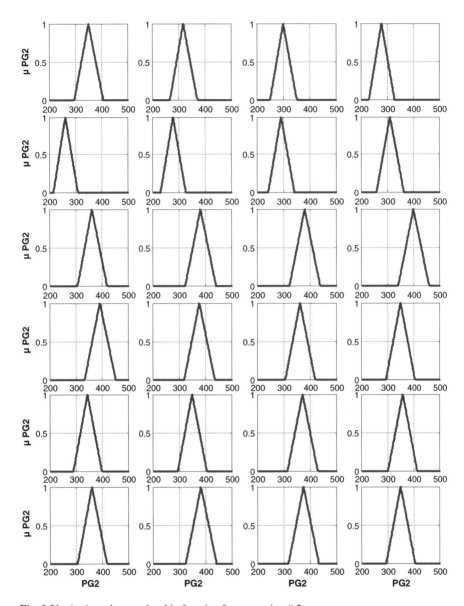

Fig. 3.31 A triangular membership function for generation # 2

156   3 Economic Operation of Electric Power Systems

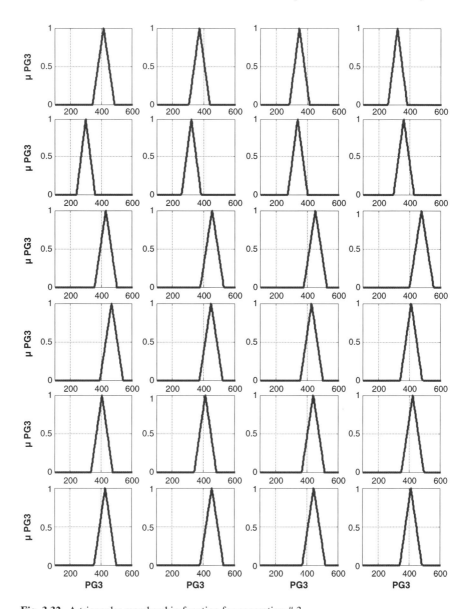

Fig. 3.32 A triangular membership function for generation # 3

3.5 Fuzzy Economical Dispatch Including Losses

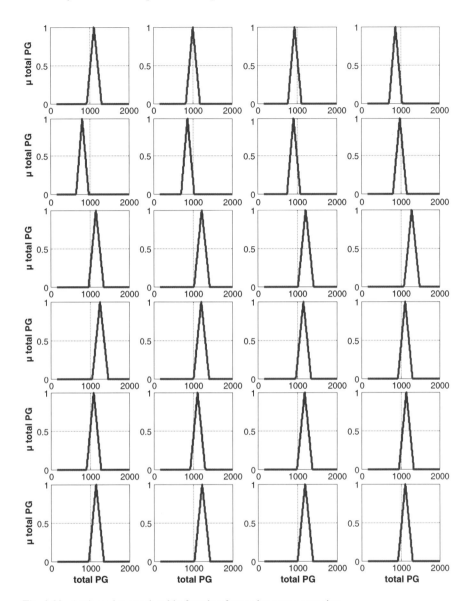

Fig. 3.33 A triangular membership function for total power generation

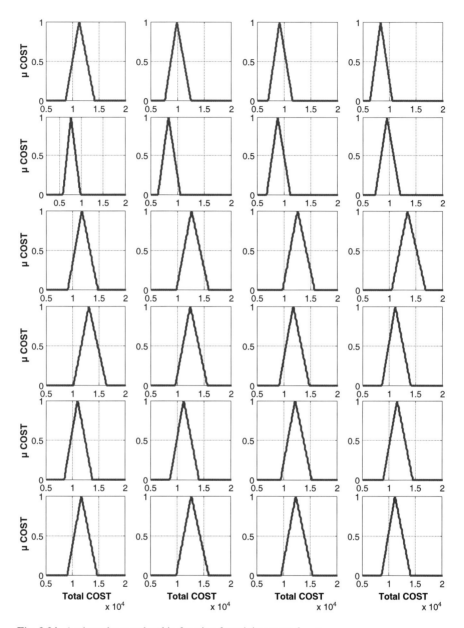

Fig. 3.34 A triangular membership function for minimum total cost

3.5 Fuzzy Economical Dispatch Including Losses 159

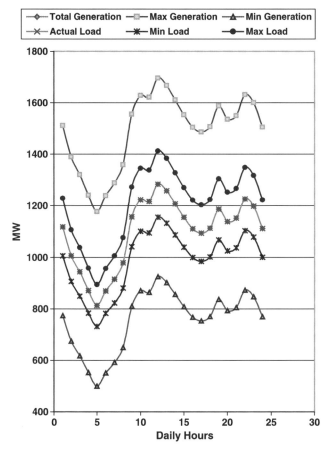

Fig. 3.35 Fuzzy total generation vs. fuzzy load for 10% deviation of (α, β, γ)

$$\sum_{i=1}^{NG} \left(\frac{\partial \tilde{P}_{Gi}}{\partial \tilde{\lambda}} \right)$$

is given as

$$\sum_{i=1}^{NG} \left(\frac{\partial \tilde{P}_{Gi}}{\partial \tilde{\lambda}} \right) = \sum_{i=1}^{NG} \left[\frac{\tilde{\gamma}_i + B_{ii} B_i}{2 \left(\tilde{\gamma}_i + \tilde{\lambda}^{[k]} B_{ii} \right)^2} \right] \quad (3.157)$$

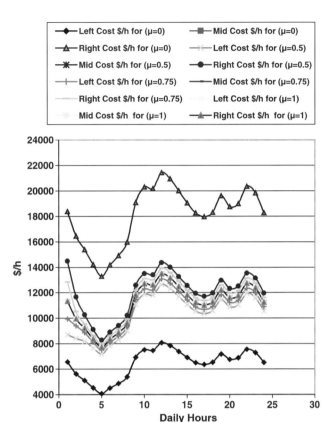

Fig. 3.36 Fuzzy minimum total cost for 10% deviation of (α, β, γ) for (0-α-cut)

Replacing the fuzzy parameters with their middle, left, and right sides in the equation we get:

$$\sum_{i=1}^{NG} \left(\frac{\partial (\bar{P}_{Gi}, L_{\tilde{P}_{Gi}}, R_{\tilde{P}_{Gi}})}{\partial (\bar{\lambda}, L_{\tilde{\lambda}}, R_{\tilde{\lambda}})} \right) = \sum_{i=1}^{NG} \left[\frac{(\bar{\gamma}_i, L_{\tilde{\gamma}_i}, R_{\tilde{\gamma}_i}) + B_{ii}(\bar{\beta}_i, L_{\tilde{\beta}_i}, R_{\tilde{\beta}_i})}{2\left((\bar{\gamma}_i, L_{\tilde{\gamma}_i}, R_{\tilde{\gamma}_i}) + (\bar{\lambda}, L_{\tilde{\lambda}}, R_{\tilde{\lambda}})^{[k]} B_{ii}\right)^2} \right] \quad (3.158)$$

The middle crisp value becomes (from Table 2.4):

$$\sum_{i=1}^{NG} \left(\frac{\partial \bar{P}_{Gi}}{\partial \bar{\lambda}_i} \right) = \sum_{i=1}^{NG} \left[\frac{\bar{\gamma}_i + B_{ii}\bar{\beta}_i}{2\left(\bar{\gamma}_i + \bar{\lambda}_i^{[k]} B_{ii}\right)^2} \right] \quad (3.159)$$

3.5 Fuzzy Economical Dispatch Including Losses

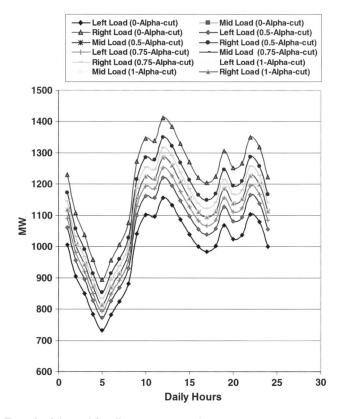

Fig. 3.37 Fuzzy load demand for all α-cut representation

The left side of the power generation becomes:

$$\sum_{i=1}^{NG} \left(\frac{\partial L_{\tilde{P}_{Gi}}}{\partial L_{\lambda i}} \right) = \sum_{i=1}^{NG} \left[\frac{L_{\tilde{\gamma}_i} + B_{ii} L_{\tilde{\beta}_i}}{2 \left(R_{\tilde{\gamma}_i} + R_{\tilde{\lambda}}^{[k]} B_{ii} \right)^2} \right] \quad (3.160)$$

The right side of the power generation becomes:

$$\sum_{i=1}^{NG} \left(\frac{\partial R_{\tilde{P}_{Gi}}}{\partial R_{\tilde{\lambda}}} \right) = \sum_{i=1}^{NG} \left[\frac{R_{\tilde{\gamma}_i} + B_{ii} R_{\tilde{\beta}_i}}{2 \left(L_{\tilde{\gamma}_i} + L_{\tilde{\lambda}}^{[k]} B_{ii} \right)^2} \right] \quad (3.161)$$

Sine $\Delta\lambda^{(k)}$ denotes the increment of change in the incremental cost is equal to:

$$\Delta\lambda^{(k)} = \frac{\Delta \tilde{P}^{[k]}}{\sum \left(\frac{dP_{Gi}}{d\lambda} \right)^{[k]}} \quad (3.162)$$

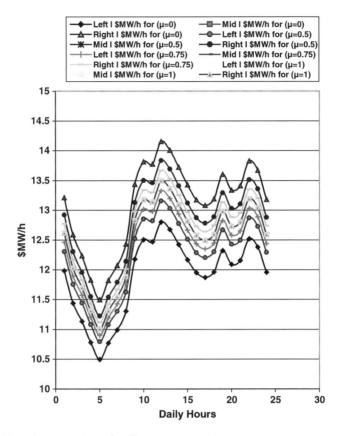

Fig. 3.38 Fuzzy incremental cost for all α-cut representation

Replacing the fuzzy parameters with their middle, left, and right values in Eq. 3.21 we get:

$$\Delta(\bar{\lambda}, L_{\bar{\lambda}}, R_{\bar{\lambda}})^{(k)} = \frac{\Delta(\bar{P}_{Gi}, L_{\tilde{P}_{Gi}}, R_{\tilde{P}_{Gi}})^{[k]}}{\sum \left(\frac{d(\bar{P}_{Gi}, L_{\tilde{P}_{Gi}}, R_{\tilde{P}_{Gi}})_i}{d(\bar{\lambda}, L_{\bar{\lambda}}, R_{\bar{\lambda}})} \right)^{[k]}} \quad (3.163)$$

The middle or crisp value from Table 2.4 will be

$$\Delta(\bar{\lambda}_i) = \frac{\Delta(\bar{P}_{Gi})}{\sum_{i=1}^{NG} \left[\frac{\bar{\gamma}_i + B_{ii}\bar{\beta}_i}{2\left(\bar{\gamma}_i + \bar{\lambda}_i^{[k]} B_{ii}\right)^2} \right]} \quad (3.164)$$

3.5 Fuzzy Economical Dispatch Including Losses

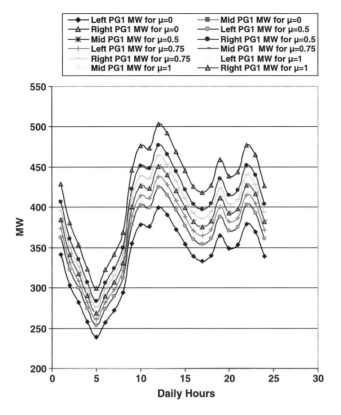

Fig. 3.39 Fuzzy power generation of unit #1 for all α-cut representation

The left side becomes

$$\Delta(L_{\tilde{\lambda}_i}) = \frac{\Delta(L_{\tilde{P}_{Gi}})}{\sum_{i=1}^{NG} \left[\frac{R_{\tilde{\gamma}_i} + B_{ii} R_{\tilde{\beta}_i}}{2\left(L_{\tilde{\gamma}_i} + L_{\tilde{\lambda}}^{[k]} B_{ii}\right)^2} \right]} \quad (3.165)$$

The right side becomes

$$\Delta(R_{\tilde{\lambda}_i}) = \frac{\Delta(R_{\tilde{P}_{Gi}})}{\sum_{i=1}^{NG} \left[\frac{L_{\tilde{\gamma}_i} + B_{ii} L_{\tilde{\beta}_i}}{2\left(R_{\tilde{\gamma}_i} + R_{\tilde{\lambda}}^{[k]} B_{ii}\right)^2} \right]} \quad (3.166)$$

Then calculate the new value of the incremental cost

$$\lambda^{(k+1)} = \lambda^{(k)} + \Delta\lambda^{(k)} \quad (3.167)$$

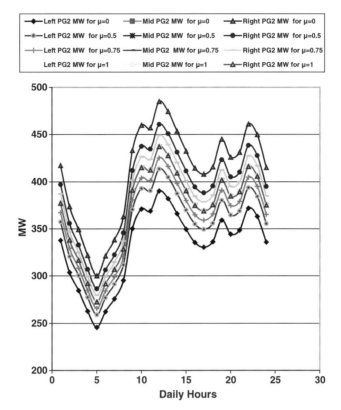

Fig. 3.40 Fuzzy power generation of unit #2 for all α-cut representation

Substituting the middle, left, and right sides into Eq. 3.26 we get:

$$(\bar{\lambda}, L_{\bar{\lambda}}, R_{\bar{\lambda}})^{(k+1)} = (\bar{\lambda}, L_{\bar{\lambda}}, R_{\bar{\lambda}})^{(k)} + \Delta(\bar{\lambda}, L_{\bar{\lambda}}, R_{\bar{\lambda}})^{(k)} \qquad (3.168)$$

If the value of $\Delta\lambda^{(k)}$ is very small then the iteration is stopped and the power generation, the power losses, and the total cost of all units are calculated. If it is not small then the iteration continues until a convergence is achieved.

3.5.2 Solution Algorithm

The iterative technique is used with a complete (ED) problem when the power losses are included in the system to find the optimal solution. In this method the initial guess of the incremental cost can be calculated for the middle, left, and right sides from Eqs. 3.130, 3.131, and 3.132. Assuming that the power losses are small and can be ignored, the iterative method will find the best equal incremental cost value.

3.5 Fuzzy Economical Dispatch Including Losses

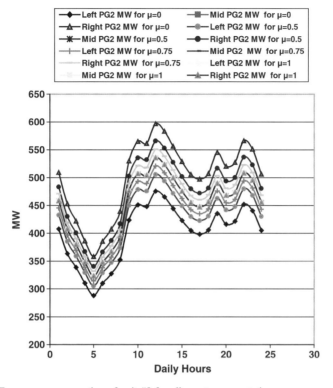

Fig. 3.41 Fuzzy power generation of unit #3 for all α-cut representation

If this value does not give the optimal solution then the iterative program repeats the process until a solution is found. The power generation equation has to be modified to take into account the power losses in the network when power losses are no longer neglected and contribute to the system performance.

3.5.3 Simulated Example

The same simulated example of Sect. 3.4.4 is used to calculate the optimal minimum cost values of the three units committed to the network. The B_{ii} loss coefficients for this example are

$$B_{ii}(pu) = \begin{bmatrix} 0.0218 & 0 & 0 \\ 0 & 0.0228 & 0 \\ 0 & 0 & 0.0179 \end{bmatrix}.$$

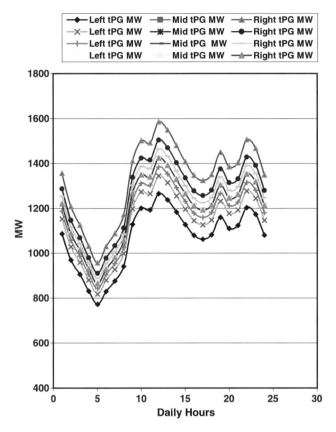

Fig. 3.42 Fuzzy total power generation for all α-cut representation

Evaluating the results obtained using the program based on the flowchart, the following observations are noted.

1. In the example presented the optimal solution was found after eight to ten iterations for each fuzzy load for 24 h.
2. The results shown in the Tables 3.20–3.22 and graphs, Figs. 3.44–3.53 satisfy the constraints set to obtain a minimum solution to the objective function.
3. Different values of α, β, and γ are tested to examine the effect on the total cost value. Those values are tabulated and plotted in different figures. The maximum and minimum ranges of the total cost increase when the value of α, β, and γ increase, and decrease when those values decrease, which is the nature of the quadratic equation of the cost function.
4. Comparing the minimum total cost results obtained in the previous section with 3% deviation for α, β, and γ with the result obtained in this section, the crisp value is higher in the transmission power loss procedure than the neglected power loss case of the previous section, which proves that when considering

3.5 Fuzzy Economical Dispatch Including Losses

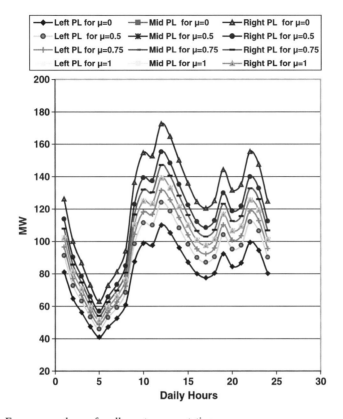

Fig. 3.43 Fuzzy power losses for all α-cut representation

power losses the overall economy of the system will be affected including the upper and lower limits of the minimum cost value. The extra cost value is a result of increased power generation to balance the equality constraint set in Eq. 3.116 to compensate the power losses in the transmission line.

5. The power losses were kept as low as possible and the variations of the hour by hour power losses were tabulated and plotted. This in fact is a great asset to the command and control center to have all this hourly information variation online.

3.5.4 Conclusion

Load conditions change from time to time. The basic objective of economic dispatch operation of power systems is "the distribution of total generation of power in the network between various regional zones; various power stations in respective zones and various units in respective power stations such that the cost of

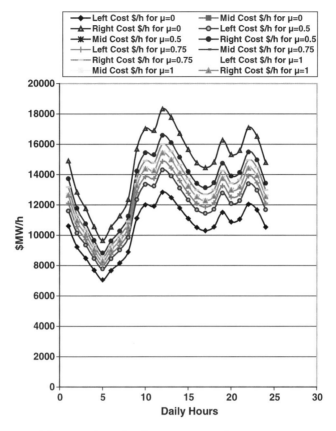

Fig. 3.44 Fuzzy minimum total cost for all α-cut representation

power delivered is a minimum." In the cost of delivered power, the cost of power generation and transmission losses should be considered. It means for every load condition, the load control center should decide the following.

(a) How much power is to be generated to meet the prevailing load condition to maintain constant frequency?
(b) How much power should each region generate?
(c) What should the exchange of power be between the regions (area)?

This aspect can be decided by the regional control center. This thesis provides all the information mentioned above. The variations of load were assumed as fuzzy, which made the output generation of each unit, the system power losses, and the total network cost become fuzzy. This fuzziness provides the load control center with valuable information, which is listed below.

1. The 10% fuzzy load deviation presented gives a range of security knowledge assessment to the load control center. Knowing the minimum and maximum

3.5 Fuzzy Economical Dispatch Including Losses

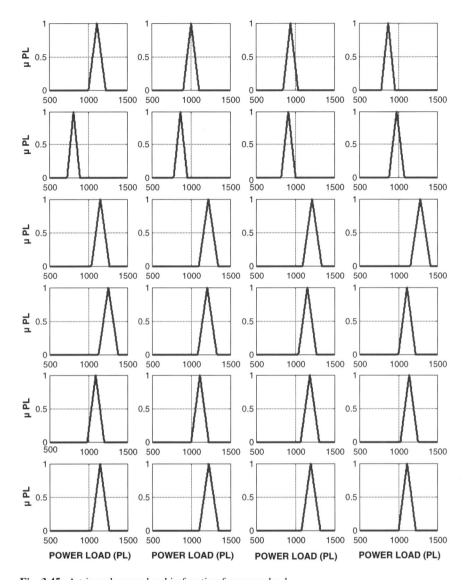

Fig. 3.45 A triangular membership function for power load

generation needed to compensate the load variation, which occurs at an hour in question, can be a great asset to the command and control engineer. If this variation cannot be supplied by the unit committed to the network, then more units can be brought in to overcome the sudden variation.

2. The maximum, minimum, and middle cost variation at each hour is calculated. This gives the company supplying the load an optimal minimum cost generation of each unit and the total cost of all units for that particular load variation at

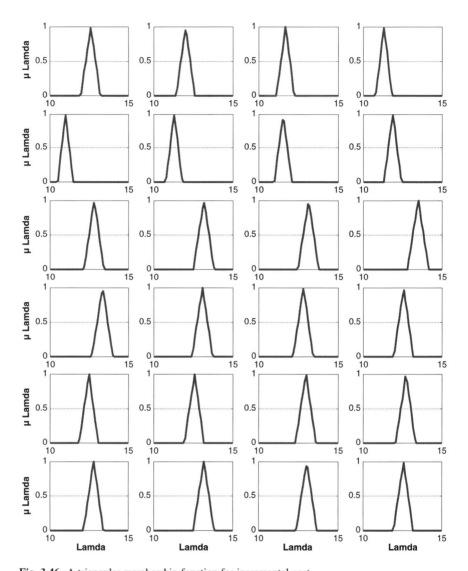

Fig. 3.46 A triangular membership function for incremental cost

the hour in question. This information is very helpful in decision making for the company supplying the load to the consumer. The company can decide whether to supply it if it is not costly or buy it from another company interconnected with the network.

3. The variation of the cost function parameters shown in Tables 3.20, 3.21, and 3.22 can affect the overall performance of the network including the total cost. This means the companies have to be very careful in choosing the unit that has

3.5 Fuzzy Economical Dispatch Including Losses

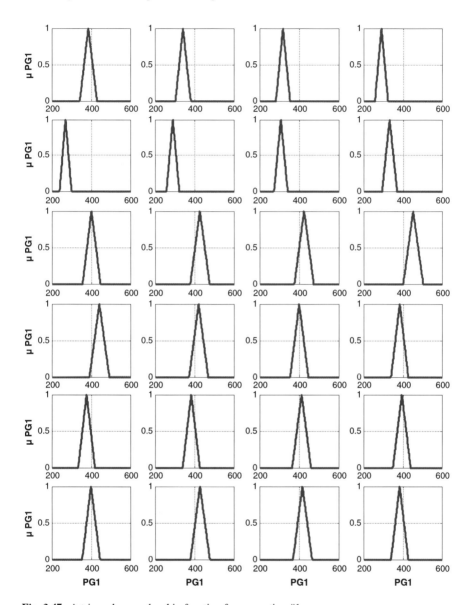

Fig. 3.47 A triangular membership function for generation #1

the best cost function parameters to commit to the network to reduce the maximum and minimum cost values.

4. Power loss information is very helpful to the substation control room where the reactive power flow is minimized through transmission lines by compensation to minimize line losses and to maintain a stable voltage level.

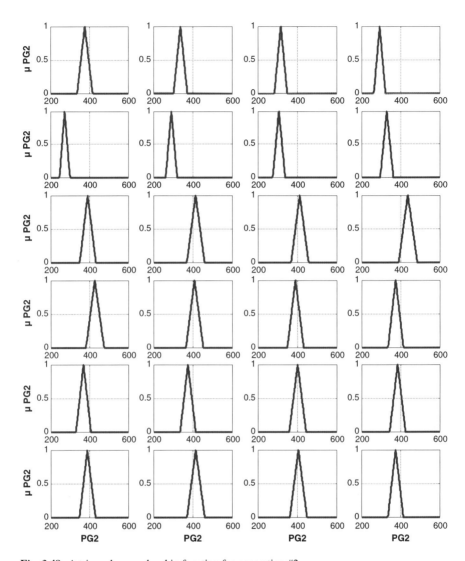

Fig. 3.48 A triangular membership function for generation #2

3.5 Fuzzy Economical Dispatch Including Losses

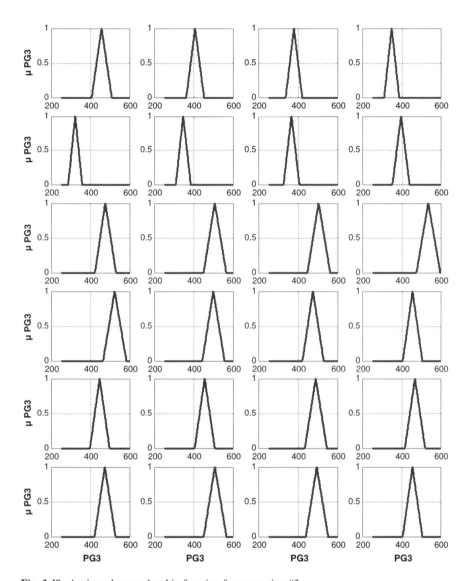

Fig. 3.49 A triangular membership function for generation #3

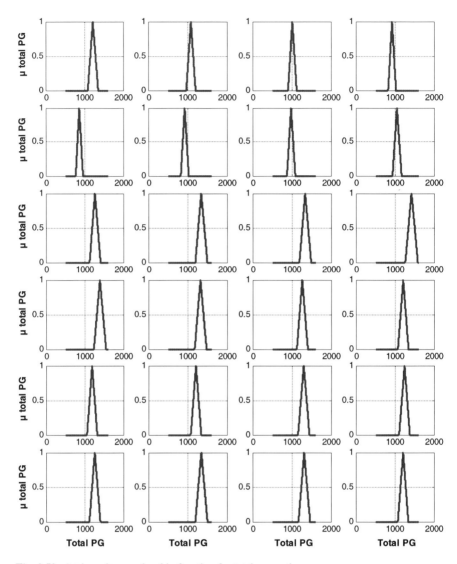

Fig. 3.50 A triangular membership function for total generation

3.5 Fuzzy Economical Dispatch Including Losses

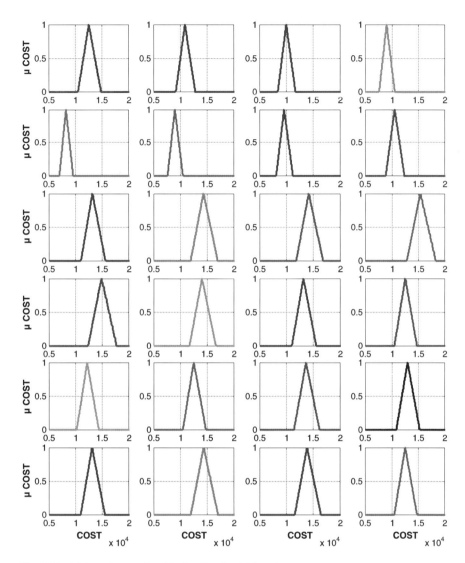

Fig. 3.51 A triangular membership function for total cost

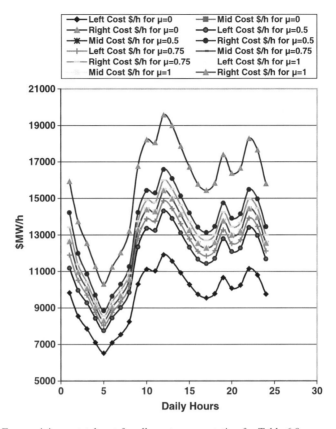

Fig. 3.52 Fuzzy minimum total cost for all α-cut representation for Table 6.9

3.5 Fuzzy Economical Dispatch Including Losses 177

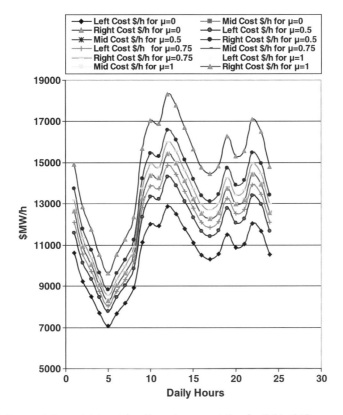

Fig. 3.53 Fuzzy minimum total cost for all α-cut representation for Table 6.10

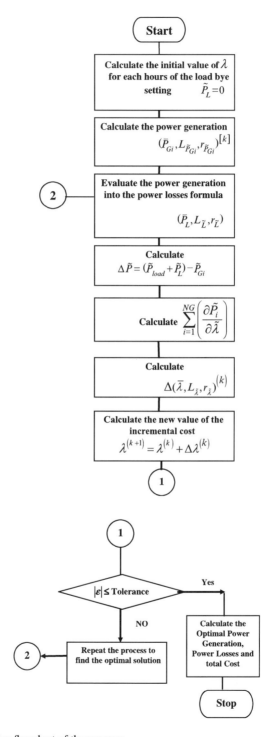

Chart 3.1 Iterative flowchart of the program

3.5 Fuzzy Economical Dispatch Including Losses

Table 3.20 Membership function of total cost for (0, 0.5, 0.75, 1) α-cut representation for model "A" weekdays with 10% deviation for (P_D) and 3% for (α, β, γ)

Membership function	$\mu_C = 0$			$\mu_C = 0.5$			$\mu_C = 0.75$			$\mu_C = 1$		
Daily hours	Left cost ($/h)	Mid cost ($/h)	Right cost ($/h)	Left cost ($/h)	Mid cost ($/h)	Right cost ($/h)	Left cost ($/h)	Mid cost ($/h)	Right cost ($/h)	Left cost ($/h)	Mid cost ($/h)	Right cost ($/h)
1	10,605	12,635	14,904	11,592	12,635	13,738	12,106	12,635	13,179	12,635	12,635	12,635
2	9,232.6	10,943	12,843	10,142	10,943	11,781	10,558	10,943	11,336	10,943	10,943	10,943
3	8,496	10,041	11,750	9,355	10,041	10,754	9,709	10,041	10,380	10,041	10,041	10,041
4	7,685.5	9,051	10,555	8,463	9,051	9,659	8,762	9,051	9,345	9,051	9,051	9,051
5	7,061	8,292	9,643	7,771	8,292	8,828	8,033	8,292	8,554	8,292	8,292	8,292
6	7,669.4	9,032	10,532	8,457	9,032	9,624	8,744	9,032	9,324	9,032	9,032	9,032
7	8,169.3	9,642	11,267	9,021	9,642	10,283	9,330	9,642	9,959	9,642	9,642	9,642
8	8,905.2	10,541	12,356	9,848	10,541	11,258	10,192	10,541	10,896	10,541	10,541	10,541
9	11,119	13,271	15,682	12,349	13,271	14,230	12,806	13,271	13,745	13,271	13,271	13,271
10	12,007	14,375	17,038	13,356	14,375	15,436	13,860	14,375	14,900	14,375	14,375	14,375
11	11,912	14,256	16,892	13,248	14,256	15,306	13,747	14,256	14,776	14,256	14,256	14,256
12	12,850	15,425	18,331	14,313	15,425	16,586	14,863	15,425	15,999	15,425	15,425	15,425
13	12,486	14,972	17,774	13,901	14,972	16,090	14,431	14,972	15,525	14,972	14,972	14,972
14	11,790	14,105	16,705	13,110	14,105	15,140	13,602	14,105	14,617	14,105	14,105	14,105
15	11,094	13,240	15,644	12,321	13,240	14,195	12,776	13,240	13,713	13,240	13,240	13,240
16	10,516	12,525	14,769	11,667	12,525	13,415	12,092	12,525	12,966	12,525	12,525	12,525
17	10,307	12,267	14,454	11,431	12,267	13,134	11,845	12,267	12,696	12,267	12,267	12,267
18	10,550	12,567	14,821	11,706	12,567	13,461	12,133	12,567	13,010	12,567	12,567	12,567
19	11,503	13,748	16,267	12,785	13,748	14,750	13,261	13,748	14,244	13,748	13,748	13,748
20	10,880	12,976	15,321	12,080	12,976	13,907	12,523	12,976	13,437	12,976	12,976	12,976
21	11,053	13,190	15,583	12,275	13,190	14,140	12,728	13,190	13,660	13,190	13,190	13,190
22	12,044	14,421	17,094	13,398	14,421	15,486	13,904	14,421	14,948	14,421	14,421	14,421
23	11,663	13,947	16,512	12,967	13,947	14,968	13,452	13,947	14,452	13,947	13,947	13,947
24	10,531	12,543	14,792	11,685	12,543	13,435	12,110	12,543	12,985	12,543	12,543	12,543

Table 3.21 Membership function of total cost for (0, 0.5, 0.75, 1) α-cut representation for model "A" weekdays with 10% deviation for (P_D, α) and 10% for (β, γ)

Membership function	$\mu_C = 0$			$\mu_C = 0.5$			$\mu_C = 0.75$			$\mu_C = 1$		
Daily hours	Left cost (\$/h)	Mid cost (\$/h)	Right cost (\$/h)	Left cost (\$/h)	Mid cost (\$/h)	Right cost (\$/h)	Left cost (\$/h)	Mid cost (\$/h)	Right cost (\$/h)	Left cost (\$/h)	Mid cost (\$/h)	Right cost (\$/h)
1	9,824.5	12,635	15,925	11,178	12,635	14,212	11,892	12,635	13,408	12,635	12,635	12,635
2	8,554.9	10,943	13,720	9,963	10,943	11,985	10,511	10,943	11,385	10,943	10,943	10,943
3	7,854.2	10,041	12,529	9,273	10,041	10,848	9,698	10,041	10,391	10,041	10,041	10,041
4	7,104.8	9,051	11,262	8,426	9,051	9,701	8,760	9,051	9,348	9,051	9,051	9,051
5	6,527.3	8,292	10,293	7,754	8,292	8,847	8,032	8,292	8,554	8,292	8,292	8,292
6	7,090	9,032	11,237	8,448	9,032	9,635	8,744	9,032	9,324	9,032	9,032	9,032
7	7,552.1	9,642	12,017	9,016	9,642	10,289	9,330	9,642	9,959	9,642	9,642	9,642
8	8,252	10,541	13,199	9,846	10,541	11,261	10,192	10,541	10,896	10,541	10,541	10,541
9	10,299	13,271	16,757	12,347	13,271	14,232	12,806	13,271	13,745	13,271	13,271	13,271
10	11,120	14,375	18,207	13,355	14,375	15,437	13,860	14,375	14,900	14,375	14,375	14,375
11	11,032	14,256	18,051	13,248	14,256	15,307	13,747	14,256	14,776	14,256	14,256	14,256
12	11,919	15,425	19,567	14,312	15,425	16,586	14,863	15,425	15,999	15,425	15,425	15,425
13	11,563	14,972	18,994	13,901	14,972	16,090	14,431	14,972	15,525	14,972	14,972	14,972
14	10,919	14,105	17,851	13,110	14,105	15,140	13,602	14,105	14,617	14,105	14,105	14,105
15	10,276	13,240	16,716	12,321	13,240	14,195	12,776	13,240	13,713	13,240	13,240	13,240
16	9,742	12,525	15,780	11,667	12,525	13,415	12,092	12,525	12,966	12,525	12,525	12,525
17	9,548.9	12,267	15,443	11,431	12,267	13,134	11,845	12,267	12,696	12,267	12,267	12,267
18	9,773.8	12,567	15,836	11,706	12,567	13,461	12,133	12,567	13,010	12,567	12,567	12,567
19	10,654	13,748	17,383	12,785	13,748	14,750	13,261	13,748	14,244	13,748	13,748	13,748
20	10,079	12,976	16,370	12,080	12,976	13,907	12,523	12,976	13,437	12,976	12,976	12,976
21	10,239	13,190	16,651	12,275	13,190	14,140	12,728	13,190	13,660	13,190	13,190	13,190
22	11,154	14,421	18,267	13,398	14,421	15,486	13,904	14,421	14,948	14,421	14,421	14,421
23	10,803	13,947	17,645	12,967	13,947	14,968	13,452	13,947	14,452	13,947	13,947	13,947
24	9,756.1	12,543	15,805	11,685	12,543	13,435	12,110	12,543	12,985	12,543	12,543	12,543

3.5 Fuzzy Economical Dispatch Including Losses 181

Table 3.22 Membership function of total cost for (0, 0.5, 0.75, 1) α-cut representation for model "A" weekdays with 2% deviation for (P_D, α) and 3% for (β, γ)

Membership function	$\mu_C = 0$			$\mu_C = 0.5$			$\mu_C = 0.75$			$\mu_C = 1$		
Daily hours	Left cost ($/h)	Mid cost ($/h)	Right cost ($/h)	Left cost ($/h)	Mid cost ($/h)	Right cost ($/h)	Left cost ($/h)	Mid cost ($/h)	Right cost ($/h)	Left cost ($/h)	Mid cost ($/h)	Right cost ($/h)
1	10,610	12,635	14,899	11,594	12,635	13,736	12,107	12,635	13,178	12,635	12,635	12,635
2	9,238	10,943	12,838	10,143	10,943	11,779	10,558	10,943	11,335	10,943	10,943	10,943
3	8,501	10,041	11,745	9,356	10,041	10,754	9,709	10,041	10,380	10,041	10,041	10,041
4	7,691	9,051	10,550	8,464	9,051	9,659	8,762	9,051	9,345	9,051	9,051	9,051
5	7,066	8,292	9,637	7,772	8,292	8,827	8,033	8,292	8,554	8,292	8,292	8,292
6	7,675	9,032	10,527	8,457	9,032	9,624	8,744	9,032	9,324	9,032	9,032	9,032
7	8,174	9,642	11,262	9,021	9,642	10,283	9,330	9,642	9,959	9,642	9,642	9,642
8	8,910	10,541	12,351	9,848	10,541	11,258	10,192	10,541	10,896	10,541	10,541	10,541
9	11,124	13,271	15,677	12,349	13,271	14,230	12,806	13,271	13,745	13,271	13,271	13,271
10	12,012	14,375	17,032	13,356	14,375	15,436	13,860	14,375	14,900	14,375	14375	14,375
11	11,917	14,256	16,887	13,248	14,256	15,306	13,747	14,256	14,776	14,256	14,256	14,256
12	12,856	15,425	18,326	14,313	15,425	16,586	14,863	15,425	15,999	15,425	15,425	15,425
13	12,492	14,972	17,769	13,901	14,972	16,090	14,431	14,972	15,525	14,972	14,972	14,972
14	11,795	14,105	16,700	13,110	14,105	15,140	13,602	14,105	14,617	14,105	14,105	14,105
15	11,099	13,240	15,639	12,321	13,240	14,195	12,776	13,240	13,713	13,240	13,240	13,240
16	10,521	12,525	14,764	11,667	12,525	13,415	12,092	12,525	12,966	12,525	12,525	12,525
17	10,312	12,267	14,449	11,431	12,267	13,134	11,845	12,267	12,696	12,267	12,267	12,267
18	10,556	12,567	14,816	11,706	12,567	13,461	12,133	12,567	13,010	12,567	12,567	12,567
19	11,508	13,748	16,262	12,785	13,748	14,750	13,261	13,748	14,244	13,748	13,748	13,748
20	10,886	12,976	15,315	12,080	12,976	13,907	12,523	12,976	13,437	12,976	12,976	12,976
21	11,058	13,190	15,578	12,275	13,190	14,140	12,728	13,190	13,660	13,190	13,190	13,190
22	12,049	14,421	17,089	13,398	14,421	15,486	13,904	14,421	14,948	14,421	14,421	14,421
23	11,669	13,947	16,507	12,967	13,947	14,968	13,452	13,947	14,452	13,947	13,947	13,947
24	10,537	12,543	14,787	11,685	12,543	13,435	12,110	12,543	12,985	12,543	12,543	12,543

Appendix A.1

For a critical reactor, the reactivity ρ_i is zero. Solving for x_i from (3.12) we obtain

$$x_i = \frac{\rho_{0i}}{\alpha_{xi}} - \frac{\rho_{ci}}{\alpha_{xi}} - \frac{\alpha_{di}}{\alpha_{xi}} P_{Ni} \tag{A.1.1}$$

From (3.10), assuming an initial condition $I_i(t_o) = I_{0i}$,

$$I_i(t) = I_{0i} e^{-\lambda_I(t-t_o)} + \frac{\gamma_I}{e_i G} \int_0^t e^{-\lambda_I(t-\tau)} P_{Ni}(\tau) d\tau \tag{A.1.2}$$

Differentiating (A.1.1) and substituting both (A.1.1) and (A.1.2) in Eq. 3.11 gives

$$\overset{o}{P}_{Ni} = \left(-\lambda_x + \frac{\Gamma_x \rho_{0i}}{e_i \sum_{fi} G \alpha_{di}} - \frac{\gamma_x \alpha_{xi}}{e_i G \alpha_{di}} \right) P_{Ni}$$

$$- \frac{\Gamma_x}{e_i \sum_{fi} G} P_{Ni}^2 - \frac{\Gamma_x}{e_i \sum_{fi} G \alpha_{di}} P_{Ci} P_{Ni}$$

$$+ \left(\frac{\lambda_x \rho_{0i}}{\alpha_{di}} - \frac{\lambda_I \alpha_{xi}}{\alpha_{di}} I_0 e^{-\lambda_I(t-t_o)} \right) - \frac{\lambda_X}{\alpha_{di}} \rho_i$$

$$- \frac{\overset{o}{\rho}_{Ci}}{\alpha_{di}} - \frac{\lambda_I \alpha_{xi} \gamma_I}{\alpha_{di} e_i G} \int_0^t e^{-\lambda_I(t-\tau)} P_{Ni}(\tau) d\tau \tag{A.1.3}$$

Appendix A.2

The following relations are used in the derivation of the objective functional (3.24).

(a) $2 \int_0^T \mu(t) v(t) \overset{o}{v}(t) dt = \mu(T) v^2(T) - \mu_0 v_0^2 - \int_0^T v^2(t) \overset{o}{\mu}(t) dt$

(b) $\int_0^T \mu(t) V(t-\tau) dt = \int_{-\tau}^0 \mu(t-\tau) V(t) dt + \int_o^T \tilde{\phi}(t,T,\tau) \mu(t+\tau) V(t) dt$

where

$$\tilde{\phi}(t,T,\tau) = \begin{cases} 1, & t \leq T-\tau \\ 0, & t > T-\tau \end{cases}$$

(c) $\int_0^T \int_0^t \mu(t) e^{-\lambda_i(t-\lambda)} V(\tau) d\tau dt = \int_o^T \int_0^t V(t) e^{\lambda_I(t-\tau)} \mu(\tau) d\tau dt$

References

1. Kiefer, W.M., Koncel, E.F.: Scheduling generations on systems with fossil and nuclear units. Trans. Am. Nucl. Soc. **13**, 768 (1970)
2. Hoskins, R.E., Rees, F.J.: Power systems optimization approach to nuclear fuel management. Trans. Am. Nucl. Soc. **13**, 768 (1970)
3. Grossman, L.M., Reinking, A.G.: Fuel management and load optimization of nuclear units in electric systems. Trans. Am. Nucl. Soc. **20**, 391 (1985)
4. Chou, Q.B.: Characteristics and maneuverability of Canada nuclear power stations operated for base-load and load following generation. IEEE Trans. Power Appar. Syst. **PAS-94**(3), 792–801 (1975)
5. El-Wakil, M.M.: Nuclear Power Engineering. McGraw-Hill, New York (1962)
6. Yasukawa, S.: An analysis of continuous reactor refueling. Nucl. Sci. Eng. **24**, 253–260 (1966)
7. Millar, C.H.: Fuel management in Canada reactors. Trans. Am. Nucl. Soc. **20**, 350 (1975)
8. El-Hawary, M.E., Christensen, G.S.: Optimal Economic Operation of Electric Power Systems. Academic, New York (1979)
9. Porter, W.A.: Modern Foundations of Systems Engineering. Macmillan, New York (1966)
10. Hamilton, E.P., Lamont, I.W.: An improved short term hydro-thermal coordination model. Paper No, A77 518-4. Institute of Electrical and Electronics Engineers Summer Power Meeting, Mexico City, 1977
11. Isbin, H.S.: Introductory Nuclear Reactor Theory. Reinhold, New York (1963)
12. Shamaly, A., et al.: A transformation for necessary optimality conditions for systems with polynomial nonlinearities. IEEE Trans. Autom. Control **AC-24**, 983–985 (1979)
13. Mahmoud, M.S.: Multilevel systems control and applications: a survey. IEEE Trans. Syst. Man Cybern. **7**(3), 125–143 (1977)
14. Nieva, R., Christensen, G.S., El-Hawary, M.E.: Functional optimization of nuclear-hydro-thermal systems. In: Proceedings, CEC, Toronto, 1978
15. Nieva, R., Christensen, G.S., El-Hawary, M.E.: Optimum load scheduling of nuclear-hydro-thermal power systems. Optim. Theor. Appl. **35**(2), 261–275 (1981)
16. Chowdhury, B.H., Rahman, S.: A review of recent advances in economic dispatch. IEEE Trans. Power Syst. **5**(4), 1248–1259 (1990)
17. Song, Y.H., Chou, C.S.V.: Advanced engineered-conditioning genetic approach to P economic dispatch. IEE Proc. Gener. Transm. Distrib. **144**(3), 285–292 (1997)
18. Walsh, M.P., O'Malley, M.J.: Augmented Hopfield network for unit commitment and economic dispatch. IEEE Trans. Power Syst. **12**(4), 1765–1774 (1997)
19. Yalcinoz, T., Short, M.J.: Large-scale economic dispatch using an improved Hopfield network. IEE Proc. Gener. Transm. Distrib. **144**(2), 181–185 (1997)
20. Song, Y.H., Wang, G.S., Wang, P.Y., Johns, A.T.: Environmental/economic dispatch using fuzzy logic controlled genetic algorithms. IEE Proc. Gener. Transm. Distrib. **144**(4), 377–382 (1997)
21. Grudinin, N.: Combined quadratic-separable programming OPF algorithm economic dispatch and security control. IEEE Trans. Power Syst. **12**(4), 1682–1688 (1997)
22. Yalcinoz, T., Short, M.J.: Neural networks approach for solving economic dispatch with transmission capacity constraints. IEEE Trans. Power Syst. **13**(2), 307–313 (1998)
23. Wang, K.P., Yuryevich, J.: Evolutionary-programming-based algorithm for environmentally constrained economic dispatch. IEEE Trans. Power Syst. **13**(2), 301–306 (1998)
24. Fan, J.Y., Zhang, L.: Real-time economic dispatch with line flow and emission constraints using quadratic programming. IEEE Trans. Power Syst. **13**(2), 320–325 (1998). doi:320
25. Xia, Q., Song, Y.H., Zhang, B., Kang, C., Xiang, N.: Dynamic queuing approach to power system short term and security dispatch. IEEE Trans. Power Syst. **13**(2), 280–285 (1998)
26. Irisarri, G., Kimball, L.M., Clements, K.A., Bagchi, A., Davis, P.W.: Economic dispatch with network and ramping constraints interior point methods. IEEE Trans. Power Syst. **13**(1), 236–242 (1998)

27. Das, D.B., Patvardhan, C.: New multi-objective stochastic search technique for economic load dispatch. IEE Proc Gener. Transm. Distrib. **145**(6), 747–752 (1998)
28. Yalcinoz, T., Short, M.J., Cory, B.J.: Security dispatch using the Hopfield neural network. IEE Gener. Transm. Distrib. **146**(5), 465–470 (1999)
29. Liang, R.-H.: A neural-based re-dispatch approach to dynamic generation allocation. IEEE Trans. Power Syst. **14**(4), 1388–1393 (1999)
30. Rudolf, A., Bayrleithner, R.: A genetic algorithm for solving the unit commitment problem hydro-thermal power system. IEEE Trans. Power Syst. **14**(4), 1460–1468 (1999)
31. Bakirtzis, A.G., Zoumas, C.E.: Lamda of Lagrangian relaxation solution to unit commitment problem. IEE Proc. Gener. Transm. Distrib. **147**(2), 131–136 (2000)
32. Jaber, R.A., Coonick, A.H., Cory, B.J.: A homogeneous linear programming algorithm for the security constrained economic dispatch problem. IEEE Trans. Power Syst. **15**(3), 930–936 (2000)
33. Ching-Tzong, S.U., Chien-Tung, L.: New approach with a Hopfield modeling framework to economic dispatch. IEEE Trans. Power Syst. **15**(2), 541–545 (2000)
34. Jaber, R.A., Coonick, A.H.: Homogeneous interior point method for constrained power scheduling. IEE Proc. Gener. Trans. Distrib. **147**(4), 239–244 (2000)
35. Whei-Min, L., Fu-Seng, C., Ming-Tong, T.: Non-convex economic dispatch by integrated artificial intelligence. IEEE Trans. Power Syst. **16**(2), 307–311 (2001)
36. Yalcinoz, T., Altun, H.: Power economic dispatch using a hybrid genetic algorithm. IEEE Power Eng. Rev. **21**(3), 59–60 (2001)
37. Aldridge, J., McKee, S., McDonald, J.R., Galloway, S.J., Dahal, K.P., Brad Macqueen, J.F.: Knowledge-based genetic algorithm for unit commitment. IEE Proc. Gener. Trans. Distrib. **148**(2), 146–152 (2001)
38. Han, X.S., Gooi, H.B., Kirschen, D.S.: Dynamic economic dispatch: feasible and optimal solution. IEEE Trans. Power Syst. **16**(1), 22–28 (2001)
39. El-Hawary, M.E.: Electric Power Applications of Fuzzy Systems. IEEE Press, New York (1998)
40. Song, Y.H., Chou, C.S.V.: Advanced engineered-conditioning genetic approach to P economic dispatch. IEE Proc. Gener. Trans. Distrib. **144**(3), 285–292 (1997)
41. Yalcinoz, T., Short, M.J.: Large-scale economic dispatch using an improved Hopfield network. IEE Proc. Gener. Transm. Distrib. **144**(2), 181–185 (1997)
42. Song, Y.H., Wang, G.S., Wang, P.Y., Johns, A.T.: Environmental/economic dispatch using fuzzy logic controlled genetic algorithms. IEE Proc. Gener. Trans. Distrib. **144**(4), 377–382 (1997)
43. Yalcinoz, T., Short, M.J.: Neural networks approach for solving economic dispatch with transmission capacity constraints. IEEE Trans. Power Syst. **13**(2), 307–313 (1998)
44. Das, D.B., Patvardhan, C.: New multi-objective stochastic search technique for economic load dispatch. IEE Proc. Gener. Transm. Distrib. **145**(6), 747–752 (1998)
45. Ching-Tzong, S.U., Chien-Tung, L.: New approach with a Hopfield modeling framework to economic dispatch. IEEE Trans. Power Syst. **15**(2), 541–545 (2000)
46. Jaber, R.A., Coonick, A.H.: Homogeneous interior point method for constrained power scheduling. IEE Proc. Gener. Trans. Distrib. **147**(4), 239–244 (2000)

Chapter 4
Economic Dispatch (ED) and Unit Commitment Problems (UCP): Formulation and Solution Algorithms

Objectives The objectives of this chapter are:

- Formulating the objectives function for ED and UCP
- Studying the system and unit constraints
- Proposing rules for generating solutions
- Generating an initial solution
- Explaining an algorithm for the economic dispatch problem
- Applying the simulated annealing algorithm to solve the problems
- Comparing simulated annealing with other simulated annealing algorithms
- Offering numerical results for the simulated annealing algorithm

4.1 Introduction

The unit commitment problem (UCP) is the problem of selecting the generating units to be in service during a scheduling period and for how long. The committed units must meet the system load and reserve requirements at minimum operating cost, subject to a variety of constraints. The economic dispatch problem (EDP) deals with the optimal allocation of the load demand among the running units while satisfying the power balance equations and unit operating limits [1].

The solution of the UCP using artificial intelligence techniques requires three major steps:

- A problem statement or system modeling
- Rules for generating trial solutions
- An efficient algorithm for solving the EDP

4.2 Problem Statement

Modeling of power system components affecting the economic operation of the system is an important step when solving the UCP. The degree of detail in component modeling varies with the desired accuracy and the nature of the problem under study. The basic components of a power system include generating power stations, transformer, transmission network, and system load.

This section is concerned with thermal generating unit scheduling. Hence it is assumed that the network is capable of transmitting the power generated to the load centers without losses or network failures. This means that the network is assumed to be perfectly reliable. Consequently, the following basic engineering assumptions are made [1–3].

- The network interchange between the system under study and other systems is fixed.
- The load demand is not affected by adding or removing generating units.
- The operating cost of a generating unit is assumed to be composed of three components: start-up cost, spinning (no load) cost, and production (loading) cost.

In the UCP under consideration, one is interested in a solution that minimizes the total operating cost during the scheduling time horizon while several constraints are satisfied [4]. The objective function and the constraints of the UCP are described in the following sections.

4.3 Rules for Generating Trial Solutions

The cornerstone in solving combinatorial optimization problems is to have good rules for generating feasible trial solutions starting from an existing feasible solution. The trial solutions (neighbors) should be randomly generated, feasible, and span the entire problem solution space as much as possible. Because of the constraints in the UCP, this is not a simple matter. The major contributions of this section are the implementation of new rules to generate randomly feasible solutions faster [5].

4.4 The Economic Dispatch Problem

The economic dispatch problem is an essential problem when solving the UCP. Once a trial solution is generated, the corresponding operating cost of this solution is calculated by solving the EDP. Consequently, using an efficient and fast algorithm for solving the EDP improves the quality of the UCP solution, and thereby the performance of the overall UCP algorithm. An efficient algorithm for solving the

4.5 The Objective Function

EDP is called the linear complementary (discussed later) and is based on Kuhn–Tucker conditions. The application of the linear complementary algorithm to solve the EDP is original. Our investigation showed that the results obtained by this algorithm are more accurate than those obtained using an IMSL quadratic programming routine.

4.5 The Objective Function

4.5.1 The Production Cost

The major component of the operating cost, for thermal and nuclear units, is the power production cost of the committed units. The production cost is mainly the cost of fuel input per hour, and maintenance and labor contribute only to a small extent. Conventionally the unit production cost is expressed as a quadratic function of the unit output power as follows.

$$F_{it}(P_{it}) = A_i P^2{}_{it} + B_i P_{it} + C_i \text{ \$/h} \tag{4.1}$$

Where

$F_{it}(P_{it})$ = The fuel cost of unit i at time t
P_{it} = the MW generated from unit i at time t
A_i, B_i, and C_i are the constants of production cost

4.5.2 The Start-Up Cost

The second component of the operating cost is the startup cost. The startup cost is attributed to the amount of energy consumed to bring the unit online. The startup cost depends upon the down time of the unit. This can vary from maximum value, when the unit is started from cold state, to a much smaller value, where the unit has recently been turned off.

Calculation of the startup cost depends also on the treatment method for the thermal unit during downtime periods. There are two methods for unit treatment during the off hours: the cooling method and the banking method. The former allows the boiler of the unit to cool down and then reheat back up to the operating temperature when recommitted online. In the latter method, the boiler operating temperature is maintained during the off time using an additional amount of energy.

The cooling method is used in this section due to its practicability when applied to real power systems. In this work, the startup cost, for a unit i at time t, based on the cooling method, is considered in a more general form as follows [6].

$$ST_{it} = So_i[1 - D_i \exp(-Toff_i/Tdown_i)] + E_i \; \$ \tag{4.2}$$

Accordingly, the overall operating cost of the generating units in the scheduling time horizon (i.e., objective function of the UCP) is

$$F_T = \sum_{t=1}^{T}\sum_{i=1}^{N}(U_{it}F_{it}(P_{it}) + V_{it}ST_{it} + W_{it}SH_{it}) \; \$ \tag{4.3}$$

4.6 The Constraints

The unit commitment problem is subject to many constraints depending on the nature of the power system under study. The constraints taken into consideration may be classified into two main groups: system constraints and unit constraints.

4.6.1 System Constraints

The system constraints, sometimes called coupling constraints, also include two categories: the load demand and the spinning reserve constraints.

4.6.1.1 Load Demand Constraints

The load demand constraint is the most important constraint in the UCP. It basically means that the generated power from all committed units must meet the system load demand. This is formulated in the so-called active power balance equation:

$$\sum_{i=1}^{N} U_{it}P_{it} = PD_t; \; \mathbf{1 \leq t \leq T} \tag{4.4}$$

4.6.1.2 Spinning Reserve Constraint

The spinning (operating) reserve is the total amount of generation capacity available from all units synchronized (spinning) on the system minus the present load demand. It is important to determine the suitable allocation of the spinning reserve from two points of view: the reliability requirements and the economical aspects.

There are various methods for determining the spinning reserve [1, 4, 7, 8]:

- The reserve is computed as a percentage of the forecasted load demand.
- It is determined that the system can make up for a loss of the highest rating unit in a given period of time.

4.6 The Constraints

- The reserve requirements are determined as a function of system reliability, which is evaluated on a probabilistic basis.

Here, the reserve is computed as a given prespecified amount which is a percentage of the forecasted load demand; that is,

$$\sum_{i=1}^{N} U_{it} P\max_i \geq (PD_t + R_t); \; \mathbf{1 \leq t \leq T} \tag{4.5}$$

4.6.2 Unit Constraints

The constraints on the generating units (sometimes called local constraints) are described as follows.

4.6.2.1 Generation Limits

The generation limits represent the minimum loading limit below which it is not economical to load the unit, and the maximum loading limit above which the unit should not be loaded.

$$U_{it} P\min_i \leq P_{it} \leq P\max_i U_{it}; \; \mathbf{1 \leq t \leq T, \; 1 \leq i \leq N} \tag{4.6}$$

4.6.2.2 Minimum Up/Down Time

If the unit is running, it cannot be turned off before a certain minimum amount of time elapses. If the unit is also down, it cannot be recommitted before a certain time elapses.

$$\begin{array}{l} T_{off\,i} \geq T_{down\,i} \\ T_{on\,i} \geq T_{up\,i} \end{array}; \; \mathbf{1 \leq i \leq N} \tag{4.7}$$

These constraints could be formulated in a mathematical form as follows.

$$\sum_{l=0}^{T_{up_i}-1} U_{i,t+l} \geq V_{it} T_{up_i}; \; \mathbf{1 \leq t \leq T, \; 1 \leq i \leq N} \tag{4.8}$$

$$\sum_{l=0}^{T_{down_i}-1} (1 - U_{i,t+l}) \geq \mathbf{W_{it} T_{down_i}}; \; \mathbf{1 \leq t \leq T, \; 1 \leq i \leq N} \tag{4.9}$$

$$V_{it} \geq U_{it} - U_{i,t-1}; \; 2 \leq t \leq T, \; 1 \leq i \leq N \quad (4.10)$$

$$W_{it} \geq U_{i,t-1} - U_{it}; \; 2 \leq t \leq T, \; 1 \leq i \leq N \quad (4.11)$$

$$V_{i1} = U_{i1}; \; 1 \leq i \leq N \quad (4.12)$$

$$W_{i1} = 1 - U_{i1}; \; 1 \leq i \leq N \quad (4.13)$$

4.6.2.3 Units Initial Status Constraint

The unit status (e.g., hours of being ON or OFF) before the first hour in the proposed schedule is an important factor to determine whether its new status violates the minimum up/down constraints. Also, the initial status of the unit affects the startup cost calculations.

4.6.2.4 Crew Constraints

If the plant consists of two or more units, they cannot be turned on at the same time due to some technical conditions or manpower availability.

4.6.2.5 Unit Availability Constraint

Due to some abnormal conditions, such as forced outage or maintenance of a unit, the unit may become *unavailable*. The unit may also be forced into service to increase reliability or stability of the system, hence the unit becomes *must run* or *fixed at a certain output*. Otherwise the unit is *available*. The availability constraint specifies the unit to be in one of the following different situations: *unavailable, must run, available,* or *fixed output* (MW).

4.6.2.6 Units Derating Constraint

During the lifetime of a unit its performance could be changed due to many conditions, for example, aging factor, the environment, and so on. These conditions may cause derating of the generating unit. Consequently, the unit maximum and minimum limits are changed.

4.7 Rules for Generating Trial Solutions

One of the most important issues in solving combinatorial optimization problems is generating a trial solution as a neighbor to an existing solution. The neighbors should be randomly generated, feasible, and span as much as possible the entire problem solution space. In the course of generating feasible solutions, the most difficult constraints to satisfy are the minimum up/down times.

The proposed rules [5] applied to get a trial solution as a neighbor of an existing feasible solution are described, with the help of an example, in the following steps. The following values are assumed: $T = 12$, $Tup_i = 2$ or 4, and $Tdown_i = 1$ or 4.

Step (1): Randomly generate a unit i, $i \sim UD(1,N)$, and an hour t, $t \sim UD(1,T)$. Figure 4.1a shows the status of some unit i over a period of 12 h. The unit is ON between the periods 5 and 8 inclusive.

Step (2): If unit i at hour t is on (e.g., 5, 6, 7, or 8 in Fig. 4.1a), then go to Step (3) to consider switching it ON around time t. Otherwise, if unit i at hour t is OFF (e.g., 1, 2, 3, 4, 9, 10, 11, or 12), then go to Step (4) to consider switching it OFF around time t.

Step (3): Switching the unit i from ON to OFF

(a) Move from the hour t backward and forward in time, to find the length of the ON period. In this example if $t = 6$, then $Ton_i = 8 - 5 + 1 = 4$, and the unit is ON during hours 5,6,7,8.

(b) If $Ton_i = Tup_i$, then turn the unit OFF in all hours comprising Ton_i. In the example if $Tup_i = 4$, then switch the unit OFF at $t = 5, 6, 7, 8$ (Fig. 4.1b).

(c) If $Ton_i > Tup_i$, then generate $L \sim UD(1, Ton_i - Tup_i)$.

(d) Turn the unit OFF for the hours $t1, t1 + 1, \ldots, t1 + L - 1$, where $t1$ is the first hour at which the unit is ON.

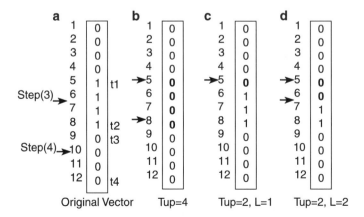

Fig. 4.1 Illustrative example for the rules of generating trial solutions

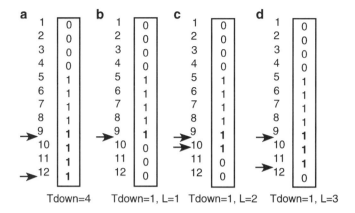

Fig. 4.2 Illustrative example for the rules of generating trial solutions

In the example if $Tup_i = 2$, then $L \sim UD\ (1, 2)$. Hence, the following two solutions are possible.

If $L = 1$, then the unit is turned OFF at $t = 5$ (Fig. 4.1c), and
If $L = 2$, the unit is turned OFF at $t = 5,6$ (Fig. 4.1d).

Step (4): Switching the unit i from OFF to ON

(a) Move from the hour t backward and forward in time, to find the length of the OFF period. In the example if $t = 10$, then $\mathbf{Toff_i} = 12 - 9 + 1 = 4$, the unit is OFF during hours 9 through 12.
(b) If $\mathbf{Toff_i} = \mathrm{Tdown}_i$, then turn the unit ON in all hours of $\mathbf{Toff_i}$. In the example if $\mathrm{Tdown}_i = 4$, then switch the unit ON at $t = 9$ through 12 (Fig. 4.2a).
(c) If $\mathbf{Toff_i} > \mathrm{Tdown}_i$, then generate $L \sim UD\ (1, \mathbf{Toff_i}\text{-}\mathrm{Tdown}_i)$.
(d) Turn the unit ON for the hours $t3, t3 + 1, t3 + 2, \ldots, t3 + L - 1$, where $t3$ is the first hour at which the unit is OFF.

In the example if $\mathrm{Tdown}_i = 1$, then $L \sim UD\ (1, 3)$. Hence, the following three solutions are possible.

If $L = 1$, then the unit is turned ON at $t = 9$ (Fig. 4.2b),
If $L = 2$, the unit is turned ON at $t = 9, 10$ (Fig. 4.2c)
If $L = 3$, the unit is turned ON at $t = 9, 10, 11$ (Fig. 4.2d).

Step (5): Check for reserve constraints

Check the reserve constraints satisfaction for the changed time periods in Steps (3) and (4). If it is satisfied, then the obtained trial solution is feasible; otherwise go to Step (1) (Fig. 4.2).

4.8 Generating an Initial Solution

Solving the UCP using any combinatorial optimization algorithm requires a starting feasible schedule. The generated starting solution must be randomly generated and feasible. The following algorithm is used for finding this starting solution [5].

Step (1): Set $U = V = P = 0$, $t = 1$.
Step (2): Do the following substeps.

    (a) Randomly generate a unit i, $i \sim UD(1,N)$.
    (b) If the unit i at hour t is OFF ($U_{it} = 0$), then go to Step (3). Otherwise go to Step (2a) to choose another unit.

Step (3): Follow the procedure in Step (4), in Sect. 4.8, to consider switching this unit ON starting from hour t.
Step (4): If $t = T$, go to Step (5), otherwise set $t = t + 1$ and go to Step (2).
Step (5): Check the reserve constraints for all hours. Repeat Steps (2) and (3) for the hours at which the constraints are not satisfied.

4.9 An Algorithm for the Economic Dispatch Problem

The EDP is the heart of any algorithm used to solve the UCP. To get the objective function of a given trial solution we have to solve the EDP. The accuracy and speed of convergence for the selected routine to solve the EDP is crucial in the efficiency of the overall UCP algorithm.

Because the production cost of the UCP, formulated in Sect. 4.6, is a quadratic function, the EDP is solved using a quadratic programming routine. In this section we present an efficient algorithm for solving the EDP. The method is based on Kuhn–Tucker conditions and is called the *linear complementary algorithm* [9].

The application of the linear complementary algorithm to solve the EDP is new. It is an efficient and fast algorithm for solving quadratic programming problems. In this algorithm the Kuhn–Tucker conditions are solved as a linear program problem in a tableau form.

In brief, the EDP for 1 h in the scheduling time horizon could be formulated as the minimization of the summation of production costs of the committed units in this hour subjected to the load demand and unit limits constraints as follows.

$$\text{Minimize} \sum_{i=1}^{N} F_{it}(P_{it}) = A_i P^2{}_{it} + B_i P_{it} + C_i \ \$/HR \quad (4.14)$$

Subject to:

$$\sum_{i=1}^{N} P_{it} = PD_t; \ 1 \leq t \leq T \quad (4.15)$$

and

$$\text{Pmin}_i \leq P_{it} \leq \text{Pmax}_i; \ 1 \leq t \leq T, 1 \leq i \leq N \tag{4.16}$$

The detailed reformulation of the EDP as a linear complementary problem is described in the following section.

4.9.1 The Economic Dispatch Problem in a Linear Complementary Form

The economic dispatch problem is generally a nonlinear programming problem. If the production cost functions of the generating units are taken in a quadratic form, then the problem can be formulated as a quadratic programming problem. Accordingly the linear complementary algorithm could be used to solve the Kuhn–Tucker conditions of these problems as proved in Appendix A.

The original formulation of the EDP, as described in Eqs. 4.14, 4.15, and 4.16, could be written, for a single time period, in a simple form as follows.

$$\text{Minimize} \sum_{i=1}^{N} F_i(p_i) = A_i p_i^2 + B_i p_i + C_i \ \$/\text{HR} \tag{4.17}$$

Subject to:

$$\sum_{i=1}^{N} p_i = PD \tag{4.18}$$

and

$$m_i \leq p_i \leq x_i; \ 1 \leq i \leq N \tag{4.19}$$

where m_i and x_i are the lower and upper limits of unit i, respectively.

For a system of N generating units, the number of constraints is $2N$ inequality constraints and one equality constraint. These constraints can be reduced to only $N + 1$ constraint; hence the tableau size and the computation effort will also be reduced. The reduction is done by defining new variables to cancel one of the sides (upper or lower) of the inequality constraints as follows.

Let

$$p'_i = p_i - m_i \tag{4.20}$$

4.9 An Algorithm for the Economic Dispatch Problem

then

$$p_i = p'_i + m_i, \quad (4.21)$$

substituting form (4.21) in (4.17) through (4.19) the EDP problem is reformulated as follows.

$$\text{Minimize} \sum_{i=1}^{N} \mathbf{F_i}(p'_i) = \mathbf{A'_i} p'^2_i + \mathbf{B'_i} p'_i + \mathbf{C'_i} \, \$/h \quad (4.22)$$

Subject to:

$$\sum_{i=1}^{N} p'_i = \mathbf{PD'} \quad (4.23)$$

$$p'_i \leq x'_i; \ 1 \leq i \leq N \quad (4.24)$$

$$p'_i \geq 0; \ 1 \leq i \leq N \quad (4.25)$$

where

$$x'_i = x_i - m_i \quad (4.26)$$

$$PD' = PD - \sum_{i=1}^{N} m_i \quad (4.27)$$

$$A'_i = A_i \quad (4.28)$$

$$B'_i = 2m_i A_i + B_i \quad (4.29)$$

$$C'_i = A_i m'^2_i + B_i m + C_i \quad (4.30)$$

Accordingly, the number of constraints are one equality constraint (Eq. 4.23) and N inequality constraints (Eq. 4.24), in addition to the nonnegativity constraints of the new N variables p'_is.

Now an analogy between the reduced formulation of the EDP and the quadratic programming formulation, could be stated as follows.

$$A' \Leftrightarrow H \quad (4.31)$$

$$B' \Leftrightarrow c \quad (4.32)$$

$$x' \Leftrightarrow b \quad (4.33)$$

$$\text{Unit matrix} \Leftrightarrow A \tag{4.34}$$

Using the problem formulation (2.22) to (2.25) and the analogous equations (4.31, 4.32, 4.33, 4.34), the EDP is solved using the linear complementary algorithm described earlier in Chap. 2.

4.9.2 Tableau Size for the Economic Dispatch Problem

Considering the modified formulation of the EDP, the number of constraints is reduced by N, where N is the number of variables or the number of committed generating units. Consequently, the tableau size of the EDP in the linear complementary formulation is as follows.

Let N be the number of variables (generating units outputs).

Inasmuch as we have N constraints as the upper limits on the generating units, and one constraint of the load demand, then $L = N + N + 1 = 2N + 1$.

The tableau size is then $L \times (2L + 2) = (2N + 1) \times (4N + 4)$.

Accordingly, the tableau size for our solved examples are:

For the 10-unit example, $N = 10$, the tableau size is 21×44.
For the 26-unit example, $N = 26$, the tableau size is 53×108.
For the 24-unit example, $N = 24$, the tableau size is 49×100.

In the following section, we describe the details of the proposed simulated annealing algorithm (SAA) as applied to the UCP.

4.10 The Simulated Annealing Algorithm (SAA) for Solving UCP

In solving the UCP, two types of variables need to be determined. The binary unit status variables U and V and the continuous unit output power variables P. The problem can then be decomposed into two subproblems: a combinatorial optimization problem in U and V and a nonlinear optimization problem in P. The SAA is used to solve the combinatorial optimization problem and a quadratic programming routine is used to solve the nonlinear optimization problem [5].

The main steps behind the SAA are given as follows.

Step (0): Initialize all variables (U,V,P) and set iteration counter $k = 0$.
Step (1): Find an initial feasible solution (U_c^K, V_c^K) randomly.
Step (2): Calculate the total operating cost, $\mathbf{F_c^k}$, as the sum of $\mathbf{F_{it}}$ and $\mathbf{S_{it}}$ in two steps:

Solve the economic dispatch problem.
Calculate the startup cost.

4.10 The Simulated Annealing Algorithm (SAA) for Solving UCP

Step (3): Determine the initial temperature Cp^k that results in a high probability of accepting any solution.

Step (4): If equilibrium is achieved go to Step (7). Otherwise repeat Steps (5) and (6) for the same temperature Cp^k until the equilibrium criterion is satisfied.

Step (5): Find a trial solution (U_t^K, V_t^K), a neighbor to (U_c^K, V_c^K), with objective function value $\mathbf{F_t^K}$.

Step (6): Perform the acceptance test; then go to Step (7). If $\mathbf{F^k}_t \leq \mathbf{F_c}^k$ or $\exp[(\mathbf{F_c}^k - \mathbf{F_t}^k)/\mathbf{Cp}] \geq \mathbf{U(0,1)}$ then accept the trial solution and let $(U_c^K, V_c^K) = (U_t^K, V_t^K)$; otherwise reject the trial solution.

Step (7): If the stopping criterion is satisfied then stop. Else decrease the temperature to Cp^{k+1} according to the cooling schedule, set $k = k + 1$, and go to Step (4).

In the following section, we present other SAA reported in the literature, and make a comparison between them and the SAA presented in this section.

4.10.1 Comparison with Other SAA in the Literature

There is only one application of the SAA to the UCP available in the literature [10]. This is referred to here as SAA-67. There are four major differences between SAA-67 and the presented algorithm [5].

In SAA-67, the starting solution is obtained using a priority list method which could be considered as a suboptimal solution, whereas in the proposed algorithm we start with a randomly generated solution that may be far from the optimal one.

- There is no rule for selecting the initial temperature in SAA-67. In the proposed algorithm the initial value of the temperature is determined by applying the heating process until a prespecified value of acceptance ratio (typical values used 0.8–0.95) is reached.
- In SAA-67, the trial solutions may not be feasible and a penalty term is used for constraint violation. In the proposed algorithm, all trial solutions are feasible which results in considerable savings of CPU time.
- Kirk's cooling schedule is used in SAA-67. The temperature is decreased by multiplying the initial temperature by a constant between 0 and 1 raised to the iteration number. In the proposed algorithm, the polynomial-time cooling schedule is used which is based on the statistics calculation during the search.

A comparison of the results obtained by the two algorithms and other methods is presented in the next section.

4.10.2 Numerical Examples

Based on the SAA and the presented rules for generating random trial solutions, a computer model has been implemented [5]. The model offers different choices for finding trial solutions, for example, completely random, semirandom, and a mix of both with certain probability. The two schedules of the Kirk and polynomial-time are implemented and compared.

To compare our results with SAA-67, we implemented the algorithm described in [10] and use the parameter settings (initial, decrement, and final temperature) as described in the reference.

Three examples from the literature are solved. The first two examples contain ten generating units, whereas the third contains 26 units. The scheduling time horizon for all cases is 24 h. The full data of the three examples are given as follows.

Example 1 [11].

The data for this example are given as follows.

1. Daily load demand

HR	Load (MW)	Spinning reserve (MW)
1	1,025	85
2	1,000	85
3	900	65
4	850	55
5	1,025	85
6	1,400	110
7	1,970	165
8	2,400	190
9	2,850	210
10	3,150	230
11	3,300	250
12	3,400	275
13	3,275	240
14	2,950	210
15	2,700	200
16	2,550	195
17	2,725	200
18	3,200	220
19	3,300	250
20	2,900	210
21	2,125	170
22	1,650	130
23	1,300	100
24	1,150	90

4.10 The Simulated Annealing Algorithm (SAA) for Solving UCP

2. Production cost function parameters

Unit	A ($/MW^2.HR)	B ($/MW.HR)	C ($/HR)
1	0.00113	9.023	820
2	0.0016	7.654	400
3	0.00147	8.752	600
4	0.0015	8.431	420
5	0.00234	9.223	540
6	0.00515	7.054	175
7	0.00131	9.121	600
8	0.00171	7.762	400
9	0.00128	8.162	725
10	0.00452	8.149	200

3. Minimum and maximum output limits of units

Unit	P_{min} (MW)	P_{max} (MW)
1	300	1,000
2	130	400
3	165	600
4	130	420
5	225	700
6	50	200
7	250	750
8	110	375
9	275	850
10	75	250

4. Minimum up/down times and startup cost parameters

Unit	T_{up}	T_{down}	ST_o	B_1	B_2	B_o
1	5	4	2,050	1	0.25	825
2	3	2	1,460	1	0.333	650
3	2	4	2,100	1	0.25	950
4	1	3	1,480	1	0.25	650
5	4	5	2,100	1	0.333	900
6	2	2	1,360	1	0.5	750
7	3	4	2,300	1	0.25	950
8	1	3	1,370	1	0.333	550
9	4	3	2,200	1	0.25	950
10	2	1	1,180	1	0.5	625

Example 2 [6].

1. Daily load demand

HR	Load (MW)
1	1,459
2	1,372
3	1,299
4	1,285
5	1,271
6	1,314
7	1,372
8	1,314
9	1,271
10	1,242
11	1,197
12	1,182
13	1,154
14	1,138
15	1,124
16	1,095
17	1,066
18	1,037
19	993
20	978
21	963
22	1,022
23	1,081
24	1,459

Spinning reserve = 10%.

2. Production cost function parameters

Unit	A ($/MW^2.HR)	B ($/MW.HR)	C ($/HR)
1	0.0051	1.4	15
2	0.00396	1.5	25
3	0.00393	1.35	40
4	0.00382	1.4	32
5	0.00212	1.54	29
6	0.00261	1.35	72
7	0.00127	1.3954	105
8	0.00135	1.3285	100
9	0.00289	1.2643	49
10	0.00148	1.2136	82

4.10 The Simulated Annealing Algorithm (SAA) for Solving UCP

3. Minimum and maximum output limits of units

Unit	P_{min} (MW)	P_{max} (MW)
1	15	60
2	20	80
3	30	100
4	25	120
5	50	150
6	75	280
7	250	520
8	50	150
9	120	320
10	75	200

4. Minimum up/down times and startup cost parameters

Unit	T_{up}	T_{down}	ST_o	B_1	B_2	B_o
1	2	5	85	0.588	0.2	0
2	2	5	101	0.594	0.2	0
3	2	5	114	0.57	0.2	0
4	2	5	94	0.65	0.18	0
5	2	5	113	0.639	0.18	0
6	2	5	176	0.568	0.15	0
7	2	5	267	0.749	0.09	0
8	2	5	282	0.749	0.09	0
9	2	5	187	0.617	0.13	0
10	2	5	227	0.641	0.11	0

Example 3 [12, 13].

1. Daily load demand

HR	Load (MW)
1	1,820
2	1,800
3	1,720
4	1,700
5	1,750
6	1,910
7	2,050
8	2,400
9	2,600
10	2,600
11	2,620
12	2,580

(continued)

HR	Load (MW)
13	2,590
14	2,570
15	2,500
16	2,350
17	2,390
18	2,480
19	2,580
20	2,620
21	2,600
22	2,480
23	2,150
24	1,900

2. Production cost function parameters

Unit	A ($/MW².HR)	B ($/MW.HR)	C ($/HR)
1	0	25.547	24.389
2	0	25.675	24.411
3	0	25.803	24.638
4	0	25.932	24.76
5	0	26.061	24.888
6	0	37.551	117.755
7	0	37.664	118.108
8	0	37.777	118.458
9	0	37.89	118.821
10	0	13.327	81.136
11	0	13.354	81.298
12	0	13.38	81.464
13	0	13.407	81.626
14	0	18	217.895
15	0	18.1	218.335
16	0	18.2	218.775
17	0	10.695	142.735
18	0	10.715	143.029
19	0	10.737	143.318
20	0	10.758	143.597
21	0	23	259.171
22	0	23.1	259.649
23	0	23.2	260.176
24	0	10.862	177.057
25	0	5.492	202.5
26	0	5.503	202.91

4.10 The Simulated Annealing Algorithm (SAA) for Solving UCP

3. Minimum and maximum output limits of units

Unit	P_{min} (MW)	P_{max} (MW)
1	2.4	12
2	2.4	12
3	2.4	12
4	2.4	12
5	2.4	12
6	4	20
7	4	20
8	4	20
9	4	20
10	15.2	76
11	15.2	76
12	15.2	76
13	15.2	76
14	25	100
15	25	100
16	25	100
17	54.25	155
18	54.25	155
19	54.25	155
20	54.25	155
21	68.95	197
22	68.95	197
23	68.95	197
24	140	350
25	140	350
26	140	350

4. Minimum up/down times and startup cost parameters

Unit	T_{up}	T_{down}	ST_o	B_1	B_2	B_o
1	0	0	0	0	0	0
2	0	0	0	0	0	0
3	0	0	0	0	0	0
4	0	0	0	0	0	0
5	0	0	0	0	0	0
6	0	0	30	0	0	0
7	0	0	30	0	0	0
8	0	0	30	0	0	0
9	0	0	30	0	0	0
10	3	2	80	0	0	0
11	3	2	80	0	0	0
12	3	2	80	0	0	0
13	3	2	80	0	0	0

(continued)

Unit	T_{up}	T_{down}	ST_o	B_1	B_2	B_o
14	4	2	100	0	0	0
15	4	2	100	0	0	0
16	4	2	100	0	0	0
17	5	3	200	0	0	0
18	5	3	200	0	0	0
19	5	3	200	0	0	0
20	5	3	200	0	0	0
21	5	4	300	0	0	0
22	5	4	300	0	0	0
23	5	4	300	0	0	0
24	8	5	500	0	0	0
25	8	5	800	0	0	0
26	8	5	800	0	0	0

Practical Example 4

1. Daily load demand

HR	Load (MW)
1	2,580
2	2,500
3	2,400
4	2,340
5	2,300
6	2,120
7	2,270
8	2,420
9	2,580
10	2,630
11	2,700
12	2,640
13	2,700
14	2,740
15	2,750
16	2,740
17	2,680
18	2,900
19	2,840
20	2,850
21	2,880
22	2,870
23	2,800
24	2,720

Spinning reserve = 400 MW.

4.10 The Simulated Annealing Algorithm (SAA) for Solving UCP

2. Unit data

Unit No.	A ($/MW².HR)	B ($/MW.HR)	C ($/HR)	P_{min} (MW)	P_{max} (MW)	T_{up} (HR)	T_{down} (HR)
1	7.62E−03	13.728	605.779	250	625	8	8
2	7.62E−03	13.728	605.779	250	625	8	8
3	1.17E−02	14.346	1,186.087	180	400	8	8
4	1.22E−02	14.020	1,235.064	190	400	8	8
5	1.22E−02	14.020	1,235.064	190	400	8	8
6	1.22E−02	14.020	1,235.064	190	400	8	8
7	3.29E−02	15.240	671.691	33	75	4	2
8	3.26E−02	15.266	671.186	33	77	4	2
9	3.24E−02	15.278	670.939	33	78	4	2
10	4.95E−02	14.087	383.963	33	69	4	2
11	4.90E−02	14.126	383.200	33	67	4	2
12	4.55E−02	14.410	377.669	20	66	4	2
13	4.46E−02	14.483	376.284	15	68	4	2
14	4.46E−02	14.483	376.291	15	68	4	2
15	4.49E−02	14.468	376.461	15	69	4	2
16	1.07E−02	16.793	633.426	20	81	4	2
17	1.08E−02	16.782	633.584	20	82	4	2
18	1.06E−02	16.795	633.408	20	81	4	2
19	4.60E−02	13.710	317.410	15	79	4	2
20	4.60E−02	13.710	317.410	15	79	4	2
21	4.60E−02	13.710	317.410	10	54	4	2
22	4.60E−02	13.710	317.417	15	54	4	2
23	4.60E−02	13.709	317.426	15	61	4	2
24	4.60E−02	13.709	317.426	15	61	4	2

Example #1 [11] is solved by Lagrangian relaxation (LR), Example #2 [6] is solved by integer programming (IP), and Example #3 [12, 13] is solved by expert systems (ES).

The polynomial-time cooling schedule is used in all of the examples. A number of tests on the performance of the developed SAA have been carried out on the three examples to find the most suitable cooling schedule parameter settings. The following parameters for the polynomial-time cooling schedule have been chosen after running a number of simulations: chain length = 150, $\varepsilon = 0.00001$, $\delta = 0.3$, initial acceptance ratio, $x = 0.95$, and the maximum number of iterations = 5,000.

Table 4.1 presents the comparison between the results of Examples 1 and 2, solved by LR, IP, and our SAA. The results show the improvement achieved by our algorithm over both IP and LR results.

Table 4.2 shows the comparison between the results obtained for Examples 1 through 3 solved by SAA-67 and our SAA. It is obvious that our SAA achieves a considerable reduction in the operating costs for the three examples.

Detailed results for Example 1 are given in Tables 4.3 and 4.4. Table 4.3 shows the load sharing among the committed units in the 24-h domain. Table 4.4 gives the

Table 4.1 Comparison among our proposed SAA, the LR, and the IP

	Example	LR [11]	IP [6]	Developed SAA
Total cost ($)	1	540,895	–	536,622
Total cost ($)	2	–	60,667	59,512
% Saving	1	0	–	0.79
% Saving	2	–	0	1.9

Table 4.2 Comparison between our proposed SAA and the SAA-67

	Example	SAA-67	Developed SAA
Total cost ($)	1	538,803	536,622
Total cost ($)	2	59,512	59,512
Total cost ($)	3	663,833	662,664
% Cost saving	1	0.38	0.79
% Cost saving	2	1.9	1.9
% Cost saving	3	0	0.17

Table 4.3 Power sharing (MW) of Example 1

	Unit number[a]							
HR	2	3	4	6	7	8	9	10
1	400.0	0	0	185.0	0	350.3	0	89.7
2	395.4	0	0	181.1	0	338.4	0	85.2
3	355.4	0	0	168.7	0	301.0	0	75.0
4	333.1	0	0	161.8	0	280.1	0	75.0
5	400.0	0	0	185.0	0	350.3	0	89.7
6	400.0	0	0	191.9	0	371.1	339.4	97.6
7	400.0	0	343.0	200.0	0	375.0	507.0	145.0
8	400.0	0	420.0	200.0	311.5	375.0	693.5	0.0
9	400.0	0	420.0	200.0	420.6	375.0	805.0	229.4
10	400.0	444.6	420.0	200.0	358.1	375.0	741.1	211.3
11	400.0	486.3	420.0	200.0	404.9	375.0	789.0	224.9
12	400.0	514.1	420.0	200.0	436.1	375.0	820.9	233.9
13	400.0	479.4	420.0	200.0	397.1	375.0	781.0	222.6
14	400.0	389.0	420.0	200.0	295.6	375.0	677.2	193.2
15	400.0	310.1	410.8	200.0	250.0	375.0	586.6	167.5
16	400.0	266.6	368.3	200.0	250.0	375.0	536.7	153.4
17	400.0	317.3	417.9	200.0	250.0	375.0	594.9	169.9
18	400.0	458.5	420.0	200.0	373.7	375.0	757.0	215.8
19	400.0	486.3	420.0	200.0	404.9	375.0	789.0	224.9
20	400.0	375.1	420.0	200.0	280.0	375.0	661.2	188.7
21	400.0	0	305.0	200.0	250.0	375.0	462.6	132.4
22	383.5	0	150.0	177.4	250.0	327.2	280.9	81.0
23	241.9	0	130.0	133.4	250.0	194.7	275.0	75.0
24	175.1	0	130.0	112.7	250.0	132.3	275.0	75.0

[a] Units 1 and 5 are OFF all hours

Table 4.4 Load demand and hourly costs of Example 1

HR	Load (MW)	ED-cost ($)	ST-cost ($)	T-cost ($)
1	1.03E+03	9.67E+03	0.00E+00	9.67E+03
2	1.00E+03	9.45E+03	0.00E+00	9.45E+03
3	9.00E+02	8.56E+03	0.00E+00	8.56E+03
4	8.50E+02	8.12E+03	0.00E+00	8.12E+03
5	1.03E+ 03	9.67E+03	0.00E+00	9.67E+03
6	1.40E+03	1.36E+04	9.50E+02	1.46E+04
7	1.97E+03	1.92E+04	6.50E+02	1.99E+04
8	2.40E+03	2.39E+04	9.50E+02	2.48E+04
9	2.85E+03	2.84E+04	6.25E+02	2.90E+04
10	3.15E+03	3.17E+04	9.50E+02	3.27E+04
11	3.30E+03	3.32E+04	0.00E+00	3.32E+04
12	3.40E+03	3.42E+04	0.00E+00	3.42E+04
13	3.28E+03	3.30E+04	0.00E+00	3.30E+04
14	2.95E+03	2.97E+04	0.00E+00	2.97E+04
15	2.70E+03	2.73E+04	0.00E+00	2.73E+04
16	2.55E+03	2.58E+04	0.00E+00	2.58E+04
17	2.73E+03	2.75E+04	0.00E+00	2.75E+04
18	3.20E+03	3.22E+04	0.00E+00	3.22E+04
19	3.30E+03	3.32E+04	0.00E+00	3.32E+04
20	2.90E+03	2.92E+04	0.00E+00	2.92E+04
21	2.13E+03	2.13E+04	0.00E+00	2.13E+04
22	1.65E+03	1.70E+04	0.00E+00	1.70E+04
23	1.30E+03	1.39E+04	0.00E+00	1.39E+04
24	1.15E+03	1.27E+04	0.00E+00	1.27E+04

hourly load demand, and the corresponding economic dispatch, startup, and total operating costs. Similar detailed results for Examples 2 and 3 are also shown in Tables 4.5, 4.6, 4.7, 4.8 and 4.9.

4.11 Summary and Conclusions

This section presents three different subjects: the problem statement, rules for generating trial solutions, and a new algorithm for solving the economic dispatch problem. In the problem statement the objective function and the constraints of the UCP are formulated in a more general form. New rules for generating trial-feasible solutions are presented.

An original application of the linear complementary algorithm for solving the EDP is also discussed. An implementation of a SAA to solve the UCP is presented in this chapter. Two different cooling schedules for the SAA are implemented and compared. The detailed description of the proposed SAA is given. The comparison between SAA and SAA-67 as reported in the literature is also presented. The computational results along with a comparison with the previously published classical optimization methods show the effectiveness of the proposed SAA.

Table 4.5 Power sharing (MW) of Example 2

HR	Unit number									
	1	2	3	4	5	6	7	8	9	10
1	60.00	80.00	100.00	120.00	150.00	189.99	372.59	0.00	186.42	200.00
2	60.00	80.00	100.00	109.98	150.00	170.54	332.62	0.00	168.85	200.00
3	60.00	80.00	100.00	99.72	146.67	155.53	301.77	0.00	155.30	200.00
4	60.00	80.00	100.00	98.10	143.74	153.15	296.87	0.00	153.14	200.00
5	60.00	80.00	100.00	96.47	140.80	150.77	291.97	0.00	150.99	200.00
6	60.00	80.00	100.00	101.47	149.82	158.09	307.02	0.00	157.60	200.00
7	60.00	80.00	100.00	109.98	150.00	170.54	332.62	0.00	168.85	200.00
8	60.00	80.00	100.00	101.47	149.82	158.09	307.02	0.00	157.60	200.00
9	60.00	80.00	100.00	96.47	140.80	150.77	291.97	0.00	150.99	200.00
10	60.00	80.00	100.00	117.59	150.00	0.00	355.50	0.00	178.91	200.00
11	60.00	80.00	100.00	109.14	150.00	0.00	330.11	0.00	167.75	200.00
12	60.00	80.00	100.00	106.33	150.00	0.00	321.64	0.00	164.03	200.00
13	60.00	80.00	100.00	101.20	149.33	0.00	306.22	0.00	157.25	200.00
14	60.00	80.00	100.00	98.96	145.29	0.00	299.47	0.00	154.28	200.00
15	60.00	80.00	100.00	97.00	141.75	0.00	293.56	0.00	151.69	200.00
16	60.00	77.69	97.37	93.62	135.68	0.00	283.42	0.00	147.23	200.00
17	60.00	74.60	94.26	90.42	129.91	0.00	273.80	0.00	143.00	200.00
18	60.00	71.52	91.15	87.23	124.15	0.00	264.18	0.00	138.78	200.00
19	60.00	66.77	86.37	82.31	115.28	0.00	250.00	0.00	132.27	200.00
20	59.82	64.41	83.99	79.86	110.88	0.00	250.00	0.00	129.04	200.00
21	58.17	62.29	81.85	77.66	106.91	0.00	250.00	0.00	126.13	200.00
22	60.00	69.92	89.54	85.57	121.17	0.00	259.20	0.00	136.59	200.00
23	60.00	76.20	95.86	92.08	132.89	0.00	278.77	0.00	145.19	200.00
24	60.00	80.00	100.00	120.00	150.00	189.99	372.59	9.00	0.00	186.42

4.12 Tabu Search (TS) Algorithm

Tabu search (TS) is a powerful optimization procedure that has been successfully applied to a number of combinatorial optimization problems [14–30]. It has the ability to avoid entrapment in local minima. TS employs a flexible memory system (in contrast to "memoryless" systems, such as simulated annealing and genetic algorithms (GA), and rigid memory systems as in branch-and-bound). Specific attention is given to the short-term memory (STM) component of TS, which has provided solutions superior to the best obtained by other methods for a variety of problems [22]. Advanced TS procedures are also used for sophisticated problems. These procedures include, in addition to the STM, intermediate-term memory (ITM), long-term memory (LTM), and strategic oscillations (SO).

In general terms, TS is an iterative improvement procedure that starts from some initial feasible solution and attempts to determine a better solution in the manner of a greatest descent algorithm. However, TS is characterized by an ability to escape local optima (which usually cause simple descent algorithms to terminate) by using

4.12 Tabu Search (TS) Algorithm

Table 4.6 Load demand and hourly costs of Example 2

HR	Load	ED-cost	ST-cost	T-cost
1	1.46E+03	3.06E+03	0.00E+00	3.06E+03
2	1.37E+03	2.86E+03	0.00E+00	2.86E+03
3	1.30E+03	2.70E+03	0.00E+00	2.70E+03
4	1.29E+03	2.67E+03	0.00E+00	2.67E+03
5	1.27E+03	2.64E+03	0.00E+00	2.64E+03
6	1.31E+03	2.73E+03	0.00E+00	2.73E+03
7	1.37E+03	2.86E+03	0.00E+00	2.86E+03
8	1.31E+03	2.73E+03	0.00E+00	2.73E+03
9	1.27E+03	2.64E+03	0.00E+00	2.64E+03
10	1.24E+03	2.57E+03	0.00E+00	2.57E+03
11	1.20E+03	2.47E+03	0.00E+00	2.47E+03
12	1.18E+03	2.44E+03	0.00E+00	2.44E+03
13	1.15E+03	2.38E+03	0.00E+00	2.38E+03
14	1.14E+03	2.34E+03	0.00E+00	2.34E+03
15	1.12E+03	2.31E+03	0.00E+00	2.31E+03
16	1.10E+03	2.25E+03	0.00E+00	2.25E+03
17	1.07E+03	2.19E+03	0.00E+00	2.19E+03
18	1.04E+03	2.13E+03	0.00E+00	2.13E+03
19	9.93E+02	2.04E+03	0.00E+00	2.04E+03
20	9.78E+02	2.01E+03	0.00E+00	2.01E+03
21	9.63E+02	1.98E+03	0.00E+00	1.98E+03
22	1.02E+03	2.10E+03	0.00E+00	2.10E+03
23	1.08E+03	2.22E+03	0.00E+00	2.22E+03
24	1.46E+03	3.06E+03	1.65E+02	3.22E+03

Total operating cost = $59,512

a short-term memory of recent solutions. Moreover, TS permits backtracking to previous solutions, which may ultimately lead, via a different direction, to better solutions [23].

The two main components of a TSA are the tabu list (TL) restrictions and the aspiration level (AV) of the solution associated with these restrictions. Discussion of these terms is presented in the following sections.

4.12.1 Tabu List (TL) Restrictions

TS may be viewed as a *metaheuristic* superimposed on another heuristic. The approach undertakes to surpass local optimality by a strategy of forbidding (or, more broadly, penalizing) certain moves. The purpose of classifying certain moves as forbidden (i.e., tabu) is basically to prevent cycling. Moves that hold tabu status are generally a small fraction of those available and a move loses its tabu status to become accessible once again after a relatively short time.

Table 4.7 Power sharing (MW) of Example 3 (for units 1–13)

HR	Unit number												
	1	2	3	4	5	6	7	8	9	10	11	12	13
1	2.40	2.40	0.00	2.40	0.00	0.00	0.00	0.00	0.00	76.00	36.40	15.20	15.20
2	2.40	2.40	0.00	2.40	0.00	0.00	0.00	0.00	0.00	76.00	16.40	15.20	15.20
3	2.40	2.40	0.00	2.40	0.00	0.00	0.00	0.00	0.00	0.00	15.20	15.20	15.20
4	2.40	2.40	0.00	2.40	0.00	0.00	0.00	0.00	0.00	0.00	15.20	15.20	15.20
5	2.40	2.40	0.00	2.40	0.00	0.00	0.00	0.00	0.00	42.40	15.20	0.00	15.20
6	2.40	2.40	0.00	2.40	0.00	0.00	0.00	0.00	0.00	76.00	76.00	40.60	15.20
7	0.00	0.00	0.00	2.40	0.00	0.00	0.00	0.00	0.00	76.00	76.00	76.00	55.65
8	0.00	0.00	0.00	2.40	0.00	0.00	0.00	0.00	0.00	76.00	76.00	76.00	76.00
9	0.00	0.00	0.00	2.40	0.00	0.00	0.00	0.00	0.00	76.00	76.00	76.00	76.00
10	0.00	0.00	0.00	2.40	0.00	0.00	0.00	0.00	0.00	76.00	76.00	76.00	76.00
11	0.00	0.00	0.00	2.40	2.40	0.00	0.00	0.00	0.00	76.00	76.00	76.00	76.00
12	0.00	0.00	0.00	0.00	2.40	0.00	0.00	0.00	0.00	76.00	76.00	76.00	76.00
13	0.00	0.00	0.00	0.00	0.00	0.00	0.00	0.00	0.00	76.00	76.00	76.00	76.00
14	0.00	0.00	0.00	0.00	0.00	0.00	0.00	0.00	0.00	76.00	76.00	76.00	76.00
15	0.00	0.00	0.00	0.00	0.00	0.00	0.00	0.00	0.00	76.00	76.00	76.00	76.00
16	0.00	0.00	0.00	0.00	0.00	0.00	0.00	0.00	0.00	76.00	76.00	76.00	76.00
17	0.00	0.00	0.00	0.00	0.00	0.00	0.00	0.00	0.00	76.00	76.00	76.00	76.00
18	0.00	0.00	0.00	0.00	0.00	0.00	0.00	0.00	0.00	76.00	76.00	76.00	76.00
19	0.00	0.00	0.00	0.00	0.00	0.00	0.00	0.00	0.00	76.00	76.00	76.00	76.00
20	2.40	0.00	0.00	2.40	0.00	0.00	0.00	0.00	0.00	76.00	76.00	76.00	76.00
21	2.40	0.00	0.00	2.40	0.00	0.00	0.00	0.00	0.00	76.00	76.00	76.00	76.00
22	2.40	0.00	0.00	2.40	0.00	0.00	0.00	0.00	0.00	76.00	76.00	76.00	76.00
23	2.40	0.00	0.00	2.40	0.00	0.00	0.00	0.00	0.00	76.00	76.00	76.00	76.00
24	2.40	0.00	0.00	2.40	0.00	0.00	0.00	0.00	0.00	76.00	76.00	33.00	15.20

The choice of appropriate type of tabu restrictions "list" depends on the problem under study. The elements of the TL are determined by a function that utilizes historical information from the search process, extending up to Z iterations in the past, where Z(TL size) can be fixed or variable depending on the application or the stage of the search.

The TL restrictions could be stated directly as a given change of variables (moves) or indirectly as a set of logical relationships or linear inequalities. Usage of these two approaches depends on the size of the TL for the problem under study.

A TL is managed by recording moves in the order in which they are made. Each time a new element is added to the "bottom" of a list, the oldest element on the list is dropped from the "top". The TL is designed to ensure the elimination of cycles of length equal to the TL size. Empirically [22], TL sizes that provide good results often grow with the size of the problem and stronger restrictions are generally coupled with smaller lists.

The way to identify a good TL size for a given problem class and choice of tabu restrictions is simply to watch for the occurrence of cycling when the size is too small and the deterioration in solution quality when the size is too large (caused by

Table 4.8 Power sharing (MW) of Example 3 (for units 14–26)

HR	Unit number												
	14	15	16	17	18	19	20	21	22	23	24	25	26
1	0.00	0.00	0.00	155.00	155.00	155.00	155.00	0.00	0.00	0.00	350.00	350.00	350.00
2	0.00	0.00	0.00	155.00	155.00	155.00	155.00	0.00	0.00	0.00	350.00	350.00	350.00
3	0.00	0.00	0.00	155.00	155.00	155.00	155.00	0.00	0.00	0.00	347.20	350.00	350.00
4	0.00	0.00	0.00	155.00	155.00	155.00	155.00	0.00	0.00	0.00	327.20	350.00	350.00
5	0.00	0.00	0.00	155.00	155.00	155.00	155.00	0.00	0.00	0.00	350.00	350.00	350.00
6	25.00	0.00	0.00	155.00	155.00	155.00	155.00	0.00	0.00	0.00	350.00	350.00	350.00
7	25.00	0.00	0.00	155.00	155.00	155.00	155.00	0.00	68.95	0.00	350.00	350.00	350.00
8	100.00	100.00	85.70	155.00	155.00	155.00	155.00	0.00	68.95	68.95	350.00	350.00	350.00
9	100.00	100.00	100.00	155.00	155.00	155.00	155.00	185.70	68.95	68.95	350.00	350.00	350.00
10	100.00	100.00	100.00	155.00	155.00	155.00	155.00	185.70	68.95	68.95	350.00	350.00	350.00
11	100.00	100.00	100.00	155.00	155.00	155.00	155.00	197.00	75.25	68.95	350.00	350.00	350.00
12	100.00	100.00	100.00	155.00	155.00	155.00	155.00	165.70	68.95	68.95	350.00	350.00	350.00
13	100.00	100.00	100.00	155.00	155.00	155.00	155.00	178.10	68.95	68.95	350.00	350.00	350.00
14	100.00	100.00	100.00	155.00	155.00	155.00	155.00	158.10	68.95	68.95	350.00	350.00	350.00
15	100.00	100.00	100.00	155.00	155.00	155.00	155.00	88.10	68.95	68.95	350.00	350.00	350.00
16	100.00	44.15	25.00	155.00	155.00	155.00	155.00	68.95	68.95	68.95	350.00	350.00	350.00
17	100.00	84.15	25.00	155.00	155.00	155.00	155.00	68.95	68.95	68.95	350.00	350.00	350.00
18	100.00	100.00	99.15	155.00	155.00	155.00	155.00	68.95	68.95	68.95	350.00	350.00	350.00
19	100.00	100.00	100.00	155.00	155.00	155.00	155.00	168.10	68.95	68.95	350.00	350.00	350.00
20	100.00	100.00	100.00	155.00	155.00	155.00	155.00	197.00	75.25	68.95	350.00	350.00	350.00
21	100.00	100.00	100.00	155.00	155.00	155.00	155.00	183.30	68.95	68.95	350.00	350.00	350.00
22	100.00	100.00	94.35	155.00	155.00	155.00	155.00	68.95	68.95	68.95	350.00	350.00	350.00
23	0.00	77.25	25.00	155.00	155.00	155.00	155.00	68.95	0.00	0.00	350.00	350.00	350.00
24	0.00	25.00	0.00	155.00	155.00	155.00	155.00	0.00	0.00	0.00	350.00	350.00	350.00

Table 4.9 Load demand and hourly costs of Example 3

HR	Load	ED-cost	ST-cost	T-cost
1	1.82E+03	1.79E+04	0.00E+00	1.79E+04
2	1.80E+03	1.77E+04	0.00E+00	1.77E+04
3	1.72E+03	1.65E+04	0.00E+00	1.65E+04
4	1.70E+03	1.63E+04	0.00E+00	1.63E+04
5	1.75E+03	1.69E+04	8.00E+01	1.70E+04
6	1.91E+03	1.95E+04	1.80E+02	1.97E+04
7	2.05E+03	2.22E+04	3.00E+02	2.25E+04
8	2.40E+03	2.95E+04	5.00E+02	3.00E+04
9	2.60E+03	3.43E+04	3.00E+02	3.46E+04
10	2.60E+03	3.43E+04	0.00E+00	3.43E+04
11	2.62E+03	3.48E+04	0.00E+00	3.48E+04
12	2.58E+03	3.38E+04	0.00E+00	3.38E+04
13	2.59E+03	3.40E+04	0.00E+00	3.40E+04
14	2.57E+03	3.35E+04	0.00E+00	3.35E+04
15	2.50E+03	3.19E+04	0.00E+00	3.19E+04
16	2.35E+03	2.91E+04	0.00E+00	2.91E+04
17	2.39E+03	2.98E+04	0.00E+00	2.98E+04
18	2.48E+03	3.15E+04	0.00E+00	3.15E+04
19	2.58E+03	3.38E+04	0.00E+00	3.38E+04
20	2.62E+03	3.48E+04	0.00E+00	3.48E+04
21	2.60E+03	3.43E+04	0.00E+00	3.43E+04
22	2.48E+03	3.16E+04	0.00E+00	3.16E+04
23	2.15E+03	2.42E+04	0.00E+00	2.42E+04
24	1.90E+03	1.93E+04	0.00E+00	1.93E+04

Total operating cost = \$662,664

forbidding too many moves). The best sizes lie in an intermediate range between these extremes. In some applications a simple choice of Z in a range centered around 7 seems to be quite effective [20].

4.12.2 Aspiration Level Criteria

Another key issue of TS arises when the move under consideration has been found to be tabu. Associated with each entry in the TL there is a certain value for the evaluation function called the *aspiration level*. If appropriate aspiration criteria are satisfied, the move will still be considered admissible in spite of the tabu classification. Roughly speaking, AV criteria are designed to override tabu status if a move is "good enough" with the compatibility of the goal of preventing the solution process from cycling [20]. Different forms of aspiration criteria are available. The one we use in this study is to override the tabu status if the tabu moves yield a solution that has a better evaluation function than the one obtained earlier for the same move.

4.12.3 Stopping Criteria

There may be several possible stopping conditions for the search. In our implementation we stop the search if either of the following two conditions is satisfied.

- The number of iterations performed since the best solution last changed is greater than a prespecified maximum number of iterations.
- The maximum allowable number of iterations is reached.

4.12.4 General Tabu Search Algorithm

In applying the TSA to solve a combinatorial optimization problem, the basic idea is to choose a feasible solution at random and then get a neighbor to this solution. A move to this neighbor is performed if either it does not belong to the TL or, in the case of being in the TL, it passes the AV test. During these search procedures the best solution is always updated and stored aside until the stopping criteria is satisfied.

A general TSA, based on the STM, for combinatorial optimization problems can be described as follows.

The following notation is used in the algorithm.

X: The set of feasible solutions for a given problem
x: Current solution, $x \in X$
x'': Best solution reached
x': Best solution among a sample of trial solutions
$E(x)$: Evaluation function of solution x
$N(x)$: Set of neighborhood of $x \in X$ (trial solutions).
$S(x)$: Sample of neighborhood, of x, $S(x) \in N(x)$.
$SS(x)$: Sorted sample in ascending order according to their evaluation functions $E(x)$.

Step (0): Set the TL as empty and the AV to be zero.
Step (1): Set iteration counter $K = 0$. Select an initial solution $x \in X$, and set $x'' = x$.
Step (2): Randomly generate a set of trial solutions $S(x) \in N(x)$ (neighbor to the current solution x) and sort them in an ascending order, to obtain $SS(x)$. Let x' be the best trial solution in the sorted set $SS(x)$ (the first in the sorted set).
Step (3): If $E(x') > E(x'')$, go to Step (4); else set the best solution $x'' = x'$ and go to Step (4).
Step (4): Perform the tabu test. If x' is NOT in the TL, then accept it as a current solution, set $x = x'$, and update the TL and AV and go to Step (6); else go to Step (5).
Step (5): Perform the AV test. If satisfied, then override the tabu state. Set $x = x'$, update the AV, and go to Step (7); else go to Step (6).

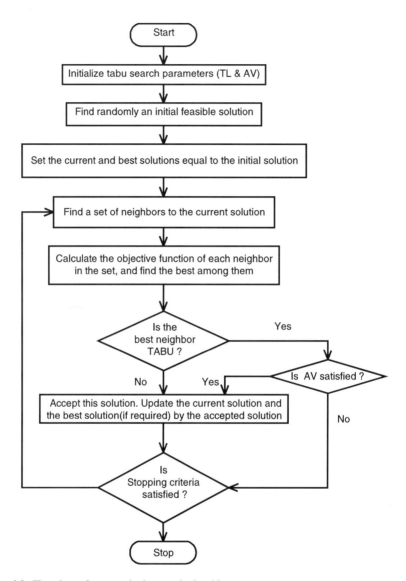

Fig. 4.3 Flowchart of a general tabu search algorithm

Step (6): If the end of the $SS(x)$ is reached, go to Step (7); otherwise, let x' be the next solution in the $SS(x)$ and go to Step (3).

Step (7): Perform termination test. If the stopping criterion is satisfied then stop; else set $K = K + 1$ and go to Step (2).

The main steps of the TSA are also shown in the flowchart of Fig. 4.3.

In the following section we describe the details of the general TSA as applied to the UCP.

4.12.5 Tabu Search Algorithm for Unit Commitment

As previously explained, the UCP can be decomposed into two subproblems, a combinatorial optimization problem in U and V and a nonlinear optimization problem in P. TS is used to solve combinatorial optimization and nonlinear optimization is solved via a quadratic programming routine. In this section a STSA, based on the STM approach, is introduced. The proposed algorithm contains three major steps.

- First, the generation of randomly feasible trial solutions
- Second, the calculation of the objective functions of the given solution by solving the EDP
- Third, the application of the TS procedures to accept or reject the solution at hand

The details of the STSA as applied to the UCP are given in the following steps.

Step (0): Initialize all variables (U,V,P) to be zeros and set iteration counter $K = 0$.

Step (1): Generate, randomly, an initial current feasible solution (U_i^0, V_i^0).

Step (2): Calculate the total operating cost F_i^0 for this solution in two steps:
- Solve the EDP to get the output power (P_i^0) and the corresponding production cost.
- Calculate the startup cost for this schedule.

Step (3): Set the global best solution equal to the current solution, $(U_B, V_B) = (U_i^0, V_i^0)$, $F_B = F_i^0$.

Step (4): Find a set of trial solutions $S(U_i^K, V_i^K)$, that are neighbors to the current solution (U_i^K, V_i^K), with objective function values $F^k(S)$.

Step (5): Sort the set of solutions in ascending order. Let $SF^k(S)$ be the sorted values. Let (U_b^K, V_b^K) be the best trial solution in the set, with an objective function F_b.

Step (6): If $F_b \geq F_B$ go to Step (7); else update the global best solution, set $(U_B, V_B) = (U_b^K, V_b^K)$, and go to Step (7).

Step (7): If the trial solution (U_b^K, V_b^K) is NOT in the TL, then update the TL, the AV, and the current solution. Set $(U_i^K, V_i^K) = (U_b^K, V_b^K)$ and $F_i^k = F_b$ and go to Step (9); else go to Step (8).

Step (8): If the AV test is NOT satisfied go to Step (9); else override the tabu state, set $(U_i^K, V_i^K) = (U_b^K, V_b^K)$, update the AV, and go to Step (10).

Step (9): If the end of the $SF^k(S)$ is reached, go to Step (10); otherwise let (U_b^K, V_b^K) be the next solution in the $SF^k(S)$ and go to Step (6).

Step (10): Stop if the termination criterion is satisfied; else set $K = K + 1$ and go to Step (4).

In the following section, we describe some methods to create TL for the UCP.

4.12.6 Tabu List Types for UCP

The TL embodies one of the primary STM functions of the TS procedure, which it executes by recording only the Z most recent moves, where Z is the TL size. A move in the UCP is defined as switching a unit from ON to OFF or the opposite at some hours in the scheduling horizon. Because the solution matrices in the UCP (U and V) have large sizes ($T \times N$), it is worth proposing and testing different approaches to create the TL move attributes rather than recording a full solution matrix.

4.12.7 Tabu List Approach for UCP

In this section, we propose original concepts for creating the TL for the UCP. During the early stages of implementing the TSA to solve the UCP, five approaches for creating the TL restrictions were tested with the aim of selecting the best among them. In our implementation we create a separate TL for each generating unit. The generating unit tabu list (GUTL) has a dimension of $Z \times L$, where L is the recorded move attribute length. In the following, the five proposed approaches of TL types for the UCP are described.

To illustrate the implementation of the proposed different approaches of TL a numerical example is shown in Figs. 4.1a and 4.2. Figure 4.1a shows an initial trial solution of a unit and four different moves generated at random. Move 1, for example, is generated from the initial schedule by randomly selecting hour 3 ($t = 3$) as the instant of changing the schedule. Move 2 is generated from move 1 and the change starts at instant $t = 7$.

4.12.7.1 Approach (1)

In this approach each GUTL is a one-dimensional array of length Z. Each entry records a time that has been previously selected randomly to generate a trial solution for this unit, irrespective of the unit status at that time. In Fig. 4.2 the TL implementation for the example of Fig. 4.1a using this approach is shown. As mentioned, only hours at which the schedule starts to be changed are recorded, that is, hours 3, 7, 1, and 4 at the moves 1, 2, 3, and 4, respectively.

4.12.7.2 Approach (2)

The TL created in this approach will be of dimension $Z \times 2$. Each entry records the time that has been previously selected randomly to generate a trial solution for this unit, in addition to the unit status at this time. The implementation of this approach is shown in Fig. 4.2. For example, at hour 3 (move 1), the unit status is 0 and at hour 4 (move 3) the unit status is 1.

4.12.7.3 Approach (3)

In this case each GUTL contains one dimensional array of Z entries. Each entry records the number of ON periods for the respective unit (the number of ones in the column of that unit in the matrix U). In Fig. 4.2 the first entry (corresponds to move 1) of the TL using this approach shows 10, which is equivalent to the number of ON hours at that move.

4.12.7.4 Approach (4)

In this approach we record the instances at which a unit is turned ON and OFF during the scheduling horizon. These instances come in pairs. If these ON–OFF pairs occur m times during the scheduling time horizon then GUTL will have the size $Z \times (2\ m)$. Figure 4.2 shows the TL of the unit when applying this approach to the solutions generated in Fig. 4.1a. The TL entries show the starting and shutdown hours for each trial solution, respectively. For example, at move 2, the unit started at hour 1 and shut down at hour 7, then started again at hour 9 and shut down at hour 11. Hence, the entry of this move in the TL is recorded as 1 and 9 as starting hours and the shutdown hours are recorded as 7 and 11.

4.12.7.5 Approach (5)

In this approach the solution vector for each generating unit (which has 0 or 1 values) is recorded as its equivalent decimal number. Hence, the generating unit tabu list (GUTL) is a one-dimensional array of length Z. Each entry records the equivalent decimal number of a specific trial solution for that unit. By using this approach we record all information of the trial solution using minimum memory requirements. The TL implementation using this approach is shown in Fig. 4.2. As shown, the entries of the TL represent the equivalent decimal number for each binary vector of a trial solution. It is clear that this approach ensures the uniqueness of the representation of a specific trial solution.

In the following section, the comparison between the results of the aforementioned five approaches is presented.

4.12.8 Comparison Among the Different Tabu Lists Approaches

To find the most efficient approach among the five described approaches for creating the TL for the UCP, several tests were conducted. Example 1 was solved with different initial solutions and different random seeds using the five different approaches of the TL. To summarize the results, Table 4.10 shows the

Table 4.10 Comparison of the five proposed approaches of TL types using TL size of 7 (Example 1)

Approach no	1	2	3	4	5
TL dimension	$Z \times 1$	$Z \times 2$	$Z \times 1$	$Z \times 2\,m$	$Z \times 1$
Cost ($)	540,986	540,409	540,174	538,390	538,390

Table 4.11 Comparison of different TL sizes using approach 4 (Example 1)

TL size	5	7	10	20	30
Cost ($)	539,496	538,390	539,374	539,422	540,215

daily operating costs for the proposed five approaches of TL types as applied to Example 1 [11]. In all cases we started with the same initial feasible solution and the same random number seed. It is obvious that the results of approaches 4 and 5 are the best. The reason is that the attributes of the moves are fully recorded; hence the search becomes more precise which prevents cycling during the age of TL. However approach 5 requires less memory space.

4.12.9 Tabu List Size for UCP

The size of the TL determines the most suitable number of moves to be recorded. To find a suitable size of TL, values of Z between 5 and 30 have been tested. Table 4.11 shows that the best value of Z was related to the TL restrictions type, where a more restricted TL corresponds to small size and vice versa. In our implementation, the results are based on a TL of size 7, which was found to be the best TL size for attempted examples.

Different experiments were carried out on different examples to find the most suitable TL size. Table 4.11 shows the daily operating costs obtained for Example 1, using approach 4, with different TL sizes starting with the same initial solution. The best results for this example are obtained at a tabu size value of 7 as shown in the table. This is in agreement with the literature [20].

The rest of the results of this section are obtained with the TL implemented according to approach 4 and of size 7, whereas approach 5 is used in Sect. 4.8.

4.12.10 Numerical Results of the STSA

Based on the STSA, a computer program has been implemented. The program offers different choices for finding a trial solution: completely random, semirandom, and a

4.12 Tabu Search (TS) Algorithm

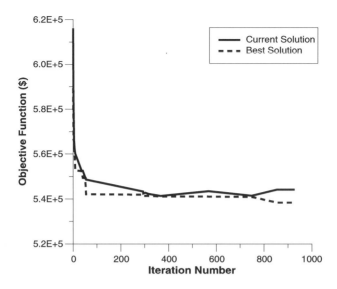

Fig. 4.4 The best and the current solutions versus iteration number for Example 1

Table 4.12 Comparison among the proposed STSA, LR, and IP

	Example	LR [11]	IP [6]	Our STSA
Total cost ($)	1	540,895	–	538,390
Total cost ($)	2	–	60,667	59,512
% Saving	1	0	–	0.46
% Saving	2	–	0	1.9
No. of iterations	1	–	–	1,924
No. of iterations	2	–	–	616

mix of both with certain probability. The previously described approaches for TL have been implemented and compared.

In order to test the model, the same examples described in the previous section are considered. To illustrate the convergence trend in the TS method, a plot of the current and best solutions with the iteration number for Example 1 [11] is given in Fig. 4.4. As shown, the current solution has no trend, because the TS criteria is to accept any non-tabu solution regardless of its objective function value. Basically this is the main idea behind the ability of the TS method to escape local minima. On the other hand, the best solution improves very fast at the beginning of the search and the improvement becomes slower at the end of the search.

Table 4.12 presents the comparison of results for Examples 1 and 2, solved by LR [11], IP [6], and the TSA. The results show the improvement achieved by the proposed STSA algorithm over both IP and LR results.

Tables 4.13, 4.14, and 4.15 show detailed results for Example 1 [11]. Table 4.13 presents the initial starting solution for the given results, which is randomly generated.

Table 4.13 Power sharing (MW) of the initial schedule (Example 1)

HR	Unit number									
	1	2	3	4	5	6	7	8	9	10
1	300.00	130.00	165.00	130.00	225.00	50.00	250.00	110.00	275.00	75.00
2	300.00	130.00	165.00	130.00	225.00	50.00	250.00	110.00	275.00	75.00
3	300.00	130.00	165.00	130.00	225.00	50.00	250.00	110.00	275.00	75.00
4	300.00	130.00	165.00	130.00	225.00	50.00	250.00	110.00	275.00	75.00
5	300.00	130.00	165.00	130.00	225.00	50.00	250.00	110.00	275.00	75.00
6	300.00	130.00	165.00	130.00	225.00	50.00	250.00	110.00	275.00	75.00
7	300.00	232.97	165.00	130.00	225.00	130.63	250.00	186.40	275.00	75.00
8	300.00	395.67	165.00	163.06	225.00	181.18	250.00	338.65	296.14	85.30
9	300.00	400.00	201.49	304.43	225.00	200.00	250.00	375.00	461.86	132.22
10	300.00	400.00	288.35	389.55	225.00	200.00	250.00	375.00	561.62	160.47
11	311.82	400.00	331.91	420.00	225.00	200.00	250.00	375.00	611.64	174.64
12	343.28	400.00	356.09	420.00	225.00	200.00	258.72	375.00	639.41	182.50
13	303.20	400.00	325.28	420.00	225.00	200.00	250.00	375.00	604.03	172.48
14	409.01	400.00	406.62	420.00	225.00	200.00	315.43	375.00	0.00	198.94
15	317.58	400.00	336.34	420.00	225.00	200.00	250.00	375.00	0.00	176.08
16	300.00	400.00	271.70	373.24	225.00	200.00	250.00	375.00	0.00	155.06
17	329.97	400.00	345.86	420.00	225.00	200.00	250.00	375.00	0.00	179.18
18	495.78	400.00	473.32	420.00	225.00	200.00	390.27	375.00	0.00	220.63
19	530.48	400.00	500.00	420.00	225.00	200.00	420.21	375.00	0.00	229.31
20	391.66	400.00	393.28	420.00	225.00	200.00	300.46	375.00	0.00	194.60
21	300.00	400.00	261.62	363.38	225.00	200.00	0.00	375.00	0.00	0.00
22	300.00	357.61	165.00	130.00	225.00	169.36	0.00	303.03	0.00	0.00
23	300.00	201.80	165.00	130.00	225.00	120.95	0.00	157.25	0.00	0.00
24	300.00	130.00	165.00	130.00	225.00	90.00	0.00	110.00	0.00	0.00

It is obvious that this initial schedule is very far from the optimal one. The cost of this initial schedule is $615,648.87, whereas that of the obtained final solution is $538,390. Table 4.14 shows the load sharing among the committed units in 24 h. Table 4.15 gives the hourly load demand and the corresponding economic dispatch costs, startup costs, and total operating cost.

In Example 2 [6], the optimal unit commitment is shown in Table 4.16 along with the load demand and the hourly operating costs. Table 4.17 gives the same results for Example 3 [62.63]. Inasmuch as the amount of reserve taken for this example in [12, 13] is not clear to us, we assumed a spinning reserve of 10%, and the corresponding total operating cost obtained is $662,583.

4.13 Advanced Tabu Search (ATS) Techniques

Advanced TS procedures are recommended for sophisticated problems [19–30]. These procedures include, in addition to short-term memory, intermediate-term memory, long-term memory, and strategic oscillation.

Table 4.14 Power sharing of the final schedule (MW) of Example 1

HR	Unit number[a]							
	2	3	4	6	7	8	9	10
1	400.00	0.00	0.00	185.04	0.00	350.26	0.00	89.70
2	395.36	0.00	0.00	181.09	0.00	338.36	0.00	85.19
3	355.38	0.00	0.00	168.67	0.00	300.95	0.00	75.00
4	333.13	0.00	0.00	161.75	0.00	280.12	0.00	75.00
5	400.00	0.00	0.00	185.04	0.00	350.26	0.00	89.70
6	400.00	270.37	0.00	200.00	0.00	375.00	0.00	154.63
7	400.00	383.56	420.00	200.00	0.00	375.00	0.00	191.44
8	400.00	295.59	396.65	200.00	0.00	375.00	569.93	162.83
9	400.00	468.07	420.00	200.00	0.00	375.00	768.01	218.92
10	400.00	444.60	420.00	200.00	358.05	375.00	741.06	211.29
11	400.00	486.30	420.00	200.00	404.89	375.00	788.95	224.86
12	400.00	514.11	420.00	200.00	436.09	375.00	820.89	233.91
13	400.00	479.35	420.00	200.00	397.09	375.00	780.96	222.60
14	400.00	388.98	420.00	200.00	295.63	375.00	677.18	193.20
15	400.00	310.07	410.84	200.00	250.00	375.00	586.56	167.54
16	400.00	266.64	368.27	200.00	250.00	375.00	536.68	153.41
17	400.00	317.31	417.93	200.00	250.00	375.00	594.87	169.89
18	400.00	458.51	420.00	200.00	373.65	375.00	757.03	215.81
19	400.00	486.30	420.00	200.00	404.89	375.00	788.95	224.86
20	400.00	375.08	420.00	200.00	280.03	375.00	661.21	188.68
21	400.00	0.00	404.87	200.00	0.00	375.00	579.57	165.56
22	400.00	0.00	0.00	200.00	0.00	375.00	524.91	150.09
23	396.46	0.00	0.00	181.43	0.00	339.39	297.13	85.58
24	377.64	0.00	0.00	175.58	0.00	321.78	275.00	0.00

[a]Units 1 and 5 are OFF all hours

4.13.1 Intermediate-Term Memory

The intermediate-term memory function is employed within TS to achieve intensification in a specified region in the solution space at some periods of the search [19–21]. ITM operates by recording and comparing features of a selected number of best trial solutions generated during a particular search period. Features that are common to all or a compelling majority of these solutions are taken to be a regional attribute of good solutions. The method then seeks new solutions that exhibit these features by correspondingly restricting or penalizing available moves during a subsequent search period. Different variants of implementation for the ITM are presented in Sect. 4.7.1.

Table 4.15 Load demand and hourly costs ($) of Example 1

HR	Load	ED-cost	ST-cost	T-cost
1	1,025.00	9,670.0	–	9,670.0
2	1,000.00	9,446.6	–	9,446.6
3	900.00	8,560.9	–	8,560.9
4	850.00	8,123.1	–	8,123.1
5	1,025.00	10,058.4	–	11,643.9
6	1,400.00	13,434.1	1,585.54	13,434.1
7	1,970.00	19,217.7	2,659.11	21,876.8
8	2,400.00	23,902.0	2,850.32	26,752.4
9	2,850.00	28,386.4	–	28,386.4
10	3,150.00	31,701.7	2,828.66	34,530.4
11	3,300.00	33,219.8	–	33,219.8
12	3,400.00	34,242.1	–	34,242.1
13	3,275.00	32,965.5	–	32,965.5
14	2,950.00	29,706.3	–	29,706.3
15	2,700.00	27,259.7	–	27,259.7
16	2,550.00	25,819.8	–	25,819.8
17	2,725.00	27,501.6	–	27,501.6
18	3,200.00	32,205.7	–	32,205.7
19	3,300.00	33,219.8	–	33,219.8
20	2,900.00	29,212.5	–	29,212.5
21	2,125.00	20,698.4	–	20,698.4
22	1,650.00	15,947.5	–	15,947.5
23	1,300.00	12,735.4	–	12,735.4
24	1,150.00	11,232.0	–	11,232.0

Total operating cost = $538,390

4.13.2 Long-Term Memory

The long-term memory function is used to perform global diversification of the search [19–21]. The function employs principles that are roughly the reverse of those for ITM. It guides the search to regions far from the best solutions examined earlier.

4.13.3 Strategic Oscillation

Strategic oscillation is a major aspect of the proposed ATSA [19–21, 25]. It allows the search to cross the feasible region in both directions to move the search into new regions and also to intensify the search in the neighborhood of the bounds. In the following section we describe the details of the proposed ATSA as applied to the UCP.

4.13 Advanced Tabu Search (ATS) Techniques

Table 4.16 Load demand and UCT of Example 2

Hour	Load	Unit number										TCOST/HR
		1	2	3	4	5	6	7	8	9	10	
1	1,459.0	1	1	1	1	1	1	1	0	1	1	.305767E+04
2	1,372.0	1	1	1	1	1	1	1	0	1	1	.285848E+04
3	1,299.0	1	1	1	1	1	1	1	0	1	1	.269791E+04
4	1,285.0	1	1	1	1	1	1	1	0	1	1	.266773E+04
5	1,271.0	1	1	1	1	1	1	1	0	1	1	.263772E+04
6	1,314.0	1	1	1	1	1	1	1	0	1	1	.273044E+04
7	1,372.0	1	1	1	1	1	1	1	0	1	1	.285848E+04
8	1,314.0	1	1	1	1	1	1	1	0	1	1	.273044E+04
9	1,271.0	1	1	1	1	1	1	1	0	1	1	.263772E+04
10	1,242.0	1	1	1	1	1	0	1	0	1	1	.257234E+04
11	1,197.0	1	1	1	1	1	0	1	0	1	1	.247036E+04
12	1,182.0	1	1	1	1	1	0	1	0	1	1	.243702E+04
13	1,154.0	1	1	1	1	1	0	1	0	1	1	.237563E+04
14	1,138.0	1	1	1	1	1	0	1	0	1	1	.234100E+04
15	1,124.0	1	1	1	1	1	0	1	0	1	1	.231092E+04
16	1,095.0	1	1	1	1	1	0	1	0	1	1	.224922E+04
17	1,066.0	1	1	1	1	1	0	1	0	1	1	.218823E+04
18	1,037.0	1	1	1	1	1	0	1	0	1	1	.212795E+04
19	993.0	1	1	1	1	1	0	1	0	1	1	.203784E+04
20	978.0	1	1	1	1	1	0	1	0	1	1	.200755E+04
21	963.0	1	1	1	1	1	0	1	0	1	1	.197752E+04
22	1,022.0	1	1	1	1	1	0	1	0	1	1	.209705E+04
23	1,081.0	1	1	1	1	1	0	1	0	1	1	.221968E+04
24	1,459.0	1	1	1	1	1	1	1	0	1	1	.322314E+04

4.13.4 ATSA for UCP

The proposed ATSA contains four major steps:

1. Applying STM procedures
2. Applying ITM procedures
3. Applying LTM procedures
4. Applying SO procedures

Figure 4.5 shows the flowchart for the proposed ATSA.

In the following subsections a description of the different components of the algorithm as applied to the UCP is presented. These components include the TL construction, and the ITM, LTM, and SO implementations.

4.13.5 Intermediate-Term Memory Implementation

The ITM is used to intensify the search in a specified region. Two different approaches are used to achieve this function

Table 4.17 Load demand and UCT of Example 3

		Unit number[a]															
Hour	Load	1	2	3	4	5	10	14	15	16	17	18	19	20	21	22	23	TCOST/HR
1	1,820.0	1	1	0	1	0	1	0	0	0	1	1	1	1	0	0	0	.179457E+05
2	1,800.0	1	1	0	1	0	1	0	0	0	1	1	1	1	0	0	0	.176786E+05
3	1,720.0	1	1	0	1	0	0	0	0	0	1	1	1	1	0	0	0	.165382E+05
4	1,700.0	1	1	0	1	0	0	0	0	0	1	1	1	1	0	0	0	.163210E+05
5	1,750.0	1	1	0	1	0	1	0	0	0	1	1	1	1	0	0	0	.169300E+05
6	1,910.0	1	1	0	1	0	1	1	0	0	1	1	1	1	0	0	0	.196623E+05
7	2,050.0	0	0	0	1	0	1	1	0	0	1	1	1	1	0	1	0	.224789E+05
8	2,400.0	0	0	0	1	0	1	1	1	1	1	1	1	1	0	1	1	.299684E+05
9	2,600.0	0	0	0	1	0	1	1	1	1	1	1	1	1	1	1	1	.345589E+05
10	2,600.0	0	0	0	1	0	1	1	1	1	1	1	1	1	1	1	1	.342589E+05
11	2,620.0	0	0	0	1	1	1	1	1	1	1	1	1	1	1	1	1	.347518E+05
12	2,580.0	0	0	0	0	1	1	1	1	1	1	1	1	1	1	1	1	.337994E+05
13	2,590.0	0	0	0	0	0	1	1	1	1	1	1	1	1	1	1	1	.339972E+05
14	2,570.0	0	0	0	0	0	1	1	1	1	1	1	1	1	1	1	1	.335372E+05
15	2,500.0	0	0	0	0	0	1	1	1	1	1	1	1	1	1	1	1	.319272E+05
16	2,350.0	0	0	0	0	0	1	1	1	1	1	1	1	1	1	1	1	.291108E+05
17	2,390.0	0	0	0	0	0	1	1	1	1	1	1	1	1	1	1	1	.298348E+05
18	2,480.0	0	0	0	0	0	1	1	1	1	1	1	1	1	1	1	1	.314712E+05
19	2,580.0	0	0	0	0	0	1	1	1	1	1	1	1	1	1	1	1	.337672E+05
20	2,620.0	1	0	0	1	0	1	1	1	1	1	1	1	1	1	1	1	.347501E+05
21	2,600.0	1	0	0	1	0	1	1	1	1	1	1	1	1	1	1	1	.342895E+05
22	2,480.0	1	0	0	1	0	1	1	1	1	1	1	1	1	1	1	1	.315566E+05
23	2,150.0	1	0	0	1	0	1	0	1	1	1	1	1	1	0	0	0	.241525E+05
24	1,900.0	1	0	0	1	0	1	0	1	0	1	1	1	1	0	0	0	.192975E+05

[a] Units 6–9 are OFF all hours. Units 11–13, and 24–26 are ON all hours

4.13.5.1 Approach (1)

In this approach the best K-solutions ($K = 5$–20) are recorded during the STM search. These best solutions are then compared to find the units that have the same schedule in a prespecified percentage of these K-solutions (in our implementation, it is taken to be 70%). These units are then included in an intermediate-memory TL. At a particular period, according to the algorithm, the TL of the ITM is activated and the search is restricted to be in the neighborhood of the best solutions.

4.13.5.2 Approach (2)

In this approach the K-best solutions are recorded during the STM search. Then, the ITM procedure always starts with one of these best solutions at a time and performs a fixed number of iterations and repeats this for the K-solutions. The best solutions list is also updated during the search.

4.13 Advanced Tabu Search (ATS) Techniques

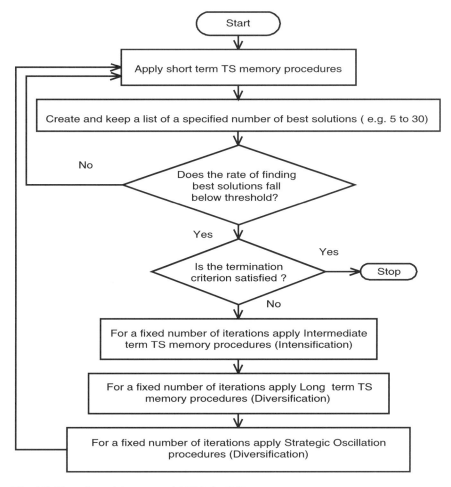

Fig. 4.5 Flow chart of the proposed ATSA for UCP

4.13.6 *Long-Term Memory Implementation*

The LTM procedure is designed to drive the search into new regions. The LTM function is activated when a local minimum is reached during the STM search. In this procedure, the search is directed to points that are far from those of the recorded best solutions. This is achieved by activating the TL of the ITM and restricting the generated trial solutions to be far from the units that were tabu in the ITM procedures.

4.13.7 Strategic Oscillation Implementation

In this procedure a specified number of moves are performed beyond the feasible boundary in a given direction before permitting a return to the feasible region. This number of moves could be changed each time the procedure begins.

4.13.8 Numerical Results of the ATSA

Considering the proposed ATSA, a computer program has been implemented. It has been concluded that Approach 5 of constructing a TL is the most efficient one because it requires less memory space with better solution quality. Accordingly, this approach has been utilized in the implementation of the TL as part of the ATSA.

The three previously described examples from the literature [6, 11–13], are solved. Table 4.18 shows a comparison of daily operating costs for the proposed ATSA, and the STSA implemented in Sect. 4.3 for Examples 1, 2, and 3. Table 4.19 presents the comparison of results obtained in the literature for Examples 1 and 2, the STSA, and the proposed ATSA.

From the last two tables, it is obvious that using the ATSA procedures improves the solution quality of the three examples. Both the required number of iterations and the solution cost are improved when using the advanced TS procedure. This emphasizes the effectiveness of using this approach, in addition to the STM, to solve difficult problems such as the UCP.

Tables 4.20 and 4.21 show detailed results for Example 1 [11]. Table 4.20 shows the load sharing among the committed units in the 24 h. Table 4.21 gives

Table 4.18 Comparison between the proposed ATSA and the STSA

	Example	STSA	ATSA	% Saving
Total cost ($)	1	538,390	537,686	0.13
Total cost ($)	2	59,512	59,385	0.21
Total cost ($)	3	662,583	660,864	0.26
No. of iterations	1	1,924	1,235	35.8
No. of iterations	2	616	138	77.5
No. of iterations	3	3,900	2,547	34.6

Table 4.19 Comparison among the ATSA, LR, IP, and the STSA

	Example	LR [11]	IP [6]	STSA	ATSA
Total cost ($)	1	540,895	–	538,390	537,686
Total cost ($)	2	–	60,667	59,512	59,385
% Saving	1	0	–	0.46	0.59
% Saving	2	–	0	1.9	2.1

4.13 Advanced Tabu Search (ATS) Techniques

Table 4.20 Power sharing (MW) of Example 1

	Unit number[a]							
Hour	2	3	4	6	7	8	9	10
1	400.00	0.00	0.00	185.04	0.00	350.26	0.00	89.70
2	395.36	0.00	0.00	181.09	0.00	338.36	0.00	85.19
3	355.38	0.00	0.00	168.67	0.00	300.95	0.00	75.00
4	333.13	0.00	0.00	161.75	0.00	280.12	0.00	75.00
5	400.00	0.00	0.00	185.04	0.00	350.26	0.00	89.70
6	400.00	0.00	295.68	200.00	0.00	375.00	0.00	129.32
7	400.00	0.00	342.97	200.00	0.00	375.00	507.02	145.01
8	400.00	295.59	396.65	200.00	0.00	375.00	569.93	162.83
9	400.00	468.07	420.00	200.00	0.00	375.00	768.01	218.92
10	400.00	444.60	420.00	200.00	358.05	375.00	741.06	211.29
11	400.00	486.30	420.00	200.00	404.89	375.00	788.95	224.86
12	400.00	514.11	420.00	200.00	436.09	375.00	820.89	233.91
13	400.00	479.35	420.00	200.00	397.09	375.00	780.96	222.60
14	400.00	388.98	420.00	200.00	295.63	375.00	677.18	193.20
15	400.00	310.07	410.84	200.00	250.00	375.00	586.56	167.54
16	400.00	266.64	368.27	200.00	250.00	375.00	536.68	153.41
17	400.00	317.31	417.93	200.00	250.00	375.00	594.87	169.89
18	400.00	458.51	420.00	200.00	373.65	375.00	757.03	215.81
19	400.00	486.30	420.00	200.00	404.89	375.00	788.95	224.86
20	400.00	375.08	420.00	200.00	280.03	375.00	661.21	188.68
21	400.00	0.00	404.87	200.00	0.00	375.00	579.57	165.56
22	400.00	0.00	0.00	200.00	0.00	375.00	675.00	0.00
23	400.00	0.00	0.00	191.64	0.00	370.14	338.22	0.00
24	377.64	0.00	0.00	175.58	0.00	321.78	275.00	0.00

[a] Units 1 and 5 are OFF at all hours

Table 4.21 Load demand and hourly costs ($) of Example 1

HR	Load	ED-cost	ST-cost	T-cost
1	1,025	9,670.0	–	9,670.0
2	1,000	9,446.6	–	9,446.6
3	900	8,560.9	–	8,560.9
4	850	8,123.1	–	8,123.1
5	1,025	9,670.0	–	9,670.0
6	1,400	13,434.1	1,706.0	15,140.0
7	1,970	19,217.7	2,659.1	21,876.8
8	2,400	23,815.5	2,685.1	26,500.6
9	2,850	28,253.9	–	28,253.9
10	3,150	31,701.7	3,007.6	34,709.3
11	3,300	33,219.8	–	33,219.8
12	3,400	34,242.1	–	34,242.1
13	3,275	32,965.5	–	32,965.5
14	2,950	29,706.3	–	29,706.3
15	2,700	27,259.7	–	27,259.7
16	2,550	25,819.8	–	25,819.8
17	2,725	27,501.6	–	27,501.6
18	3,200	32,205.7	–	32,205.7
19	3,300	33,219.8	–	33,219.8

(continued)

Table 4.21 (continued)

HR	Load	ED-cost	ST-cost	T-cost
20	2,900	29,212.5	–	29,212.5
21	2,125	20,698.4	–	20,698.4
22	1,650	15,878.2	–	15,878.2
23	1,300	12,572.8	–	12,572.8
24	1,150	11,232.0	–	11,232.0

Total operating cost = \$537,686

Table 4.22 Power sharing (MW) of Example 3 (units 1–13)

	Unit number												
HR	1	2	3	4	5	6	7	8	9	10	11	12	13
1	0.00	0.00	0.00	0.00	0.00	0.00	4.00	0.00	0.00	0.00	76.00	54.80	15.20
2	0.00	0.00	0.00	0.00	0.00	0.00	4.00	0.00	0.00	0.00	0.00	76.00	50.00
3	2.40	0.00	0.00	0.00	0.00	0.00	4.00	0.00	0.00	0.00	0.00	28.40	15.20
4	2.40	0.00	0.00	0.00	0.00	0.00	4.00	0.00	0.00	0.00	0.00	15.20	15.20
5	2.40	0.00	0.00	0.00	0.00	0.00	4.00	0.00	0.00	0.00	0.00	58.40	15.20
6	2.40	0.00	0.00	0.00	0.00	0.00	4.00	0.00	0.00	76.00	76.00	66.40	15.20
7	0.00	0.00	2.40	0.00	0.00	0.00	4.00	0.00	0.00	76.00	76.00	76.00	76.00
8	0.00	0.00	2.40	0.00	0.00	0.00	4.00	0.00	0.00	76.00	76.00	76.00	76.00
9	0.00	0.00	2.40	0.00	0.00	0.00	4.00	0.00	0.00	76.00	76.00	76.00	76.00
10	0.00	0.00	2.40	0.00	0.00	0.00	4.00	0.00	0.00	76.00	76.00	76.00	76.00
11	0.00	0.00	2.40	0.00	0.00	0.00	4.00	0.00	0.00	76.00	76.00	76.00	76.00
12	0.00	0.00	2.40	0.00	0.00	0.00	4.00	0.00	0.00	76.00	76.00	76.00	76.00
13	0.00	0.00	2.40	0.00	0.00	0.00	4.00	0.00	0.00	76.00	76.00	76.00	76.00
14	0.00	0.00	2.40	0.00	0.00	0.00	0.00	0.00	0.00	76.00	76.00	76.00	76.00
15	0.00	0.00	0.00	0.00	0.00	0.00	0.00	0.00	0.00	76.00	76.00	76.00	76.00
16	0.00	0.00	0.00	0.00	0.00	0.00	0.00	0.00	0.00	76.00	76.00	76.00	76.00
17	0.00	0.00	0.00	0.00	0.00	0.00	0.00	0.00	0.00	76.00	76.00	76.00	76.00
18	0.00	0.00	0.00	0.00	0.00	0.00	0.00	0.00	0.00	76.00	76.00	76.00	76.00
19	0.00	0.00	0.00	0.00	0.00	0.00	0.00	0.00	0.00	76.00	76.00	76.00	76.00
20	0.00	0.00	0.00	0.00	0.00	0.00	0.00	0.00	0.00	76.00	76.00	76.00	76.00
21	0.00	0.00	0.00	0.00	0.00	0.00	0.00	0.00	0.00	76.00	76.00	76.00	76.00
22	0.00	0.00	0.00	0.00	0.00	0.00	0.00	0.00	0.00	76.00	76.00	76.00	76.00
23	0.00	0.00	0.00	0.00	0.00	0.00	0.00	0.00	0.00	76.00	76.00	76.00	64.10
24	0.00	0.00	0.00	0.00	0.00	0.00	0.00	0.00	0.00	15.20	15.20	15.20	15.20

the hourly load demand, and the corresponding economic dispatch costs, startup costs, and total operating cost. For Example 3, Tables 4.22, 4.23, and 4.24 show detailed results and the corresponding total operating cost obtained is \$660,864.8750.

4.13 Advanced Tabu Search (ATS) Techniques

Table 4.23 Power sharing (MW) of Example 3 (units 14–26)

HR	Unit number												
	14	15	16	17	18	19	20	21	22	23	24	25	26
1	0.00	0.00	0.00	155.00	155.00	155.00	155.00	0.00	0.00	0.00	350.00	350.00	350.00
2	0.00	0.00	0.00	155.00	155.00	155.00	155.00	0.00	0.00	0.00	350.00	350.00	350.00
3	0.00	0.00	0.00	155.00	155.00	155.00	155.00	0.00	0.00	0.00	350.00	350.00	350.00
4	0.00	0.00	0.00	155.00	155.00	155.00	155.00	0.00	0.00	0.00	343.20	350.00	350.00
5	0.00	0.00	0.00	155.00	155.00	155.00	155.00	0.00	0.00	0.00	350.00	350.00	350.00
6	0.00	0.00	0.00	155.00	155.00	155.00	155.00	0.00	0.00	0.00	350.00	350.00	350.00
7	0.00	0.00	69.60	155.00	155.00	155.00	155.00	0.00	0.00	0.00	350.00	350.00	350.00
8	0.00	100.00	100.00	155.00	155.00	155.00	155.00	150.65	68.95	0.00	350.00	350.00	350.00
9	100.00	100.00	100.00	155.00	155.00	155.00	155.00	197.00	122.60	0.00	350.00	350.00	350.00
10	100.00	100.00	100.00	155.00	155.00	155.00	155.00	197.00	122.60	0.00	350.00	350.00	350.00
11	100.00	100.00	100.00	155.00	155.00	155.00	155.00	197.00	142.60	0.00	350.00	350.00	350.00
12	100.00	100.00	100.00	155.00	155.00	155.00	155.00	197.00	102.60	0.00	350.00	350.00	350.00
13	100.00	100.00	100.00	155.00	155.00	155.00	155.00	197.00	112.60	0.00	350.00	350.00	350.00
14	100.00	100.00	100.00	155.00	155.00	155.00	155.00	197.00	96.60	0.00	350.00	350.00	350.00
15	100.00	100.00	100.00	155.00	155.00	155.00	155.00	157.05	68.95	0.00	350.00	350.00	350.00
16	100.00	100.00	38.10	155.00	155.00	155.00	155.00	68.95	68.95	0.00	350.00	350.00	350.00
17	100.00	100.00	78.10	155.00	155.00	155.00	155.00	68.95	68.95	0.00	350.00	350.00	350.00
18	100.00	100.00	100.00	155.00	155.00	155.00	155.00	137.05	68.95	0.00	350.00	350.00	350.00
19	100.00	100.00	100.00	155.00	155.00	155.00	155.00	197.00	109.00	0.00	350.00	350.00	350.00
20	100.00	100.00	100.00	155.00	155.00	155.00	155.00	197.00	149.00	0.00	350.00	350.00	350.00
21	100.00	100.00	100.00	155.00	155.00	155.00	155.00	197.00	129.00	0.00	350.00	350.00	350.00
22	100.00	100.00	100.00	155.00	155.00	155.00	155.00	137.05	68.95	0.00	350.00	350.00	350.00
23	0.00	25.00	25.00	155.00	155.00	155.00	155.00	68.95	68.95	0.00	350.00	350.00	350.00
24	0.00	25.00	25.00	155.00	155.00	155.00	155.00	68.95	68.95	0.00	331.30	350.00	350.00

Table 4.24 Load demand and hourly costs ($) of Example 3

HR	Load	ED-cost	ST-cost	T-cost
1	1.82E+03	1.79E+04	0.00E+00	1.79E+04
2	1.80E+03	1.76E+04	0.00E+00	1.76E+04
3	1.72E+03	1.66E+04	0.00E+00	1.66E+04
4	1.70E+03	1.63E+04	0.00E+00	1.63E+04
5	1.75E+03	1.70E+04	0.00E+00	1.70E+04
6	1.91E+03	1.93E+04	1.60E+02	1.94E+04
7	2.05E+03	2.17E+04	1.00E+02	2.18E+04
8	2.40E+03	2.98E+04	7.00E+02	3.05E+04
9	2.60E+03	3.42E+04	1.00E+02	3.43E+04
10	2.60E+03	3.42E+04	0.00E+00	3.42E+04
11	2.62E+03	3.46E+04	0.00E+00	3.46E+04
12	2.58E+03	3.37E+04	0.00E+00	3.37E+04
13	2.59E+03	3.39E+04	0.00E+00	3.39E+04
14	2.57E+03	3.33E+04	0.00E+00	3.33E+04
15	2.50E+03	3.17E+04	0.00E+00	3.17E+04
16	2.35E+03	2.85E+04	0.00E+00	2.85E+04
17	2.39E+03	2.92E+04	0.00E+00	2.92E+04
18	2.48E+03	3.12E+04	0.00E+00	3.12E+04
19	2.58E+03	3.35E+04	0.00E+00	3.35E+04
20	2.62E+03	3.44E+04	0.00E+00	3.44E+04
21	2.60E+03	3.40E+04	0.00E+00	3.40E+04
22	2.48E+03	3.12E+04	0.00E+00	3.12E+04
23	2.15E+03	2.47E+04	0.00E+00	2.47E+04
24	1.90E+03	2.14E+04	0.00E+00	2.14E+04

Total operating cost = $660,864.8750

4.14 Conclusions

In this section, the application of the TS method for the UCP is introduced for the first time. Two algorithms for the UCP are proposed and tested. The first algorithm uses the STM procedure of the TS method, and the second algorithm is based on advanced TS procedures. Different criteria for constructing the TL restrictions for the UCP are implemented and compared. Several examples are solved to test the algorithms.

The computational results of the two algorithms along with a comparison with previously published works are presented. The results showed that the algorithm based on the STM outperforms the results reported in the literature. On the other hand, both the required number of iterations and the solution cost are improved when using the advanced TS procedure in the second algorithm. This emphasizes the effectiveness of using this approach, along with the STM, to solve difficult problems such as the UCP.

4.15 Genetic Algorithms for Unit Commitment

Genetic algorithms were developed in the early 1970s [31] and have become increasingly popular in recent years in science and engineering disciplines [31–39]. GAs have been quite successfully applied to optimization problems including wire routing, scheduling, adaptive control, game playing, cognitive modeling, transportation problems, traveling salesman problems, optimal control problems, and so on.

GA is a global search technique based on the mechanics of natural selection and genetics. It is a general-purpose optimization algorithm that is distinguished from conventional optimization techniques by the use of concepts of population genetics to guide the optimization search. Instead of point-to-point search, GA searches from population to population. The advantages of GA over traditional techniques are:

- It needs only rough information of the objective function and places no restriction such as differentiability and convexity on the objective function.
- The method works with a set of solutions from one generation to the next, and not a single solution, thus making it less likely to converge on local minima.
- The solutions developed are randomly based on the probability rate of the genetic operators such as mutation and crossover; the initial solutions thus would not dictate the search direction of GA.

A major disadvantage of the GA method is that it requires tremendously high CPU time. Following are the major applications of GA in power systems.

- Planning (transmission expansion planning, capacitor placement
- Operation (voltage/reactive power control, unit commitment/economic dispatch, hydrothermal scheduling)

GAs are general-purpose search techniques based on principles inspired by the genetic and evolutionary mechanisms observed in natural systems and populations of living beings. Their basic principle is the maintenance of a population of solutions to a problem (genotypes) in the form of encoded information individuals that evolve in time. A GA for a particular problem must have the following five components.

1. A genetic representation for a potential solution to the problem
2. A way to create an initial population of potential solutions
3. An evaluation function that plays the role of the environment, rating solutions in terms of their "fitness"
4. Genetic operators that alter the composition of children
5. Values for various parameters that the GA uses (population size, probabilities of applying genetic operators, etc.)

A genetic search starts with a randomly generated initial population within which each individual is evaluated by means of a fitness function. Individuals in this and subsequent generations are duplicated or eliminated according to their fitness values. Further generations are created by applying GA operators. This eventually leads to a generation of high-performing individuals [37]. In this section an implementation of a genetic algorithm to the UCP is implemented. Several examples are solved to test the algorithm.

4.15.1 Solution Coding

GAs require the natural parameter set of the optimization problem to be coded as a finite-length string over some finite alphabet. Coding is the most important point in applying the GA to solve any optimization problem. Coding could be in a real or binary form. Coded strings of solutions are called "chromosomes." A group of these solutions (chromosomes) is called a population.

4.15.2 Fitness Function

The fitness function is the second important issue in solving optimization problems using GAs. It is often necessary to map the underlying natural objective function to a fitness function through one or more mappings.

The first mapping is done to transform the objective function into a maximization problem rather than minimization to suit the GA concepts of selecting the fittest chromosome that has the highest objective function.

A second important mapping is the scaling of the fitness function values. Scaling is an important step during the search procedures of the GA. This is done to keep appropriate levels of competition throughout a simulation. Without scaling, early on there is a tendency for a few superindividuals to dominate the selection process. Later on, when the population has largely converged, competition among population members is weaker and simulation tends to wander. Thus, scaling is a useful process to prevent both the premature convergence of the algorithm and the random improvement that may occur in the late iterations of the algorithm. There are many methods for scaling such as linear, sigma truncation, and power law scaling [31]. Linear scaling is the most commonly used. In the sigma truncation method, population variance information to preprocess raw fitness values prior to scaling is used. It is called sigma (σ) truncation because of the use of population standard deviation information; a constant is subtracted from raw fitness values as follows.

$$\mathbf{f}' = \mathbf{f} - (\mathbf{f}' - \mathbf{c}.\sigma) \tag{4.35}$$

In Eq. 4.35 the constant c is chosen as a reasonable multiple of the population standard deviation and negative results ($f' < 0$) are arbitrarily set to 0. Following sigma truncation, fitness scaling can proceed as described without the danger of negative results.

4.15.3 Genetic Algorithms Operators

There are usually three operators in a typical GA [37]. The first is the production operator which makes one or more copies of any individual that possesses a high fitness value; otherwise, the individual is eliminated from the solution pool.

The second operator is the recombination (also known as the "crossover") operator. This operator selects two individuals within the generation and a crossover site and performs a swapping operation of the string bits to the right-hand side of the crossover site of both individuals. The crossover operator serves two complementary search functions: it provides new points for further testing within the hyperplanes already represented in the population, and it introduces representatives of new hyperplanes into the population, which are not represented by either parent structure. Thus, the probability of a better performing offspring is greatly enhanced.

The third operator is the "mutation" operator. This operator acts as a background operator and is used to explore some of the unvisited points in the search space by randomly flipping a "bit" in a population of strings. Frequent application of this operator would lead to a completely random search, therefore a very low probability is usually assigned to its activation.

4.15.4 Constraint Handling (Repair Mechanism)

Constraint handling techniques for the GAs can be grouped into a few categories [33]. One way is to generate a solution without considering the constraints but to include them with penalty factors in the fitness function. This method has been used previously [5, 40–44]. Another category is based on the application of a special repair algorithm to correct any infeasible solution so generated. The third approach concentrates on the use of special representation mappings (decoders) that guarantee (or at least increase the probability of) the generation of a feasible solution or the use of problem-specific operators that preserve feasibility of the solutions.

In our implementation, we always generate solutions that satisfy the minimum up/down constraints. However, due to applying the crossover and mutation operations the load demand and/or the reserve constraints might be violated. A mechanism to restore feasibility is applied by randomly committing more units at the violated time periods and keeping the feasibility of the minimum up/down time constraints.

4.15.5 A General Genetic Algorithm

In applying the GAs to optimization problems, certain steps for simulating evolution must be performed. These are described as follows [32].

Step (1): Initialize a population of chromosomes.
Step (2): Evaluate each chromosome in the population.
Step (3): Create new chromosomes by mating current chromosomes; apply mutation and recombination as the parent chromosomes mate.
Step (4): Delete members of the population to make room for the new chromosomes.
Step (5): Evaluate the new chromosomes and insert them into the population.
Step (6): If the termination criterion is satisfied, stop and return the best chromosomes; otherwise, go to Step (3).

4.15.6 Implementation of a Genetic Algorithm to the UCP

The proposed GA implementation for the UCP differs from other conventional GA implementations in three respects [45].

- First, the UCP solution is coded using a mix of binary and decimal representations, thus saving computer memory as well as computation time of the GA search procedure.
- Second, the fitness function is based only on the total operating cost and no penalties are included.
- Third, to improve the local fine-tuning capabilities of the proposed GA, a special mutation operator based on a local search procedure is designed.

The proposed algorithm involves four major steps [45].

1. Creating an initial population by randomly generating a set of feasible solutions (chromosomes), using the rules explained in Chap. 2.
2. Evaluating each chromosome by solving the economic dispatch problem, using the algorithm explained in Chap. 2.
3. Determining the fitness function for each chromosome in the population.
4. Applying GA operators to generate new populations:

   - Copy the best solution from the current to the new population.
   - Generate new members (typically 1–10% of the population size) as neighbors to solutions in the current population, and add them to the new population.
   - Apply the crossover operator to complete the members of the new population.
   - Apply the mutation operator to the new population. The flowchart of Fig. 4.6 shows the main steps of the proposed algorithm.

   In the following sections, implementations of the different components of the proposed algorithm are presented.

4.15 Genetic Algorithms for Unit Commitment

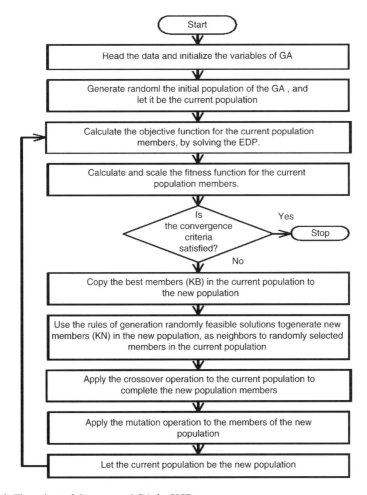

Fig. 4.6 Flow chart of the proposed GA for UCP

4.15.7 Solution Coding

Because the UCP lends itself to binary coding in which a zero denotes the OFF state and a one represents the ON state, all published works use binary coding [5, 40–44]. The UCP solution is represented by a binary matrix (U) of dimension $T \times N$ (Fig. 4.7a). A candidate solution in the GA could then be represented by a string whose length is the product of the scheduling periods and the number of generating units $T \times N$. In the GA a number of these solutions equal to the population size (NPOP) are stored. The required storage size is then equal to NPOP $\times$ T $\times$ N which is a large value even for a moderate size system.

The new proposed method for coding is based on a mix of a binary number and its equivalent decimal number. Each column vector of length T in the solution

Fig. 4.7 (a) The binary solution matrix U. (b) The equivalent decimal vector $(1 \times N)$ (one chromosome). (c) Population of size $NPOP \times N$ ($NPOP$ chromosomes)

a

HR	Unit Number							
	1	2	3	4	.	.	.	N
1	1	1	0	0	.	.	.	1
2	1	1	0	0	.	.	.	1
3	1	0	1	0	.	.	.	0
.	.	.	.	.	.	.	.	.

b

U1	U2	U3	U4	.	.	.	UN

c

23	14	45	56	.	.	.	62
34	52	72	18	.	.	.	91
.	.	.	.	.	.	.	.
51	36	46	87	.	.	.	21

matrix (which represents the operation schedule of one unit) is converted to its equivalent decimal number. The solution matrix is then converted into one row vector (chromosome) of N decimal numbers, $(U1, U2, \ldots, UN)$; each represents the schedule of one unit as shown in Fig. 4.7b. Typically the numbers $U1, U2, \ldots, UN$ are integers ranging between 0 and $2^N - 1$. Accordingly, a population of size $NPOP$ can be stored in a matrix of dimension $NPOP \times N$ as arbitrarily shown in Fig. 4.7c. Hence, the proposed method requires only $1/T$ of the storage required if a normal binary coding is used.

4.15.8 Fitness Function

Unlike the previous solutions of the UCP using GAs [5, 40–44], the fitness function is taken as the reciprocal of the total operating cost, inasmuch as we are always generating feasible solutions.

The fitness function is then scaled to prevent premature convergence. Linear scaling is used. This requires a linear relationship between the original fitness function (f) and the scaled one ($\mathbf{f_s}$) [31]:

$$\mathbf{f_s} = \mathbf{af} + \mathbf{b} \tag{4.36}$$

$$\mathbf{a} = (\mathbf{c} - \mathbf{1})\mathbf{f_{av}}/(\mathbf{f_{max}} - \mathbf{f_{min}}) \tag{4.37}$$

$$\mathbf{b} = (\mathbf{1} - \mathbf{a})\mathbf{f_{av}} \tag{4.38}$$

where c is a parameter between 1.2 and 2, and $\mathbf{f_{max}}, \mathbf{f_{min}}, \mathbf{f_{av}}$ are maximum, minimum, and average values of the original fitness functions, respectively.

4.15 Genetic Algorithms for Unit Commitment

```
         | 12 | 34 | 45 | 62 | 93 | 72 | 82 | 32 |
         | 52 | 81 | 69 | 55 | 26 | 38 | 57 | 76 |
Two Parents

         | 12 | 34 | 69 | 55 | 26 | 72 | 82 | 32 |
         | 52 | 81 | 45 | 62 | 93 | 38 | 57 | 76 |
Two children
```

Fig. 4.8 Two-point crossover example

4.15.9 Selection of Chromosomes

The selection of chromosomes for applying various GA operators is based on their scaled fitness function in accordance with the roulette wheel selection rule. The roulette wheel slots are sized according to the accumulated probabilities of reproducing each chromosome.

4.15.10 Crossover

To speed up the calculations, the crossover operation is performed between two chromosomes in their decimal form. A two-point crossover operation is used. The following steps are applied to perform the crossover operation.

- Select two parents according to the roulette wheel rule.
- Randomly select two positions in the two chromosomes.
- Exchange the bits between the two selected positions in the two parents to produce two children (Fig. 4.8).
- Decode the two children into their binary equivalents and check for reserve constraint violation.
- If the reserve constraints are not satisfied apply the repair mechanism to restore the feasibility of the produced children.

4.15.11 Mutation

The crossover operation explained in the last section is not enough to create a completely new solution. The reason is that it exchanges the schedule of units as black boxes among different chromosomes without applying any changes in the schedules of the units themselves.

Two new types of mutation operators are introduced to create changes in the units' schedules. The mutation operation is applied after reproducing all the new population members. It is done by applying the probability test to the members of the new population one by one. The mutation operation is then applied to the selected chromosome. Details of the two mutation operators are described in the following sections.

4.15.11.1 Mutation Operator (1)

The first mutation operator is implemented as follows.

1. Select a chromosome as explained before and decode it into its binary equivalent.
2. Randomly pick a unit number and a time period.
3. Apply the rules in Sect. 2.4 to reverse the status of this unit keeping the feasibility of the unit constraints related to its minimum up- and downtimes.
4. For the changed time periods, check the reserve constraints.
5. If the reserve constraints are violated, apply the proposed correction mechanism and go to the next step; otherwise go to the next step.
6. Decode the modified solution matrix from binary to decimal form and update the new population.

4.15.11.2 Mutation Operator (2)

The second mutation operator is based on a local search algorithm to perform fine-tuning on some of the chromosomes in the newly generated population. The selection of chromosomes for applying this type of mutation could be random or based on the roulette wheel method.

The local search algorithm steps are described in detail as follows.

1. Decode the selected chromosome into its binary form.
2. Sort the time periods in descending order according to the difference between the committed unit capacity and the load demand.
3. Identify the time periods at which the committed unit capacity is greater than 10% above the load plus the desired reserve. These time periods have a surplus of committed power capacity.
4. At the surplus capacity time periods, sort the committed units in ascending order according to their percentage loading.
5. Identify the units that have a percentage loading less than 20% above their minimum output limits. These units are the most cost units among the committed units in the respective time periods, inasmuch as they are lightly loaded.
6. Take the time periods, according to their order found in Step (2) and consider switching off the underloaded units one at a time, according to their order.

7. Check the feasibility of the solution obtained. If it is feasible, go to Step (8); otherwise go to Step (6).
8. Calculate the objective function of the solution obtained by solving the economic dispatch problem for the changed time periods.
9. Decode the new solution obtained to its decimal equivalent and replace the old one in the new population.

4.15.12 Adaptive GA Operators

The search for the optimal GA parameter setting is a very complex task. To achieve a good performance of the GA, an adaptive scheme to control the probability rate of performing the crossover and mutation operators is designed.

The crossover rate controls the frequency with which the crossover operator is applied. The higher the crossover rate, the more quickly new structures are introduced into the population. If the crossover rate is too high, high-performance structures are discarded faster than selection can produce improvements. If the crossover rate is too low, the search may stagnate due to the lower exploration rate. In this implementation, the crossover rate is initialized with a high value (typically between 0.6 and 0.8) and is then decreased during the search according to the convergence rate of the algorithm (decrement value is 0.01).

Mutation is a secondary search operator that increases the variability of the population. A low level of mutation prevents any given bit position from remaining forever converged to a single value in the entire population, and consequently increases the probability of entrapment at local minima. A high level of mutation yields an essentially random search, which may lead to very slow convergence. To guide the search, the mutation rate starts at a low value (between 0.2 and 0.5); it is then incremented by 0.01 as the algorithm likely converged to a local minimum.

4.15.13 Numerical Examples

In order to test the proposed algorithm, three systems are considered. Preliminary experiments have been performed on the three systems to find the most suitable GA parameter settings. These control parameters have been chosen after running a number of simulations: population size = 50, initial value of crossover rate = 0.8, decrement value of crossover = 0.01, initial value of mutation rate = 0.2, increment value of mutation = 0.01, local search mutation rate = 0.1, elite copies = 2, and the maximum number of generations = 1,000.

Different experiments were carried out to investigate the effect of the local search mutation on the results. It was found that the proposed algorithm with local search performs better than the simple GA without local search in terms of both solution quality and number of iterations.

Table 4.25 Comparison among LR, IP, and the proposed GA

	Example	LR [11]	IP [6]	GA
Total cost ($)	1	540,895	–	537,372
Total cost ($)	2	–	60,667	59,491
% Saving	1	–		0.65
% Saving	2	–		1.93
Generations no	1	–		411
Generations no	2	–		393

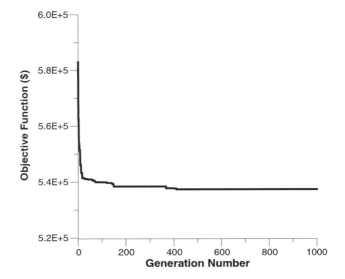

Fig. 4.9 Convergence of the proposed GA

Table 4.25 presents the comparison of results obtained in the literature (LR and IP) for Examples 1 and 2.

Figure 4.9 shows progress in the best objective function versus the generation number. The algorithm converges after about 400 generations, which is relatively fast.

Tables 4.26, 4.27, and 4.28 show detailed results for Example 1 [11]. Table 4.26 shows the load sharing among the committed units in 24 h. Table 4.27 gives the hourly load demand and the corresponding economic dispatch, startup, and total operating costs. Table 5.4 presents the final schedule of the 24 h, given in Table 4.26 in the form of its equivalent decimal numbers.

Tables 4.29, 4.30, 4.31, and 4.32 also present the detailed results for Example 3 with a total operating cost of $661,439.8

4.15 Genetic Algorithms for Unit Commitment

Table 4.26 Power sharing (MW) of Example 1

	Unit number[a]							
HR	2	3	4	6	7	8	9	10
1	400.0	0.0	0.0	185.0	0.0	350.3	0.0	89.7
2	395.4	0.0	0.0	181.1	0.0	338.4	0.0	85.2
3	355.4	0.0	0.0	168.7	0.0	301.0	0.0	75.0
4	333.1	0.0	0.0	161.8	0.0	280.1	0.0	75.0
5	400.0	0.0	0.0	185.0	0.0	350.3	0.0	89.7
6	400.0	0.0	295.7	200.0	0.0	375.0	0.0	129.3
7	400.0	0.0	343.0	200.0	0.0	375.0	507.0	145.0
8	400.0	295.6	396.7	200.0	0.0	375.0	569.9	162.8
9	400.0	468.1	420.0	200.0	0.0	375.0	768.0	218.9
10	400.0	444.6	420.0	200.0	358.1	375.0	741.1	211.3
11	400.0	486.3	420.0	200.0	404.9	375.0	789.0	224.9
12	400.0	514.1	420.0	200.0	436.1	375.0	820.9	233.9
13	400.0	479.4	420.0	200.0	397.1	375.0	781.0	222.6
14	400.0	389.0	420.0	200.0	295.6	375.0	677.2	193.2
15	400.0	310.1	410.8	200.0	250.0	375.0	586.6	167.5
16	400.0	266.6	368.3	200.0	250.0	375.0	536.7	153.4
17	400.0	317.3	417.9	200.0	250.0	375.0	594.9	169.9
18	400.0	458.5	420.0	200.0	373.7	375.0	757.0	215.8
19	400.0	486.3	420.0	200.0	404.9	375.0	789.0	224.9
20	400.0	0.0	420.0	200.0	442.2	375.0	827.2	235.7
21	400.0	0.0	404.9	200.0	0.0	375.0	579.6	165.6
22	400.0	0.0	0.0	200.0	0.0	375.0	675.0	0.0
23	400.0	0.0	0.0	191.6	0.0	370.1	338.2	0.0
24	377.6	0.0	0.0	175.6	0.0	321.8	275.0	0.0

[a] Units 1,5 are OFF all hours

Table 4.27 Load demand and hourly costs ($) of Example 1

HR	Load	ED-cost	ST-cost	T-cost
1	1,025	9,670.04	0.00	9,670.04
2	1,000	9,446.62	0.00	9,446.62
3	900	8,560.91	0.00	8,560.91
4	850	8,123.13	0.00	8,123.13
5	1,025	9,670.04	0.00	9,670.04
6	1,400	13,434.10	1,705.97	15,140.00
7	1,970	19,217.70	2,659.11	21,876.80
8	2,400	23,815.50	2,685.07	26,500.60
9	2,850	28,253.90	0.00	28,253.90
10	3,150	31,701.70	3,007.58	34,709.30
11	3,300	33,219.80	0.00	33,219.80
12	3,400	34,242.10	0.00	34,242.10
13	3,275	32,965.50	0.00	32,965.50
14	2,950	29,706.30	0.00	29,706.30
15	2,700	27,259.70	0.00	27,259.70

(continued)

Table 4.27 (continued)

HR	Load	ED-cost	ST-cost	T-cost
16	2,550	25,819.80	0.00	25,819.80
17	2,725	27,501.60	0.00	27,501.60
18	3,200	32,205.70	0.00	32,205.70
19	3,300	33,219.80	0.00	33,219.80
20	2,900	28,899.00	0.00	28,899.00
21	2,125	20,698.40	0.00	20,698.40
22	1,650	15,878.20	0.00	15,878.20
23	1,300	12,572.80	0.00	12,572.80
24	1,150	11,232.00	0.00	11,232.00

Total operating cost = \$537,371.94

Table 4.28 The UCT of Example 1 in its equivalent decimal form (best chromosome)

Unit number				
1,6	2,7	3,8	4,9	5,10
0	16777215	524160	2097120	0
16777215	1048064	16777215	16777152	2097151

Table 4.29 Power sharing (MW) of Example 3 (units 1–13)

	Unit number												
HR	1	2	3	4	5	6	7	8	9	10	11	12	13
1	0.00	0.00	0.00	0.00	0.00	0.00	0.00	0.00	4.00	0.00	76.00	54.80	15.20
2	0.00	0.00	0.00	0.00	0.00	0.00	0.00	0.00	4.00	0.00	0.00	76.00	50.00
3	2.40	0.00	0.00	0.00	0.00	0.00	0.00	0.00	4.00	0.00	0.00	28.40	15.20
4	2.40	0.00	0.00	0.00	0.00	0.00	0.00	0.00	4.00	0.00	0.00	15.20	15.20
5	2.40	0.00	0.00	0.00	0.00	0.00	0.00	0.00	4.00	0.00	0.00	58.40	15.20
6	2.40	0.00	0.00	0.00	0.00	0.00	0.00	0.00	4.00	76.00	76.00	66.40	15.20
7	0.00	0.00	2.40	0.00	0.00	0.00	0.00	0.00	4.00	76.00	76.00	76.00	76.00
8	0.00	0.00	2.40	0.00	0.00	0.00	0.00	0.00	4.00	76.00	76.00	76.00	76.00
9	0.00	0.00	2.40	0.00	0.00	0.00	0.00	0.00	4.00	76.00	76.00	76.00	76.00
10	0.00	0.00	2.40	0.00	0.00	0.00	0.00	0.00	4.00	76.00	76.00	76.00	76.00
11	0.00	0.00	2.40	0.00	0.00	0.00	0.00	0.00	4.00	76.00	76.00	76.00	76.00
12	0.00	0.00	2.40	0.00	0.00	0.00	0.00	0.00	4.00	76.00	76.00	76.00	76.00
13	0.00	0.00	2.40	0.00	0.00	0.00	0.00	0.00	4.00	76.00	76.00	76.00	76.00
14	0.00	0.00	2.40	0.00	0.00	0.00	0.00	0.00	4.00	76.00	76.00	76.00	76.00
15	0.00	0.00	0.00	0.00	0.00	0.00	0.00	0.00	4.00	76.00	76.00	76.00	76.00
16	0.00	0.00	0.00	0.00	0.00	0.00	0.00	0.00	4.00	76.00	76.00	76.00	76.00
17	0.00	0.00	0.00	0.00	0.00	0.00	0.00	0.00	0.00	76.00	76.00	76.00	76.00
18	0.00	0.00	0.00	0.00	0.00	0.00	0.00	0.00	0.00	76.00	76.00	76.00	76.00
19	0.00	0.00	0.00	0.00	0.00	0.00	0.00	0.00	0.00	76.00	76.00	76.00	76.00
20	0.00	0.00	0.00	0.00	0.00	0.00	0.00	0.00	0.00	76.00	76.00	76.00	76.00
21	0.00	0.00	0.00	0.00	0.00	0.00	0.00	0.00	0.00	76.00	76.00	76.00	76.00
22	0.00	0.00	0.00	0.00	0.00	0.00	0.00	0.00	0.00	76.00	76.00	76.00	76.00
23	0.00	0.00	0.00	0.00	0.00	0.00	0.00	0.00	0.00	76.00	76.00	76.00	64.10
24	0.00	0.00	0.00	0.00	0.00	0.00	0.00	0.00	0.00	15.20	15.20	15.20	15.20

4.15 Genetic Algorithms for Unit Commitment

Table 4.30 Power sharing (MW) of Example 3 (units 14–26)

HR	Unit number												
	14	15	16	17	18	19	20	21	22	23	24	25	26
1	0.00	0.00	0.00	155.00	155.00	155.00	155.00	0.00	0.00	0.00	350.00	350.00	350.00
2	0.00	0.00	0.00	155.00	155.00	155.00	155.00	0.00	0.00	0.00	350.00	350.00	350.00
3	0.00	0.00	0.00	155.00	155.00	155.00	155.00	0.00	0.00	0.00	350.00	350.00	350.00
4	0.00	0.00	0.00	155.00	155.00	155.00	155.00	0.00	0.00	0.00	343.20	350.00	350.00
5	0.00	0.00	0.00	155.00	155.00	155.00	155.00	0.00	0.00	0.00	350.00	350.00	350.00
6	0.00	0.00	0.00	155.00	155.00	155.00	155.00	0.00	0.00	0.00	350.00	350.00	350.00
7	0.00	0.00	69.60	155.00	155.00	155.00	155.00	0.00	0.00	0.00	350.00	350.00	350.00
8	0.00	100.00	100.00	155.00	155.00	155.00	155.65	150.65	68.95	0.00	350.00	350.00	350.0
9	100.00	100.00	100.00	155.00	155.00	155.00	155.00	197.00	122.60	0.00	350.00	350.00	350.00
10	100.00	100.00	100.00	155.00	155.00	155.00	155.00	197.00	122.60	0.00	350.00	350.00	350.00
11	100.00	100.00	100.00	155.00	155.00	155.00	155.00	197.00	142.60	0.00	350.00	350.00	350.00
12	100.00	100.00	100.00	155.00	155.00	155.00	155.00	197.00	102.60	0.00	350.00	350.00	350.00
13	100.00	100.00	100.00	155.00	155.00	155.00	155.00	197.00	112.60	0.00	350.00	350.00	350.00
14	100.00	100.00	100.00	155.00	155.00	155.00	155.00	197.00	92.60	0.00	350.00	350.00	350.00
15	100.00	100.00	100.00	155.00	155.00	155.00	155.00	153.05	68.95	0.00	350.00	350.00	350.00
16	100.00	100.00	34.10	155.00	155.00	155.00	155.00	68.95	68.95	0.00	350.00	350.00	350.00
17	100.00	100.00	78.10	155.00	155.00	155.00	155.00	68.95	68.95	0.00	350.00	350.00	350.00
18	100.00	100.00	100.00	155.00	155.00	155.00	155.00	137.05	68.95	0.00	350.00	350.00	350.00
19	100.00	100.00	100.00	155.00	155.00	155.00	155.00	197.00	109.00	0.00	350.00	350.00	350.00
20	100.00	100.00	100.00	155.00	155.00	155.00	155.00	197.00	149.00	0.00	350.00	350.00	350.00
21	100.00	100.00	100.00	155.00	155.00	155.00	155.00	197.00	129.00	0.00	350.00	350.00	350.00
22	100.00	100.00	100.00	155.00	155.00	155.00	155.00	137.05	68.95	0.00	350.00	350.00	350.00
23	0.00	25.00	25.00	155.00	155.00	155.00	155.00	68.95	68.95	0.00	350.00	350.00	350.00
24	0.00	25.00	25.00	155.00	155.00	155.00	155.00	68.95	68.95	0.00	331.30	350.00	350.00

Table 4.31 Load demand and hourly costs ($) of Example 3

HR	Load	ED-cost	ST-cost	T-cost
1	1.82E+03	1.79E+04	0.00E+00	1.79E+04
2	1.80E+03	1.76E+04	0.00E+00	1.76E+04
3	1.72E+03	1.66E+04	0.00E+00	1.66E+04
4	1.70E+03	1.63E+04	0.00E+00	1.63E+04
5	1.75E+03	1.70E+04	0.00E+00	1.70E+04
6	1.91E+03	1.93E+04	1.60E+02	1.94E+04
7	2.05E+03	2.17E+04	1.00E+02	2.18E+04
8	2.40E+03	2.98E+04	7.00E+02	3.05E+04
9	2.60E+03	3.42E+04	1.00E+02	3.43E+04
10	2.60E+03	3.42E+04	0.00E+00	3.42E+04
11	2.62E+03	3.46E+04	0.00E+00	3.46E+04
12	2.58E+03	3.37E+04	0.00E+00	3.37E+04
13	2.59E+03	3.39E+04	0.00E+00	3.39E+04
14	2.57E+03	3.35E+04	0.00E+00	3.35E+04
15	2.50E+03	3.18E+04	0.00E+00	3.18E+04
16	2.35E+03	2.87E+04	0.00E+00	2.87E+04
17	2.39E+03	2.92E+04	0.00E+00	2.92E+04
18	2.48E+03	3.12E+04	0.00E+00	3.12E+04
19	2.58E+03	3.35E+04	0.00E+00	3.35E+04
20	2.62E+03	3.44E+04	0.00E+00	3.44E+04
21	2.60E+03	3.40E+04	0.00E+00	3.40E+04
22	2.48E+03	3.12E+04	0.00E+00	3.12E+04
23	2.15E+03	2.47E+04	0.00E+00	2.47E+04
24	1.90E+03	2.14E+04	0.00E+00	2.14E+04

Total operating cost = $661,439.8125

4.15.14 Summary

In this chapter, a new implementation of a GA to solve the UCP is proposed.

The proposed new GA implementation [45] for the UCP differs from other GA implementations in three respects. First, the UCP solution is coded using a mix between binary and decimal representations. Second, the fitness function is based only on the total operating cost and no penalties are included. Third, to improve the local fine-tuning capabilities of the proposed GA, a special mutation operator based on a local search procedure is designed. The detailed description of the local search algorithm that has been used with the GA is also presented.

The computational results as well as a comparison with previously published work showed the effectiveness of the proposed new approach in saving costs and computer memory.

In the next chapter, we introduce three new hybrid algorithms for the UCP.

Table 4.32 The UCT of Example 3 in its equivalent decimal form (best chromosome)

Unit number																										
1	2	3	4	5	6	7	8	9	10	11	12	13	14	15	16	17	18	19	20	21	22	23	24	25	26
60	0	16320	0	0	0	0	0	65535	16777184	16777185	16777215	16777215	4194048	16777088	16777152	16777215	16777215	16777215	16777215	16777088	16777088	0	16777215	16777215	16777215

4.16 Hybrid Algorithms for Unit Commitment

In the above section three different algorithms, based on SA, TS, and GA methods, to solve the UCP are discussed. The effectiveness of the three methods to solve the UCP has been proved. The main features of these methods were also investigated. These methods, of course, have their own merits and drawbacks.

- SA is a Markov chain Monte Carlo method, and therefore, it is memoryless. Hence, the main problem is that SA will continue to jump up and down without noticing that the movement is confined.
- TS is characterized as a memory-based method, and therefore learning is achieved during the search process.
- GA is a global stochastic search method. The fine-tuning capability when approaching a local minimum is weak.

The competitive performance of the combinatorial optimization algorithms is still an open issue. Recently, hybrid methods have come into the picture capturing the merits of different methods and exploring them in a form of hybrid scheme. It is often proved that a hybrid scheme of some methods outperforms the performance of these methods as individuals.

In this chapter we propose three different new hybrid algorithms for the UCP. The proposed hybrid algorithms integrate the use of the previously introduced algorithms, SA, TS, and GA. The bases of the hybridization of these algorithms are completely new ideas and are applied to the UCP for the first time.

4.17 Hybrid of Simulated Annealing and Tabu Search (ST)

Inasmuch as the main feature of the TS method is to prevent cycling of solutions, we could explore this point in refining the SA algorithm. The main idea in the proposed ST algorithm is to use the TS algorithm to prevent the repeated solutions from being accepted by the SA. This will save time and improve the quality of the obtained solution.

The proposed ST algorithm [46] may be described as an SA algorithm with the TS algorithm used as a filter to reject the repeated trial solutions from being tested by the SAA. The TS method is implemented as a preprocessor step in the SAA to test a set of neighbors to the current solution. The trial solution that satisfies the tabu test is accepted. This accepted trial solution is then accepted or rejected according to the SA test. The main steps of the ST algorithm are described in the flowchart of Fig. 4.10.

4.17 Hybrid of Simulated Annealing and Tabu Search (ST) 247

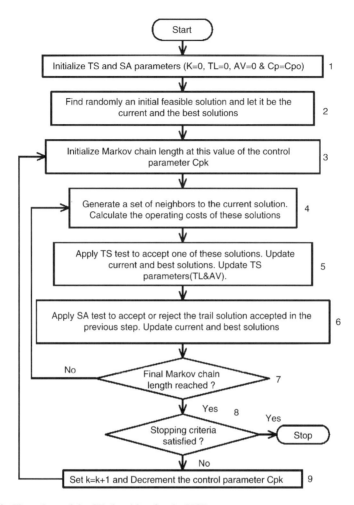

Fig. 4.10 Flow chart of the ST algorithm for the UCP

4.17.1 Tabu Search Part in the ST Algorithm

In the TS part of the ST algorithm, the STM procedures are implemented. In this implementation, the TL is created using approach 5 which was explained earlier in this chapter.

The tabu test (block 5) can be described as follows.

- Sort the set of trial solutions (neighbors) in an ascending order according to their objective functions.
- Apply the tabu acceptance test in order until one of these solutions is accepted.

- *Tabu acceptance test*: If the trial solution (U_j) is NOT tabu or tabu but satisfies the AV, then accept the trial solution as the current solution, set $U_c = U_j$ and $F_c = F_j$, and go to the SA test. Otherwise apply the test to the next solution.

4.17.2 Simulated Annealing Part in the ST Algorithm

In the SA part of the ST algorithm we used the polynomial-time cooling schedule to decrement the control parameter during the search (block 9). The SA test implemented in the ST algorithm (block 6) is described in the following steps.

Let $\mathbf{U_c}, \mathbf{F_c}$ be the SA current solution and the corresponding operating cost, respectively.

Let $\mathbf{Uj}, \mathbf{Fj}$ be the trial solution accepted by the previous tabu test and the corresponding operating cost, respectively.

SA acceptance test:

- If $\mathbf{F_j} \leq \mathbf{F_c}$, or $\exp[(\mathbf{Fc} - \mathbf{F_j})/\mathbf{Cp}] \geq U(\mathbf{0}, \mathbf{1})$, then accept the trial solution and update the current solution; set $\mathbf{U_c} = \mathbf{U_j}$ and $\mathbf{F_c} = \mathbf{F_j}$; then go to block (7).
- Otherwise reject the trial solution and set $\mathbf{U_j} = \mathbf{U_c}$ and $F_j = F_c$; then go to block (7).

4.18 Numerical Results of the ST Algorithm

Considering the proposed ST algorithm, a computer program has been implemented. The three examples, previously solved in the last chapters, are solved again using the ST algorithm [6, 11, 12].

These control parameters have been chosen after running a number of simulations: maximum number of iterations = 3,000, tabu size = 7, initial control parameter (temperature) = 5,000, chain length = 150, ε = 0.00001, and δ = 0.3.

Table 4.33 shows a comparison of the results obtained for the three examples 1, 2, and 3 as solved by the SA, the TS, and the ST algorithms. It is obvious that the ST algorithm achieves reduction in the operating costs for the three examples. Also, the

Table 4.33 Comparison among SAA, STSA, ATSA, and the ST algorithms

	Example	SAA	STSA	ATSA	ST
Total cost ($)	1	536,622	538,390	537,686	536,386
Total cost ($)	2	59,512	59,512	59,385	59,385
Total cost ($)	3	662,664	662,583	660,864	660,596
Iterations no.	1	384	1,924	1,235	625
Iterations no.	2	652	616	138	538
Iterations no.	3	2,361	3,900	2,547	2,829

4.18 Numerical Results of the ST Algorithm

Table 4.34 Comparison among the LR, IP, and ST algorithms

	Example	LR [11]	IP [6]	ST
Total cost ($)	1	540,895	–	536,386
Total cost ($)	2	–	60,667	59,380
% Saving	1	0	–	0.83
% Saving	2	–	0	2.11

Table 4.35 Power sharing (MW) of Example 1

	Unit number							
HR	2	3	4	6	7	8	9	10
1	400.00	0.00	0.00	185.04	0.00	350.26	0.00	89.70
2	395.36	0.00	0.00	181.09	0.00	338.36	0.00	85.19
3	355.38	0.00	0.00	168.67	0.00	300.95	0.00	75.00
4	333.13	0.00	0.00	161.75	0.00	280.12	0.00	75.00
5	400.00	0.00	0.00	185.04	0.00	350.26	0.00	89.70
6	400.00	0.00	295.68	200.00	0.00	375.00	0.00	129.32
7	400.00	383.56	420.00	200.00	0.00	375.00	0.00	191.44
8	400.00	295.59	396.65	200.00	0.00	375.00	569.93	162.83
9	400.00	468.07	420.00	200.00	0.00	375.00	768.01	218.92
10	400.00	444.60	420.00	200.00	358.05	375.00	741.06	211.29
11	400.00	486.30	420.00	200.00	404.89	375.00	788.95	224.86
12	400.00	514.11	420.00	200.00	436.09	375.00	820.89	233.91
13	400.00	479.35	420.00	200.00	397.09	375.00	780.96	222.60
14	400.00	388.98	420.00	200.00	295.64	375.00	677.18	193.20
15	400.00	310.07	410.84	200.00	250.00	375.00	586.56	167.54
16	400.00	266.64	368.27	200.00	250.00	375.00	536.68	153.41
17	400.00	317.31	417.93	200.00	250.00	375.00	594.87	169.89
18	400.00	458.51	420.00	200.00	373.65	375.00	757.03	215.81
19	400.00	486.30	420.00	200.00	404.89	375.00	788.95	224.86
20	400.00	375.08	420.00	200.00	280.03	375.00	661.21	188.68
21	400.00	215.96	318.62	200.00	0.00	375.00	478.49	136.93
22	400.00	217.46	320.12	200.00	0.00	375.00	0.00	137.42
23	400.00	165.00	246.88	0.00	0.00	375.00	0.00	113.12
24	396.36	165.00	163.80	0.00	0.00	339.29	0.00	85.55

number of iterations is less for Example 1, whereas for Examples 2 and 3 the ATSA converges faster.

Table 4.34 also shows a comparison of the ST algorithm results with the results of the LR and IP for Examples 1 and 2. It is obvious that significant cost savings are achieved.

Detailed results for Example 1 are given in Tables 4.35 and 4.36. Table 4.35 shows the load sharing among the committed units in the 24 h. Table 4.36 gives the hourly load demand, and the corresponding economic dispatch costs, startup costs, and total operating cost.

Detailed results for Example 3 are given in Tables 4.37, 4.38, and 4.39.

Table 4.36 Load demand and hourly costs ($) of Example 1

HR	Load	ED-cost	ST-cost	T-cost
1	1.03E+03	9.67E+03	0.00E+00	9.67E+03
2	1.00E+03	9.45E+03	0.00E+00	9.45E+03
3	9.00E+02	8.56E+03	0.00E+00	8.56E+03
4	8.50E+02	8.12E+03	0.00E+00	8.12E+03
5	1.03E+03	9.67E+03	0.00E+00	9.67E+03
6	1.40E+03	1.34E+04	1.06E+03	1.45E+04
7	1.97E+03	1.94E+04	1.63E+03	2.10E+04
8	2.40E+03	2.38E+04	1.82E+03	2.56E+04
9	2.85E+03	2.83E+04	0.00E+00	2.83E+04
10	3.15E+03	3.17E+04	2.06E+03	3.38E+04
11	3.30E+03	3.32E+04	0.00E+00	3.32E+04
12	3.40E+03	3.42E+04	0.00E+00	3.42E+04
13	3.28E+03	3.30E+04	0.00E+00	3.30E+04
14	2.95E+03	2.97E+04	0.00E+00	2.97E+04
15	2.70E+03	2.73E+04	0.00E+00	2.73E+04
16	2.55E+03	2.58E+04	0.00E+00	2.58E+04
17	2.73E+03	2.75E+04	0.00E+00	2.75E+04
18	3.20E+03	3.22E+04	0.00E+00	3.22E+04
19	3.30E+03	3.32E+04	0.00E+00	3.32E+04
20	2.90E+03	2.92E+04	0.00E+00	2.92E+04
21	2.13E+03	2.12E+04	0.00E+00	2.12E+04
22	1.65E+03	1.63E+04	0.00E+00	1.63E+04
23	1.30E+03	1.31E+04	0.00E+00	1.31E+04
24	1.15E+03	1.18E+04	0.00E+00	1.18E+04

Total operating cost = $536,386

Table 4.37 Power sharing (MW) of Example 3 (for units 1–13)

	Unit number												
HR	1	2	3	4	5	6	7	8	9	10	11	12	13
1	2.40	2.40	0.00	0.00	0.00	0.00	0.00	0.00	0.00	0.00	45.85	15.20	15.20
2	2.40	2.40	0.00	0.00	0.00	0.00	0.00	0.00	0.00	0.00	0.00	41.05	15.20
3	2.40	2.40	0.00	0.00	0.00	0.00	0.00	0.00	0.00	0.00	0.00	30.00	15.20
4	2.40	2.40	0.00	0.00	0.00	0.00	0.00	0.00	0.00	0.00	0.00	15.20	15.20
5	2.40	2.40	0.00	0.00	0.00	0.00	0.00	0.00	0.00	0.00	0.00	60.00	15.20
6	2.40	2.40	0.00	0.00	0.00	0.00	0.00	0.00	0.00	76.00	76.00	68.00	15.20
7	2.40	2.40	0.00	0.00	0.00	0.00	0.00	0.00	0.00	76.00	76.00	76.00	76.00
8	2.40	2.40	0.00	0.00	0.00	0.00	0.00	0.00	0.00	76.00	76.00	76.00	76.00
9	2.40	2.40	0.00	0.00	0.00	0.00	0.00	0.00	0.00	76.00	76.00	76.00	76.00
10	2.40	2.40	0.00	0.00	0.00	0.00	0.00	0.00	0.00	76.00	76.00	76.00	76.00
11	0.00	0.00	0.00	0.00	0.00	0.00	0.00	0.00	0.00	76.00	76.00	76.00	76.00
12	0.00	0.00	0.00	0.00	0.00	0.00	0.00	0.00	0.00	76.00	76.00	76.00	76.00
13	0.00	0.00	0.00	0.00	0.00	0.00	0.00	0.00	0.00	76.00	76.00	76.00	76.00
14	0.00	0.00	0.00	0.00	0.00	0.00	0.00	0.00	0.00	76.00	76.00	76.00	76.00
15	0.00	0.00	0.00	0.00	0.00	0.00	0.00	0.00	0.00	76.00	76.00	76.00	76.00
16	0.00	0.00	0.00	0.00	0.00	0.00	0.00	0.00	0.00	76.00	76.00	76.00	76.00

(continued)

Table 4.37 (continued)

	Unit number												
HR	1	2	3	4	5	6	7	8	9	10	11	12	13
17	0.00	0.00	0.00	0.00	0.00	0.00	0.00	0.00	0.00	76.00	76.00	76.00	76.00
18	0.00	0.00	0.00	0.00	0.00	0.00	0.00	0.00	0.00	76.00	76.00	76.00	76.00
19	0.00	0.00	0.00	0.00	0.00	0.00	0.00	0.00	0.00	76.00	76.00	76.00	76.00
20	0.00	0.00	0.00	0.00	0.00	0.00	0.00	0.00	0.00	76.00	76.00	76.00	76.00
21	0.00	0.00	0.00	0.00	0.00	0.00	0.00	0.00	0.00	76.00	76.00	76.00	76.00
22	0.00	0.00	0.00	0.00	0.00	0.00	0.00	0.00	0.00	76.00	76.00	76.00	76.00
23	0.00	0.00	0.00	0.00	0.00	0.00	0.00	0.00	0.00	76.00	76.00	76.00	64.10
24	0.00	0.00	0.00	0.00	0.00	0.00	0.00	0.00	0.00	15.20	15.20	15.20	15.20

4.19 Hybrid of Genetic Algorithms and Tabu Search

4.19.1 The Proposed Genetic Tabu (GT) Algorithm

In this section we propose a new algorithm (GT) based on integrating the use of GA and TS methods to solve the UCP. The proposed algorithm is mainly based on the GA approach. TS is incorporated in the reproduction phase of the GA to generate a number of new solutions (chromosomes). These new solutions are generated as neighbors to randomly selected solutions (chromosomes) in the current population and are added to the new population of the GA. The idea behind using TS is to ensure generating new solutions and hence to prevent the search from being trapped in a local minimum.

The major steps of the proposed GT algorithm are summarized as follows.

- Create an initial population by randomly generating a set of feasible solutions (chromosomes) using rules described earlier.
- Evaluate the population and check the convergence.
- Generate a new population from the current population by applying the GA operators.
- Use TS to create new solutions (chromosomes) and add them to the new GA population.

The details of the GT algorithm are also shown in the flowchart of Fig. 4.11.

The following two sections summarize the implementations of different components of the GT algorithm.

4.19.2 Genetic Algorithm as a Part of the GT Algorithm

GA is the basic part of the GT algorithm. GA implementation is similar to that described earlier. The implementation of the GA can be summarized as follows.

Table 4.38 Power sharing (MW) of Example 3 (for units 14–26)

HR	Unit number												
	14	15	16	17	18	19	20	21	22	23	24	25	26
1	0.00	0.00	0.00	155.00	155.00	155.00	155.00	68.95	0.00	0.00	350.00	350.00	350.00
2	0.00	0.00	0.00	155.00	155.00	155.00	155.00	68.95	0.00	0.00	350.00	350.00	350.00
3	0.00	0.00	0.00	155.00	155.00	155.00	155.00	0.00	0.00	0.00	350.00	350.00	350.00
4	0.00	0.00	0.00	155.00	155.00	155.00	155.00	0.00	0.00	0.00	344.80	350.00	350.00
5	0.00	0.00	0.00	155.00	155.00	155.00	155.00	0.00	0.00	0.00	350.00	350.00	350.00
6	0.00	0.00	0.00	155.00	155.00	155.00	155.00	0.00	0.00	0.00	350.00	350.00	350.00
7	0.00	0.00	71.20	155.00	155.00	155.00	155.00	0.00	0.00	0.00	350.00	350.00	350.00
8	0.00	100.00	100.00	155.00	155.00	155.00	155.00	152.25	68.95	0.00	350.00	350.00	350.00
9	100.00	100.00	100.00	155.00	155.00	155.00	155.00	197.00	124.20	0.00	350.00	350.00	350.00
10	100.00	100.00	100.00	155.00	155.00	155.00	155.00	197.00	124.20	0.00	350.00	350.00	350.00
11	100.00	100.00	100.00	155.00	155.00	155.00	155.00	197.00	149.00	0.00	350.00	350.00	350.00
12	100.00	100.00	100.00	155.00	155.00	155.00	155.00	197.00	109.00	0.00	350.00	350.00	350.00
13	100.00	100.00	100.00	155.00	155.00	155.00	155.00	197.00	119.00	0.00	350.00	350.00	350.00
14	100.00	100.00	100.00	155.00	155.00	155.00	155.00	197.00	99.00	0.00	350.00	350.00	350.00
15	100.00	100.00	100.00	155.00	155.00	155.00	155.00	157.05	68.95	0.00	350.00	350.00	350.00
16	100.00	100.00	38.10	155.00	155.00	155.00	155.00	68.95	68.95	0.00	350.00	350.00	350.00
17	100.00	100.00	78.10	155.00	155.00	155.00	155.00	68.95	68.95	0.00	350.00	350.00	350.00
18	100.00	100.00	100.00	155.00	155.00	155.00	155.00	137.05	68.95	0.00	350.00	350.00	350.00
19	100.00	100.00	100.00	155.00	155.00	155.00	155.00	197.00	109.00	0.00	350.00	350.00	350.00
20	100.00	100.00	100.00	155.00	155.00	155.00	155.00	197.00	149.00	0.00	350.00	350.00	350.00
21	100.00	100.00	100.00	155.00	155.00	155.00	155.00	197.00	129.00	0.00	350.00	350.00	350.00
22	100.00	100.00	100.00	155.00	155.00	155.00	155.00	137.05	68.95	0.00	350.00	350.00	350.00
23	0.00	25.00	25.00	155.00	155.00	155.00	155.00	68.95	68.95	0.00	350.00	350.00	350.00
24	0.00	25.00	25.00	155.00	155.00	155.00	155.00	68.95	68.95	0.00	331.30	350.00	350.00

4.19 Hybrid of Genetic Algorithms and Tabu Search

Table 4.39 Load demand and hourly costs ($) of Example 3

HR	Load	ED-cost	ST-cost	T-cost
1	1.82E+03	1.87E+04	0.00E+00	1.87E+04
2	1.80E+03	1.84E+04	0.00E+00	1.84E+04
3	1.72E+03	1.64E+04	0.00E+00	1.64E+04
4	1.70E+03	1.61E+04	0.00E+00	1.61E+04
5	1.75E+03	1.68E+04	0.00E+00	1.68E+04
6	1.91E+03	1.91E+04	1.60E+02	1.93E+04
7	2.05E+03	2.15E+04	1.00E+02	2.16E+04
8	2.40E+03	2.97E+04	7.00E+02	3.04E+04
9	2.60E+03	3.40E+04	1.00E+02	3.41E+04
10	2.60E+03	3.40E+04	0.00E+00	3.40E+04
11	2.62E+03	3.44E+04	0.00E+00	3.44E+04
12	2.58E+03	3.35E+04	0.00E+00	3.35E+04
13	2.59E+03	3.37E+04	0.00E+00	3.37E+04
14	2.57E+03	3.33E+04	0.00E+00	3.33E+04
15	2.50E+03	3.17E+04	0.00E+00	3.17E+04
16	2.35E+03	2.85E+04	0.00E+00	2.85E+04
17	2.39E+03	2.92E+04	0.00E+00	2.92E+04
18	2.48E+03	3.12E+04	0.00E+00	3.12E+04
19	2.58E+03	3.35E+04	0.00E+00	3.35E+04
20	2.62E+03	3.44E+04	0.00E+00	3.44E+04
21	2.60E+03	3.40E+04	0.00E+00	3.40E+04
22	2.48E+03	3.12E+04	0.00E+00	3.12E+04
23	2.15E+03	2.47E+04	0.00E+00	2.47E+04
24	1.90E+03	2.14E+04	0.00E+00	2.14E+04

Total operating cost = $660,596.75

- Creating an initial population by randomly generating a set of feasible solutions (chromosomes) using rules described earlier in the previous sections.
- The solution is coded as a mix of binary and decimal numbers.
- The fitness function is constructed from the objective function only without penalty terms.
- Reproduction operators of the GA, crossover, and mutation described earlier are used.

4.19.3 Tabu Search as a Part of the GT Algorithm

In the implementation of the GT algorithm, the TS is incorporated in the reproduction phase of the GA as a tool for escaping the local minimum and the premature convergence of the GA. TS is used to generate a prespecified number of new solutions that have not been generated before (typically 5–10% of the population size). The tabu list in the initial population is initially empty. It is then updated to

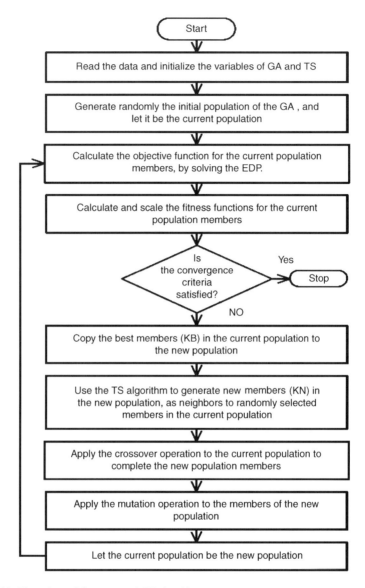

Fig. 4.11 Flow chart of the proposed GT algorithm

accept or reject the new solutions in each generation of the GA. The TS is implemented as a short-term memory algorithm. The flowchart of Fig. 4.12 shows the main steps of the TS algorithm implementation that have been used as a part of the proposed GT algorithm.

4.20 Numerical Results of the GT Algorithm

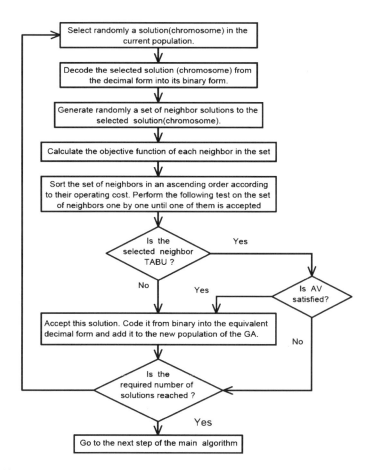

Fig. 4.12 Tabu search as a part of the GT algorithm

4.20 Numerical Results of the GT Algorithm

In order to test the proposed GT algorithm, the same three examples are considered.

A number of tests on the performance of the proposed algorithm, have been carried out to determine the most suitable GA and TS parameter settings. The following control parameters have been chosen after running a number of simulations: population size $= 50$, crossover rate $= 0.8$, mutation rate $= 0.3$, best solution copies $= 2$, maximum number of generations $= 1,000$, and tabu list size $= 7$.

Different experiments with different random number seeds were carried out to investigate the performance of the proposed algorithm. It was found that the proposed algorithm performs better than the TS algorithm and the simple GA, in

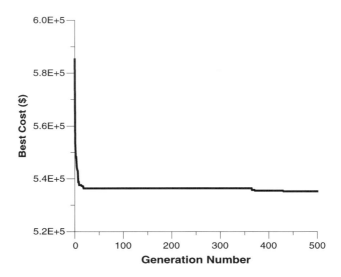

Fig. 4.13 Convergence of the proposed GT algorithm (Example 1)

Table 4.40 Comparison among the STSA, GA, and GT algorithms

	Example	STSA	GA	GT
Total cost ($)	1	538,390	537,372	535,234
Total cost ($)	2	59,512	59,491	59,380
Total cost ($)	3	662,583	661,439	660,412
Generations/iterations no.	1	1,924	411	434
Generations/iterations no.	2	616	393	513
Generations/iterations no.	3	3,900	985	623

terms of both solution quality and number of iterations. Figure 4.13 shows the convergence process of the proposed algorithm when applied to solve Example 1.

Table 4.40 shows the comparison with the results of the TS and the GA for the three examples. It is obvious that a substantial reduction in the objective function, compared to the simple GA, has been achieved, whereas the GT algorithm converges more slowly. This improvement in the objective function is due to the role of the TS in generating new and good members in each new population of the GA.

Table 4.41 presents the comparison with the results obtained in the literature (LR and IP) for Examples 1 and 2. In addition to the high percentage saving in the cost over these classical methods, the proposed algorithm has many other advantages. For instance, it may produce various solutions with the same objective function. This gives the operator the flexibility to select any of them. Also, any additional operating constraints could be easily handled without reformulating the problem.

Tables 4.42, 4.43, and 4.44 show the detailed results for Example 1 [11]. Table 4.42 shows the load sharing among the committed units in 24 h. Table 4.43 presents the final schedule of the 24 h given in Table 4.42 in the form of its

4.20 Numerical Results of the GT Algorithm

Table 4.41 Comparison among the LR, IP, and the GT algorithms

	Example	LR [11]	IP [6]	GT
Total cost ($)	1	540,895	–	535,234
Total cost ($)	2	–	60,667	59,380
% Saving	1	0	–	1.05
% Saving	2	–	0	2.12

Table 4.42 Power sharing (MW) of Example 1

	Unit number[a]							
HR	2	3	4	6	7	8	9	10
1	400.0	0.0	0.0	185.0	0.0	350.3	0.0	89.7
2	395.4	0.0	0.0	181.1	0.0	338.4	0.0	85.2
3	355.4	0.0	0.0	168.7	0.0	301.0	0.0	75.0
4	333.1	0.0	0.0	161.8	0.0	280.1	0.0	75.0
5	400.0	0.0	0.0	185.0	0.0	350.3	0.0	89.7
6	400.0	0.0	295.7	200.0	0.0	375.0	0.0	129.3
7	400.0	383.6	420.0	200.0	0.0	375.0	0.0	191.4
8	400.0	295.6	396.7	200.0	0.0	375.0	569.9	162.8
9	400.0	468.1	420.0	200.0	0.0	375.0	768.0	218.9
10	400.0	444.6	420.0	200.0	358.1	375.0	741.1	211.3
11	400.0	486.3	420.0	200.0	404.9	375.0	789.0	224.9
12	400.0	514.1	420.0	200.0	436.1	375.0	820.9	233.9
13	400.0	479.4	420.0	200.0	397.1	375.0	781.0	222.6
14	400.0	389.0	420.0	200.0	295.6	375.0	677.2	193.2
15	400.0	310.1	410.8	200.0	250.0	375.0	586.6	167.5
16	400.0	266.6	368.3	200.0	250.0	375.0	536.7	153.4
17	400.0	317.3	417.9	200.0	250.0	375.0	594.9	169.9
18	400.0	458.5	420.0	200.0	373.7	375.0	757.0	215.8
19	400.0	486.3	420.0	200.0	404.9	375.0	789.0	224.9
20	400.0	0.0	420.0	200.0	442.2	375.0	827.2	235.7
21	400.0	0.0	404.9	200.0	0.0	375.0	579.6	165.6
22	400.0	0.0	216.5	196.7	0.0	375.0	358.8	103.0
23	0.0	0.0	235.2	200.0	0.0	375.0	380.6	109.2
24	0.0	0.0	186.4	188.0	0.0	359.1	323.5	93.0

[a] Units 1,5 are OFF at all hours

Table 4.43 GA population of the best solution for Example 1

Unit number				
1,6	2,7	3,8	4,9	5,10
0	4194303	524224	16777184	0
16777215	1048064	16777215	16777088	16777215

equivalent decimal numbers. Table 4.44 gives the hourly load demand and the corresponding economic dispatch, startup, and total operating costs of the final schedule.

Tables 4.45, 4.46, and 4.47 show the detailed results for Example 3 [12].

Table 4.44 Load demand and hourly costs ($) of Example 1

HR	Load	ED-cost	ST-cost	T-cost
1	1,025	9,670.04	0.00	9,670.04
2	1,000	9,446.62	0.00	9,446.62
3	900	8,560.91	0.00	8,560.91
4	850	8,123.13	0.00	8,123.13
5	1,025	9,670.04	0.00	9,670.04
6	1,400	13,434.10	1,055.97	14,490.00
7	1,970	19,385.10	1,631.43	21,016.50
8	2,400	23,815.50	1,817.70	25,633.20
9	2,850	28,253.90	0.00	28,253.90
10	3,150	31,701.70	2,057.58	33,759.30
11	3,300	33,219.80	0.00	33,219.80
12	3,400	34,242.00	0.00	34,242.00
13	3,275	32,965.50	0.00	32,965.50
14	2,950	29,706.30	0.00	29,706.30
15	2,700	27,259.70	0.00	27,259.70
16	2,550	25,819.80	0.00	25,819.80
17	2,725	27,501.60	0.00	27,501.60
18	3,200	32,205.70	0.00	32,205.70
19	3,300	33,219.80	0.00	33,219.80
20	2,900	28,899.00	0.00	28,899.00
21	2,125	20,698.40	0.00	20,698.40
22	1,650	16,251.80	0.00	16,251.80
23	1,300	12,989.60	0.00	12,989.60
24	1,150	11,631.00	0.00	11,631.00

Total operating cost = $535,234

Table 4.45 Power sharing (MW) of Example 3 (for units 1–13)

	Unit number												
HR	1	2	3	4	5	6	7	8	9	10	11	12	13
1	0.00	2.40	0.00	0.00	0.00	0.00	4.00	0.00	0.00	0.00	76.00	52.40	15.20
2	2.40	2.40	0.00	0.00	0.00	0.00	4.00	0.00	0.00	0.00	0.00	76.00	45.20
3	2.40	0.00	0.00	0.00	0.00	0.00	4.00	0.00	0.00	0.00	0.00	28.40	15.20
4	2.40	0.00	0.00	0.00	0.00	0.00	4.00	0.00	0.00	0.00	0.00	15.20	15.20
5	2.40	0.00	0.00	0.00	0.00	0.00	4.00	0.00	0.00	0.00	0.00	58.40	15.20
6	2.40	0.00	0.00	0.00	0.00	0.00	4.00	0.00	0.00	76.00	76.00	66.40	15.20
7	0.00	0.00	0.00	0.00	0.00	0.00	4.00	0.00	0.00	76.00	76.00	76.00	76.00
8	0.00	0.00	0.00	0.00	0.00	0.00	4.00	0.00	0.00	76.00	76.00	76.00	76.00
9	0.00	0.00	0.00	0.00	0.00	0.00	4.00	0.00	0.00	76.00	76.00	76.00	76.00
10	0.00	0.00	0.00	0.00	0.00	0.00	4.00	0.00	0.00	76.00	76.00	76.00	76.00
11	0.00	0.00	0.00	0.00	0.00	0.00	4.00	0.00	0.00	76.00	76.00	76.00	76.00
12	0.00	0.00	0.00	0.00	0.00	0.00	0.00	0.00	0.00	76.00	76.00	76.00	76.00
13	0.00	0.00	0.00	0.00	0.00	0.00	0.00	0.00	0.00	76.00	76.00	76.00	76.00
14	0.00	0.00	0.00	0.00	0.00	0.00	0.00	0.00	0.00	76.00	76.00	76.00	76.00
15	0.00	0.00	0.00	0.00	0.00	0.00	0.00	0.00	0.00	76.00	76.00	76.00	76.00

(continued)

Table 4.45 (continued)

	Unit number												
HR	1	2	3	4	5	6	7	8	9	10	11	12	13
16	0.00	0.00	0.00	0.00	0.00	0.00	0.00	0.00	0.00	76.00	76.00	76.00	76.00
17	0.00	0.00	0.00	0.00	0.00	0.00	0.00	0.00	0.00	76.00	76.00	76.00	76.00
18	0.00	0.00	0.00	0.00	0.00	0.00	0.00	0.00	0.00	76.00	76.00	76.00	76.00
19	0.00	0.00	0.00	0.00	0.00	0.00	0.00	0.00	0.00	76.00	76.00	76.00	76.00
20	0.00	0.00	0.00	0.00	0.00	0.00	0.00	0.00	0.00	76.00	76.00	76.00	76.00
21	0.00	0.00	0.00	0.00	0.00	0.00	0.00	0.00	0.00	76.00	76.00	76.00	76.00
22	0.00	0.00	0.00	0.00	0.00	0.00	0.00	0.00	0.00	76.00	76.00	76.00	76.00
23	0.00	0.00	0.00	0.00	0.00	0.00	0.00	0.00	0.00	76.00	76.00	76.00	64.10
24	0.00	0.00	0.00	0.00	0.00	0.00	0.00	0.00	0.00	15.20	15.20	15.20	15.20

4.21 Hybrid of Genetic Algorithms, Simulated Annealing, and Tabu Search

This section presents a new algorithm (GST) based on integrating the GA, TS, and SA methods to solve the UCP. The proposed GST algorithm could be considered as a further improvement to the GT algorithm implemented earlier.

The core of the proposed algorithm is based on the GA. The TS is used to generate new population members in the reproduction phase of the GA. Moreover, the SA method is adopted to improve the convergence of the GA by testing the population members of the GA after each generation. The SA test allows the acceptance of any solution at the beginning of the search, and only good solutions will have a higher probability of acceptance as the generation number increases. The effect of using the SA is to accelerate the convergence of the GA and also increase the fine-tuning capability of the GA when approaching a local minimum.

The major steps of the algorithm are summarized as follows [47].

- Create an initial population by randomly generating a set of feasible solutions, and initialize the current solution of the SA algorithm.
- Apply GA operators to generate new population members.
- Use the TS algorithm to generate some members in the new population (typically 5–10% of the population size), as neighbors to the randomly selected solutions in the current population.
- Apply the SA algorithm to test all the members of the new population.

Figure 4.14 shows the flowchart of the proposed GTS algorithm. The following sections summarize the implementations of different components of the proposed GTS algorithm.

Table 4.46 Power sharing (MW) of Example 3 (for units 14–26)

HR	Unit number												
	14	15	16	17	18	19	20	21	22	23	24	25	26
1	0.00	0.00	0.00	155.00	155.00	155.00	155.00	0.00	0.00	0.00	350.00	350.00	350.00
2	0.00	0.00	0.00	155.00	155.00	155.00	155.00	0.00	0.00	0.00	350.00	350.00	350.00
3	0.00	0.00	0.00	155.00	155.00	155.00	155.00	0.00	0.00	0.00	350.00	350.00	350.00
4	0.00	0.00	0.00	155.00	155.00	155.00	155.00	0.00	0.00	0.00	343.20	350.00	350.00
5	0.00	0.00	0.00	155.00	155.00	155.00	155.00	0.00	0.00	0.00	350.00	350.00	350.00
6	0.00	0.00	0.00	155.00	155.00	155.00	155.00	0.00	0.00	0.00	350.00	350.00	350.00
7	0.00	0.00	72.00	155.00	155.00	155.00	155.00	0.00	0.00	0.00	350.00	350.00	350.00
8	0.00	100.00	100.00	155.00	155.00	155.00	155.00	153.05	68.95	0.00	350.00	350.00	350.00
9	100.00	100.00	100.00	155.00	155.00	155.00	155.00	197.00	125.00	0.00	350.00	350.00	350.00
10	100.00	100.00	100.00	155.00	155.00	155.00	155.00	197.00	125.00	0.00	350.00	350.00	350.00
11	100.00	100.00	100.00	155.00	155.00	155.00	155.00	197.00	145.00	0.00	350.00	350.00	350.00
12	100.00	100.00	100.00	155.00	155.00	155.00	155.00	197.00	109.00	0.00	350.00	350.00	350.00
13	100.00	100.00	100.00	155.00	155.00	155.00	155.00	197.00	119.00	0.00	350.00	350.00	350.00
14	100.00	100.00	100.00	155.00	155.00	155.00	155.00	197.00	99.00	0.00	350.00	350.00	350.00
15	100.00	100.00	100.00	155.00	155.00	155.00	155.00	157.05	68.95	0.00	350.00	350.00	350.00
16	100.00	100.00	38.10	155.00	155.00	155.00	155.00	68.95	68.95	0.00	350.00	350.00	350.00
17	100.00	100.00	78.10	155.00	155.00	155.00	155.00	68.95	68.95	0.00	350.00	350.00	350.00
18	100.00	100.00	100.00	155.00	155.00	155.00	155.00	137.05	68.95	0.00	350.00	350.00	350.00
19	100.00	100.00	100.00	155.00	155.00	155.00	155.00	197.00	109.00	0.00	350.00	350.00	350.00
20	100.00	100.00	100.00	155.00	155.00	155.00	155.00	197.00	149.00	0.00	350.00	350.00	350.00
21	100.00	100.00	100.00	155.00	155.00	155.00	155.00	197.00	129.00	0.00	350.00	350.00	350.00
22	100.00	100.00	100.00	155.00	155.00	155.00	155.00	137.05	68.95	0.00	350.00	350.00	350.00
23	0.00	25.00	25.00	155.00	155.00	155.00	155.00	68.95	68.95	0.00	350.00	350.00	350.00
24	0.00	25.00	25.00	155.00	155.00	155.00	155.00	68.95	68.95	0.00	331.30	350.00	350.00

Table 4.47 Load demand and hourly costs ($) of Example 3

HR	Load	ED-cost	ST-cost	T-cost
1	1.82E+03	1.80E+04	0.00E+00	1.80E+04
2	1.80E+03	1.77E+04	0.00E+00	1.77E+04
3	1.72E+03	1.66E+04	0.00E+00	1.66E+04
4	1.70E+03	1.63E+04	0.00E+00	1.63E+04
5	1.75E+03	1.70E+04	0.00E+00	1.70E+04
6	1.91E+03	1.93E+04	1.60E+02	1.94E+04
7	2.05E+03	2.16E+04	1.00E+02	2.17E+04
8	2.40E+03	2.98E+04	7.00E+02	3.05E+04
9	2.60E+03	3.41E+04	1.00E+02	3.42E+04
10	2.60E+03	3.41E+04	0.00E+00	3.41E+04
11	2.62E+03	3.46E+04	0.00E+00	3.46E+04
12	2.58E+03	3.35E+04	0.00E+00	3.35E+04
13	2.59E+03	3.37E+04	0.00E+00	3.37E+04
14	2.57E+03	3.33E+04	0.00E+00	3.33E+04
15	2.50E+03	3.17E+04	0.00E+00	3.17E+04
16	2.35E+03	2.85E+04	0.00E+00	2.85E+04
17	2.39E+03	2.92E+04	0.00E+00	2.92E+04
18	2.48E+03	3.12E+04	0.00E+00	3.12E+04
19	2.58E+03	3.35E+04	0.00E+00	3.35E+04
20	2.62E+03	3.44E+04	0.00E+00	3.44E+04
21	2.60E+03	3.40E+04	0.00E+00	3.40E+04
22	2.48E+03	3.12E+04	0.00E+00	3.12E+04
23	2.15E+03	2.47E+04	0.00E+00	2.47E+04
24	1.90E+03	2.14E+04	0.00E+00	2.14E+04

Total operating cost = $660,412.4375

4.21.1 Genetic Algorithm as a Part of the GST Algorithm

The implementations of the GA in the GST algorithm are exactly the same as those described earlier.

4.21.2 Tabu Search Part of the GST Algorithm

The implementations of the TS part in the GST are also the same as those described earlier.

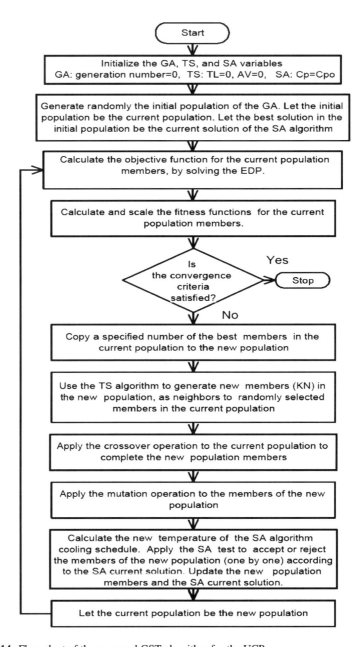

Fig. 4.14 Flow chart of the proposed GST algorithm for the UCP

4.21.3 Simulated Annealing as a Part of the GST Algorithm

After creating a new population of the GA, the SA test is applied to the members of this population, one by one. The steps of the SA algorithm as applied at the *k*th generation of the proposed GST algorithm are described as follows.

Let $\mathbf{U}_c, \mathbf{F}_c$ be the SA current solution and the corresponding operating cost, respectively.

Let $\mathbf{U}_j, \mathbf{F}_j$ be the *j*th solution (chromosome) in a given population and the corresponding operating cost, respectively.

Step (1): Calculate the new temperature $\mathbf{Cp}^k = \mathbf{Cp}^o(\beta)^k$, where $0 < \beta < 1$.

Step (2): At the same calculated temperature $c_p{}^k$ apply the following acceptance test for the population members of the GA one by one.

Step (3): *Acceptance test*: If $\mathbf{F}_j \leq \mathbf{F}_c$, or $\exp[(\mathbf{F}_c - \mathbf{F}_j)/\mathbf{Cp}] \geq \mathbf{U}(0,1)$, then accept the population member, and update the current solution; set $\mathbf{U}_c = \mathbf{U}_j$ and $\mathbf{F}_c = \mathbf{F}_j$; then go to Step (4). Otherwise, reject the population member and set $\mathbf{U}_j = \mathbf{U}_c$ and $\mathbf{F}_j = \mathbf{F}_c$; then go to Step (4).

Step (4): If all the population members are tested go to the next step in the main algorithm; otherwise go to Step (2).

4.22 Numerical Results of the GST Algorithm

For the purpose of testing the hybrid GST algorithm, the same three examples from the literature [6, 11, 12] are considered. These control parameters have been chosen after running a number of simulations: population size = 50, crossover rate = 0.8, mutation rate = 0.3, elite copies = 2, maximum number of generations = 1,000, tabu size = 7, and initial control parameter (temperature) = 5,000, $\beta = 0.9$.

Figure 4.15 shows the convergence speed of the GST algorithm when Example 1 is solved.

Different experiments were carried out to evaluate the results obtained by the proposed GST algorithm and those obtained from the individual algorithms. Table 4.48 shows the results of this comparison for the three examples. The superiority of the GST algorithm is obvious: it performs better than each of the individual algorithms in terms of both solution quality and number of generations.

Table 4.49 presents the comparison of results obtained in the literature (LR and IP) for Examples 1 and 2, and the proposed GST algorithm.

Tables 4.50, 4.51, and 4.52 show detailed results for Example 1 [11]. Table 4.50 shows the load sharing among the committed units in the 24 h. Table 4.50 presents the final schedule of the 24-h period, given in Table 4.51, in the form of its equivalent decimal numbers. Table 4.52 gives the hourly load demand and the corresponding economic dispatch, startup, and total operating costs. Tables 4.53, 4.54, and 4.55 show detailed results for Example 3 [12].

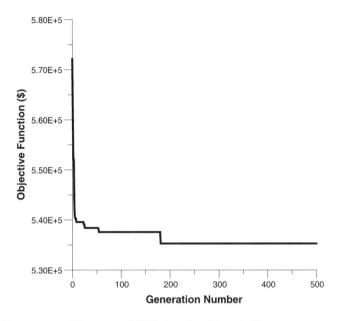

Fig. 4.15 Convergence of the proposed GST algorithm (Example 1)

Table 4.48 Comparison among the SAA, STSA, GA, and GST algorithms

	Example	SA	STSA	GA	GST
Total cost ($)	1	536,622	538,390	537,372	535,271
Total cost ($)	2	59,385	59,512	59,491	59,380
Total cost ($)	3	662,664	662,583	661,439	660,151
Generations/iterations no.	1	384	1,924	411	181
Generations/iterations no.	2	652	616	393	180
Generations/iterations no.	3	2,361	3,900	985	762

Table 4.49 Comparison among the LR, IP, and GST algorithms

	Example	LR	IP	GST
Total cost ($)	1	540,895	–	535,271
Total cost ($)	2	–	60,667	59,380
% Saving	1	0	–	1.04
% Saving	2	–	0	2.12

4.22 Numerical Results of the GST Algorithm

Table 4.50 Power sharing (MW) of Example 1

HR	Unit number[a]							
	2	3	4	6	7	8	9	10
1	400	0	0	185.04	0	350.26	0	89.7
2	395.36	0	0	181.09	0	338.36	0	85.19
3	355.38	0	0	168.67	0	300.95	0	75
4	333.13	0	0	161.75	0	280.12	0	75
5	400	0	0	185.04	0	350.26	0	89.7
6	400	0	295.68	200	0	375	0	129.32
7	400	383.56	420	200	0	375	0	191.44
8	400	295.59	396.65	200	0	375	569.93	162.83
9	400	468.07	420	200	0	375	768.01	218.92
10	400	444.6	420	200	358.05	375	741.06	211.29
11	400	486.3	420	200	404.89	375	788.95	224.86
12	400	514.11	420	200	436.09	375	820.89	233.91
13	400	479.35	420	200	397.09	375	780.96	222.6
14	400	388.98	420	200	295.63	375	677.18	193.2
15	400	310.07	410.84	200	250	375	586.56	167.54
16	400	266.64	368.27	200	250	375	536.68	153.41
17	400	317.31	417.93	200	250	375	594.87	169.89
18	400	458.51	420	200	373.65	375	757.03	215.81
19	400	486.3	420	200	404.89	375	788.95	224.86
20	400	491.88	0	200	411.1	375	795.35	226.67
21	400	344.77	0	200	0	375	626.41	178.82
22	400	459.02	0	200	0	375	0	215.98
23	400	194.91	0	200	0	375	0	130.09
24	389.59	165	0	179.3	0	332.96	0	83.15

[a] Units 1,5 are OFF at all hours

Table 4.51 The best population in the GA for Example 1

Unit number				
1,6	2,7	3,8	4,9	5,10
0	16777215	16777152	16777184	0
4194303	1048064	16777215	2097024	16777215

Table 4.52 Load demand and hourly costs ($) of Example 1

HR	Load	ED-cost	ST-cost	T-cost
1	1,025	9,670.0	–	9,670.0
2	1,000	9,446.6	–	9,446.6
3	900	8,560.9	–	8,560.9
4	850	8,123.1	–	8,123.1
5	1,025	9,670.0	–	9,670.0
6	1,400	13,434.1	1,056.0	14,490.0
7	1,970	19,385.1	1,631.4	21,016.5
8	2,400	23,815.5	1,817.7	25,633.2

(continued)

Table 4.52 (continued)

HR	Load	ED-cost	ST-cost	T-cost
9	2,850	28,253.9	–	28,253.9
10	3,150	31,701.7	2,057.6	33,759.3
11	3,300	33,219.8	–	33,219.8
12	3,400	34,242.1	–	34,242.1
13	3,275	32,965.5	–	32,965.5
14	2,950	29,706.3	–	29,706.3
15	2,700	27,259.7	–	27,259.7
16	2,550	25,819.8	–	25,819.8
17	2,725	27,501.6	–	27,501.6
18	3,200	32,205.7	–	32,205.7
19	3,300	33,219.8	–	33,219.8
20	2,900	29,198.0	–	29,198.0
21	2,125	20,994.5	–	20,994.5
22	1,650	16,158.6	–	16,158.6
23	1,300	12,758.9	–	12,758.9
24	1,150	11,397.1	–	11,397.1

Total operating cost = $535,270.94

Table 4.53 Power sharing (MW) of Example 3 (for units 1–13)

	Unit number												
HR	1	2	3	4	5	6	7	8	9	10	11	12	13
1	2.40	2.40	0.00	0.00	0.00	4.00	0.00	0.00	0.00	0.00	76.00	50.00	15.20
2	2.40	2.40	0.00	0.00	0.00	4.00	0.00	0.00	0.00	0.00	0.00	76.00	45.20
3	0.00	2.40	0.00	0.00	0.00	4.00	0.00	0.00	0.00	0.00	0.00	28.40	15.20
4	0.00	2.40	0.00	0.00	0.00	4.00	0.00	0.00	0.00	0.00	0.00	15.20	15.20
5	0.00	2.40	0.00	0.00	0.00	4.00	0.00	0.00	0.00	0.00	0.00	58.40	15.20
6	0.00	2.40	0.00	0.00	0.00	4.00	0.00	0.00	0.00	76.00	76.00	66.40	15.20
7	0.00	2.40	2.40	0.00	0.00	4.00	0.00	0.00	0.00	76.00	76.00	76.00	76.00
8	0.00	2.40	2.40	0.00	0.00	0.00	0.00	0.00	0.00	76.00	76.00	76.00	76.00
9	0.00	2.40	2.40	0.00	0.00	0.00	0.00	0.00	0.00	76.00	76.00	76.00	76.00
10	0.00	2.40	2.40	0.00	0.00	0.00	0.00	0.00	0.00	76.00	76.00	76.00	76.00
11	0.00	0.00	2.40	0.00	0.00	0.00	0.00	0.00	0.00	76.00	76.00	76.00	76.00
12	0.00	0.00	2.40	0.00	0.00	0.00	0.00	0.00	0.00	76.00	76.00	76.00	76.00
13	0.00	0.00	2.40	0.00	0.00	0.00	0.00	0.00	0.00	76.00	76.00	76.00	76.00
14	0.00	0.00	2.40	0.00	0.00	0.00	0.00	0.00	0.00	76.00	76.00	76.00	76.00
15	0.00	0.00	0.00	0.00	0.00	0.00	0.00	0.00	0.00	76.00	76.00	76.00	76.00
16	0.00	0.00	0.00	0.00	0.00	0.00	0.00	0.00	0.00	76.00	76.00	76.00	76.00
17	0.00	0.00	0.00	0.00	0.00	0.00	0.00	0.00	0.00	76.00	76.00	76.00	76.00
18	0.00	0.00	0.00	0.00	0.00	0.00	0.00	0.00	0.00	76.00	76.00	76.00	76.00
19	0.00	0.00	0.00	0.00	0.00	0.00	0.00	0.00	0.00	76.00	76.00	76.00	76.00
20	0.00	0.00	0.00	0.00	0.00	0.00	0.00	0.00	0.00	76.00	76.00	76.00	76.00
21	0.00	0.00	0.00	0.00	0.00	0.00	0.00	0.00	0.00	76.00	76.00	76.00	76.00
22	0.00	0.00	0.00	0.00	0.00	0.00	0.00	0.00	0.00	76.00	76.00	76.00	76.00
23	0.00	0.00	0.00	0.00	0.00	0.00	0.00	0.00	0.00	76.00	76.00	76.00	64.10
24	0.00	0.00	0.00	0.00	0.00	0.00	0.00	0.00	0.00	15.20	15.20	15.20	15.20

4.22 Numerical Results of the GST Algorithm

Table 4.54 Power sharing (MW) of Example 3 (for units 14–26)

HR	14	15	16	17	18	19	20	21	22	23	24	25	26
1	0.00	0.00	0.00	155.00	155.00	155.00	155.00	0.00	0.00	0.00	350.00	350.00	350.00
2	0.00	0.00	0.00	155.00	155.00	155.00	155.00	0.00	0.00	0.00	350.00	350.00	350.00
3	0.00	0.00	0.00	155.00	155.00	155.00	155.00	0.00	0.00	0.00	350.00	350.00	350.00
4	0.00	0.00	0.00	155.00	155.00	155.00	155.00	0.00	0.00	0.00	343.20	350.00	350.00
5	0.00	0.00	0.00	155.00	155.00	155.00	155.00	0.00	0.00	0.00	350.00	350.00	350.00
6	0.00	0.00	0.00	155.00	155.00	155.00	155.00	0.00	0.00	0.00	350.00	350.00	350.00
7	0.00	0.00	67.20	155.00	155.00	155.00	155.00	0.00	0.00	0.00	350.00	350.00	350.00
8	0.00	100.00	100.00	155.00	155.00	155.00	155.00	152.25	68.95	0.00	350.00	350.00	350.00
9	100.00	100.00	100.00	155.00	155.00	155.00	155.00	197.00	124.20	0.00	350.00	350.00	350.00
10	100.00	100.00	100.00	155.00	155.00	155.00	155.00	197.00	124.20	0.00	350.00	350.00	350.00
11	100.00	100.00	100.00	155.00	155.00	155.00	155.00	197.00	146.60	0.00	350.00	350.00	350.00
12	100.00	100.00	100.00	155.00	155.00	155.00	155.00	197.00	106.60	0.00	350.00	350.00	350.00
13	100.00	100.00	100.00	155.00	155.00	155.00	155.00	197.00	116.60	0.00	350.00	350.00	350.00
14	100.00	100.00	100.00	155.00	155.00	155.00	155.00	197.00	96.60	0.00	350.00	350.00	350.00
15	100.00	100.00	100.00	155.00	155.00	155.00	155.00	157.05	68.95	0.00	350.00	350.00	350.00
16	100.00	100.00	38.10	155.00	155.00	155.00	155.00	68.95	68.95	0.00	350.00	350.00	350.00
17	100.00	100.00	78.10	155.00	155.00	155.00	155.00	68.95	68.95	0.00	350.00	350.00	350.00
18	100.00	100.00	100.00	155.00	155.00	155.00	155.00	137.05	68.95	0.00	350.00	350.00	350.00
19	100.00	100.00	100.00	155.00	155.00	155.00	155.00	197.00	109.00	0.00	350.00	350.00	350.00
20	100.00	100.00	100.00	155.00	155.00	155.00	155.00	197.00	149.00	0.00	350.00	350.00	350.00
21	100.00	100.00	100.00	155.00	155.00	155.00	155.00	197.00	129.00	0.00	350.00	350.00	350.00
22	100.00	100.00	100.00	155.00	155.00	155.00	155.00	137.05	68.95	0.00	350.00	350.00	350.00
23	0.00	25.00	25.00	155.00	155.00	155.00	155.00	68.95	68.95	0.00	350.00	350.00	350.00
24	0.00	25.00	25.00	155.00	155.00	155.00	155.00	68.95	68.95	0.00	331.30	350.00	350.00

Table 4.55 Load demand and hourly costs ($) of Example 3

HR	Load	ED-cost	ST-cost	T-cost
1	1.82E+03	1.80E+04	0.00E+00	1.80E+04
2	1.80E+03	1.77E+04	0.00E+00	1.77E+04
3	1.72E+03	1.66E+04	0.00E+00	1.66E+04
4	1.70E+03	1.63E+04	0.00E+00	1.63E+04
5	1.75E+03	1.70E+04	0.00E+00	1.70E+04
6	1.91E+03	1.93E+04	1.60E+02	1.94E+04
7	2.05E+03	2.17E+04	1.00E+02	2.18E+04
8	2.40E+03	2.97E+04	7.00E+02	3.04E+04
9	2.60E+03	3.40E+04	1.00E+02	3.41E+04
10	2.60E+03	3.40E+04	0.00E+00	3.40E+04
11	2.62E+03	3.45E+04	0.00E+00	3.45E+04
12	2.58E+03	3.35E+04	0.00E+00	3.35E+04
13	2.59E+03	3.38E+04	0.00E+00	3.38E+04
14	2.57E+03	3.33E+04	0.00E+00	3.33E+04
15	2.50E+03	3.17E+04	0.00E+00	3.17E+04
16	2.35E+03	2.85E+04	0.00E+00	2.85E+04
17	2.39E+03	2.92E+04	0.00E+00	2.92E+04
18	2.48E+03	3.12E+04	0.00E+00	3.12E+04
19	2.58E+03	3.35E+04	0.00E+00	3.35E+04
20	2.62E+03	3.44E+04	0.00E+00	3.44E+04
21	2.60E+03	3.40E+04	0.00E+00	3.40E+04
22	2.48E+03	3.12E+04	0.00E+00	3.12E+04
23	2.15E+03	2.47E+04	0.00E+00	2.47E+04
24	1.90E+03	2.14E+04	0.00E+00	2.14E+04

Total operating cost = $660,151.25

4.23 Summary

In this chapter, three AI-based novel hybrid algorithms for the UCP are proposed. The hybrid algorithms integrate the use of the previously introduced algorithms: SA, TS, and GA. The ideas of the hybridization of these algorithms are original and are applied to the UCP for the first time. Various details of implementation have also been discussed.

In the first algorithm [46], the main features of the SA and the TS methods are integrated. The TS test is embedded in the SA algorithm to create a memory that prevents cycling of the solutions accepted by the SA. A significant cost saving has been achieved over the individual TS and SA methods.

The second hybrid algorithm is based on integrating the use of GA and TS methods to solve the UCP. The proposed algorithm is mainly based on the GA approach. The TS is used to induce new population members in the reproduction phase of the GA.

A third new hybrid algorithm that integrates the main features of the three algorithms, GA, TS, and SA, is also proposed. The algorithm is mainly structured around the GA, and the TS is used to generate new members in the GA population.

The SA algorithm is used to test all the GA members after each reproduction of a new population.

Among the three hybrid algorithms, it is found that the overall performance of the GT algorithm is superior. The GT algorithm converges faster and gives a better quality solution. In the next chapter, a comparison of the seven proposed algorithms and the other methods reported in the literature is detailed.

4.24 Comparisons of the Algorithms for the Unit Commitment Problem

In the previous sections, seven AI-based algorithms were proposed for solving the UCP.

The two proposed algorithms are presented and implemented. These two algorithms are

- A simulated annealing algorithm
- A genetic algorithm

The other five algorithms are original and are applied for the first time to solve the UCP. These algorithms are

- A simple tabu search algorithm
- An advanced tabu search algorithm
- A hybrid of the simulated annealing and tabu search algorithms
- A hybrid of genetic and tabu search algorithms
- A hybrid of genetic, simulated annealing, and tabu search algorithms

4.24.1 Results of Example 1

Table 4.56 and Figs. 4.16 and 4.17 show the daily operating costs and the number of iterations (or generations for the GA-based algorithms) of the seven proposed algorithms, and the LR and the SAA-67 methods as well, for Example 1.

Generally speaking, all the proposed algorithms outperform the results reported in the literature using the LR and the SAA-67 methods for the same example. The daily cost saving amounts range between 2,505 (using STSA) and 5,661 (using GT), which is equivalent to a percentage saving of 0.46–1.05, respectively.

Comparing the results of the proposed algorithms, for Example 1, it is clear that the GT algorithm is the best as far as quality of solution and convergence speed are concerned. Considering the algorithms based on a single technique (e.g., SA, STSA, ATSA, and GA), we conclude that the SA performance is the best among these algorithms as far as the objective function value and the number of iterations required for convergence are concerned. It can also be concluded that the results of

Table 4.56 Comparison of different algorithms for Example 1

Algorithm	Daily operating cost ($)	Amount of daily saving	% Saving	No. of iterations or generations
LR [11]	540,895	0	0	–
SAA [10]	538,803	2,092	0.38	821
SAA	536,622	4,273	0.79	384
STSA	538,390	2,505	0.46	1,924
ATSA	537,686	3,209	0.59	1,235
GA	537,372	3,523	0.65	411
ST	536,386	4,509	0.83	625
GT	535,234	5,661	1.05	434
GST	535,271	5,624	1.04	181

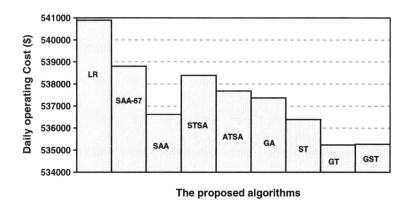

Fig. 4.16 Operating costs of different proposed algorithms for Example 1

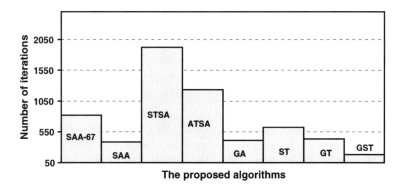

Fig. 4.17 Operating costs of different proposed algorithms for Example 2

Table 4.57 Comparison of different algorithms for Example 2

Algorithm	Daily operating cost ($)	Amount of daily saving	% Saving	No. of iterations or generations
IP [6]	60,667	0	0	–
SAA [10]	59,512	1,155	1.9	945
SAA	59,512	1,155	1.9	652
STSA	59,512	1,155	1.9	616
ATSA	59,385	1,282	2.1	138
GA	59,491	1,176	1.93	393
ST	59,385	1,282	2.11	538
GT	59,380	1,287	2.12	513
GST	59,380	1,287	2.12	180

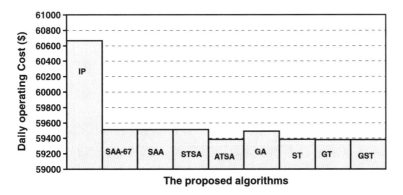

Fig. 4.18 Operating costs of different proposed algorithms for Example 2

the hybrid algorithms (e.g., ST, GT, and GST) are better than the results of the algorithms based on a single method. Among the proposed hybrid algorithms, it is also found that the GT algorithm provides a better quality of solution and also a faster speed of convergence.

4.24.2 Results of Example 2

Table 4.57 and Figs. 4.18 and 4.19 show the daily operating costs and the number of iterations (or generations of the GA-based algorithms) of the seven proposed algorithms, and the IP and SAA-67 methods as well, for Example 2.

As shown, the results of our proposed algorithms are better than the results reported in the literature using the IP method for this example. Compared with the IP results, the daily cost saving amount ranges between 1,155 (by SA and STSA) and 1,287 (by GT and GST) which is equivalent to a percentage saving of 1.9–2.12, respectively.

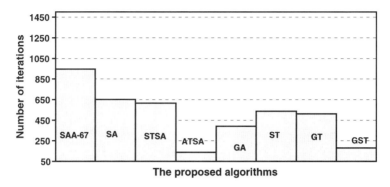

Fig. 4.19 Number of iterations of different proposed algorithms for Example 2

Among the results of all the proposed algorithms, for Example 2, it is clear that the GT and GST algorithms are the best. Considering the algorithms based on a single method (e.g., SA, STSA, ATSA, and GA), although there is not much difference among them, the ATSA gives slightly better results than the others as far as the objective function value and the number of iterations required for convergence are concerned. It can also be concluded that the results of the hybrid algorithms (e.g., ST, GT, and GST) are better than the results of the algorithms based on a single method. Among the proposed hybrid algorithms, it is also found that the GST algorithm is the best in terms of the quality of solution and the speed of convergence.

4.24.3 Results of Example 3

Table 4.12 and Figs. 4.5 and 4.6 present the daily operating costs and the number of iterations (or generations of the GA-based algorithms) of the seven proposed algorithms, and of the SAA-67 method, for Example 3.

In this example, the proposed algorithms also give better results than those of the SAA-67. The daily cost saving amount, compared to the SAA-67 results, ranges between 1,169 (by SA) and 3,682 (by GST) which is equivalent to a percentage saving of 0.17 and 0.55, respectively.

Among all the proposed algorithms, for Example 3, it is obvious that the GST algorithm provides the best quality of solution, whereas the GT algorithm converges faster. Considering the algorithms based on a single method (e.g., SA, STSA, ATSA, and GA), based on the results of this example, it is noted that the ATSA performance is the best among these algorithms as far as the objective function value is concerned, and the GA converges faster. It can also be concluded that the results of the hybrid algorithms (e.g., ST, GT, and GST are better than the results of the algorithms based on a single method. As far as the proposed hybrid algorithms are concerned, it is also found that the GST algorithm is the best in the quality of solution and the GT convergence is better (Figs. 4.20 and 4.21, Table 4.58).

4.24 Comparisons of the Algorithms for the Unit Commitment Problem

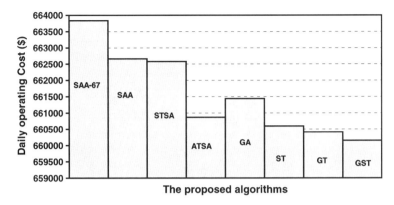

Fig. 4.20 Operating costs of different proposed algorithms for Example 3

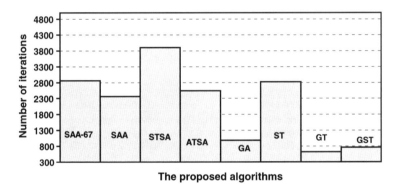

Fig. 4.21 Number of iterations of different proposed algorithms for Example 3

Table 4.58 Comparison of different algorithms for Example 3

Algorithm	Daily operating cost ($)	Amount of daily saving	% Saving	No. of iterations or generations
SAA [10]	663,833	0	0	2,864
SAA	662,664	1,169	0.17	2,361
STSA	662,583	1,250	0.19	3,900
ATSA	660,864	2,969	0.45	2,547
GA	661,439	2,394	0.36	985
ST	660,596	3,237	0.48	2,829
GT	660,412	3,421	0.51	623
GST	660,151	3,682	0.55	762

4.24.4 Summary

This chapter is intended for the comparison of the results of our proposed algorithms and those of other methods reported in the literature. Three systems extracted from the literature are considered. The effectiveness of our proposed algorithms is demonstrated.

It can be concluded that all the proposed algorithms outperform the results reported in the literature using the LR , IP, and SA methods. Considering the algorithms based on a single technique (e.g., SAA, STSA, ATSA, and GA), we conclude that the performance of both the SAA and the GA are the best. It can also be concluded that the results of the hybrid algorithms (e.g., ST, GT, and GST) are superior to the results of the algorithms based on a single method. Among the proposed hybrid algorithms, it is also found that the GST algorithm provides a better quality of solution and also a faster speed of convergence.

References

1. Wood, A.J., Wollenberg, B.F.: Power Generation, Operation, and Control. John Wiley & Sons, New York (1984)
2. El-Hawary, M.E., Christensen, G.S.: Optimal Economic Operation of Electric Power Systems. Academic Press, Inc, New York (1979)
3. Nagrath, J., Koathari, D.P.: Modern Power System Analysis. TATA McGraw Hill Publishing Co, New Delhi (1987)
4. Mantawy,H.: Optimal scheduling of thermal generation in electric power systems. A master thesis, Ain Shams University, Cairo, Egypt (1988)
5. Kazarilis, S.A., Bakirtzis, A.G., Petridis, V.: A genetic algorithm solution to the unit commitment problem. IEEE Trans. Power Syst. **11**(1), 83–91 (1996)
6. Hamam, K., Hamam, Y.M., Hindi, K.S., Brameller, A.: Unit commitment of thermal generation. IEE Proc. C **127**(1), 3–8 (1980)
7. Baldick, R.: The generalized unit commitment problem. IEEE Trans. Power Syst. **10**(1), 465–475 (1995)
8. Lee, F.L., Huang, J., Adapa, R.: Multi-area unit commitment via sequential method and a DC power flow network model. IEEE Trans. Power Syst. **9**(1), 279–287 (1994)
9. Bazaraa, M.S., Shetty, C.M.: Nonlinear Programming. Theory and Algorithms. John Wiley & Sons, New York (1979)
10. Ma, X., El-Keib, A.A., Smith, R.E., Ma, H.: A genetic algorithm based approach to thermal unit commitment of electric power systems. Electr. Power Syst. Res. **34**, 29–36 (1995)
11. Bard, J.F.: Short-term scheduling of thermal-electric generators using Lagrangian relaxation. Oper. Res. **36**(5), 756–766 (1988)
12. Li, S., Shahidehpour, S.M., Wang, C.: Promoting the application of expert systems in short-term unit commitment. IEEE Trans. Power Syst. **3**(1), 286–292 (1993)
13. Sendaula, M.H., Biswas, S.K., Eltom, A., Parten, C., Kazibwe, W.: Simultaneous solution of unit commitment and dispatch problems using artificial neural networks. Electr. Power Energy Syst. **15**(3), 193–199 (1993)
14. Glover, F., Greenberg, H.J.: New approach for heuristic search: a bilateral linkage with artificial intelligence. Eur. J. Oper. Res. **39**, 119–130 (1989)

15. Glover, F.: Future paths for integer programming and links to artificial intelligence. Comput. Oper. Res. **13**(5), 533–549 (1986)
16. Glover, F.: Tabu search-part I. ORSA J. Comput. **1**(3), 190–206 (1989)
17. Glover, F.: Artificial intelligence, heuristic frameworks and tabu search. Manag. Decis. Econ. **11**, 365–375 (1990)
18. Glover, F.: Tabu search-part II. ORSA J. Comput. **2**(1), 4–32 (1990)
19. Bland, J.A., Dawson, G.P.: Tabu search and design optimization. Compu. Aided Des. **23**(3), 195–201 (1991)
20. Glover, F.: A user's guide to tabu search. Ann. Oper. Res. **41**, 3–28 (1993)
21. Laguna, M., Glover, F.: Integrating target analysis and tabu search for improved scheduling systems. Expert Syst. Appl. **6**, 287–297 (1993)
22. Kelly, J.P., Olden, B.L., Assad, A.A.: Large-scale controlled rounding using tabu search with strategic oscillation. Ann. Oper. Res. **41**, 69–84 (1993)
23. Barnes, J.W., Laguna, M.: A tabu search experience in production scheduling. Ann. Oper. Res. **41**, 141–156 (1993)
24. Charest, M., Ferland, J.A.: Preventive maintenance scheduling of power generating units. Ann. Oper. Res. **41**, 185–206 (1993)
25. Daniels, R.L., Mazzola, J.B.: A tabu search heuristic for the flexible-resource flow shop scheduling problem. Ann. Oper. Res. **41**, 207–230 (1993)
26. Amico, M.D., Trubian, M.: Applying tabu search to the jop-shop scheduling problem. Ann. Oper. Res. **41**, 231–252 (1993)
27. Mooney, E.L., Rardin, R.L.: Tabu search for a class of scheduling problems. Ann. Oper. Res. **41**, 253–278 (1993)
28. Goldberg, D.E.: Genetic Algorithms in Search, Optimization and Machine Learning. Addison Wesely, Reading, Mass (1989)
29. Davis, L. (ed.): Handbook of Genetic Algorithms. Van Nostrand, New York (1991)
30. Michalewicz, Z. (ed.): Genetic Algorithms + Data Structures = Evolution Programs. Springer-Verlag, Berlin/Heidelberg/New York (1992)
31. Grefenstette, J.J.: Optimization of control parameters for genetic algorithms. IEEE Trans. Syst. Man Cybern. **16**(1), 122–128 (1986)
32. Grefenstette, J. J., Baker, J. E.: How genetic algorithm work: a critical look at implicit parallelism. In: The Proceedings of the Third International Conference on Genetic Algorithms, Morgan Kaufmann Publishers, San Mateo, California (1989)
33. Buckles, B. P., Petry, F. E., Kuester, R. L.: Schema survival rates and heuristic search in genetic algorithms. In: Proceedings of Tools for AI, IEEE Computer Society Press, Los Alamitos, pp. 322–327 (1990)
34. Awadh, B., Sepehri, N., Hawaleshka, O.: A computer-aided process planning model based on genetic algorithms. Comput. Oper. Res. **22**(8), 841–856 (1995)
35. Goldberg, D.E., Deb, K., Clark, J.H.: Genetic algorithms, noise, and the sizing of populations. Complex Syst. **6**, 333–362 (1992)
36. Homaifar, A., Guan, S., Liepins, G.E.: Schema analysis of the traveling salesman problem using genetic algorithms. Complex Syst. **6**, 533–552 (1992)
37. Mantawy, H., Abdel-Magid Y. L., Selim, S. Z., Salah, M. A.: An improved simulated annealing algorithm for unit commitment-application to Sceco-East. In: Proceedings of 3rd International Conference on Intelligent Applications in Communications and Power Systems, IACPS'97, UAE, pp. 133–139 (1997)
38. Turgeon, A.: Optimal scheduling of thermal generating units. IEEE Trans. Autom. Control **Ac-23**(6), 1000–1005 (1978)
39. Cohen, A., Yoshimura, M.: A branch-and-bound algorithm for unit commitment. IEEE Trans. PAS **PAS-102**(2), 444–451 (1983)
40. Yang, P.-C., Yang, H.-T., Huang, C.-L.: Solving the unit commitment problem with a genetic algorithm through a constraint satisfaction technique. Electr. Power Syst. Res. **37**, 55–65 (1996)

41. Orero, S.O., Irving, M.R.: A genetic algorithm for generators scheduling in power systems. Electr. Power Energy Syst. **18**(1), 19–26 (1996)
42. Sheble, G.B., Maifeld, T.T., Birttig, K., Fahd, G., Fukurozaki-Coppinger, S.: Unit commitment by genetic algorithm with penalty methods and a comparison of Lagrangian search and genetic algorithm-economic dispatch example. Electr. Power Energy Syst. **18**(6), 339–346 (1996)
43. Mantawy, H., Abdel-Magid, YL., Selim, SZ.: A genetic algorithm with local search for unit commitment. In: Accepted for presentation in the International Conference on Intelligent Systems Applications to Power Systems, ISAP'97, Korea (1997)
44. Waight, J.G., Albuyeh, F., Bose, A.: Scheduling of generation and reserve margin using dynamic and linear programming. IEEE Trans. Power Appar. Syst. **PAS-100**(5), 2226–2230 (1981)
45. Sheble, G.B., Grigsby, L.: Decision analysis solution of the unit commitment problem. Electr. Power Syst. Res. **11**, 85–93 (1986)
46. Cerny, V.: Thermodynamical approach to the traveling salesman problem: an efficient simulation algorithm. J. Optim. Theory Appl. **45**(1) (1985)
47. Kirkpatrick, S., Gelatt, C.D., Vecchi, M.P.: Optimization by simulated annealing. Science **220**, 671–680 (1983)
48. Billintion, R.: Power System Reliability Evaluation. Gordon and Breach Science Publishers, New York/London/Paris (1980)
49. Gillets, B.E.: Introduction to Operations Research. McGraw Hill, Inc, New York, NY (1976)
50. Larson, R.E.: State Incremental Dynamic Programming. Elsevier Pub. Co, California (1968)
51. Sheble, G.B., Fahd, G.N.: Unit commitment literature synopsis. IEEE Trans. Power Syst. **9**(1), 128–135 (1994)
52. Lowery, P.G.: Generating unit commitment by dynamic programming. IEEE Trans. PAS **85**(5), 422–426 (1966)
53. Guy, J.D.: Security constrained unit commitment. IEEE Trans. PAS **90**(3), 1385–1389 (1971)
54. Ayoub, A.K., Patton, A.D.: Optimal thermal generating unit commitment. IEEE Trans. PAS **PAS-90**(4), 1752–1756 (1971)
55. Pang, C.K., Chen, H.C.: Optimal short-term thermal unit commitment. IEEE Trans. PAS **PAS-102**(1), 1336–1346 (1976)
56. Pang, C.K., Sheble, G.B., Albuyeh, F.: Evaluation of dynamic programming based methods and multiple area representation for thermal unit commitments. IEEE Trans. PAS **PAS-100**(3), 1212–1218 (1981)
57. Van De Bosch, P.P., Honderd, G.: A solution of the unit commitment problem via decomposition and dynamic programming. IEEE Trans. Power Appar. Syst. **PAS-194**(7), 1684–1690 (1985)
58. Bond, S.D., Fox, B.: Optimal thermal unit scheduling using improved dynamic programming algorithms. IEE Proc. C **133**(1), 1–5 (1986)
59. Kusic, G.L., Putnam, H.A.: Dispatch and unit commitment including commonly owned units. IEEE Trans. Power Appar. Syst. **PAS-104**(9), 2408–2412 (1985)
60. Snyder, W.L., Powell, H.D., Rayburn, J.C.: Dynamic programming approach to unit commitment. IEEE Trans. Power Syst. **PWRS-2**(2), 339–350 (1987)
61. Hobbs, W.G., Warner, G.H., Sheble, G.B.: An enhanced dynamic programming approach for unit commitment. IEEE Trans. Power Syst. **3**(3), 1201–1205 (1988)
62. Hsu, Y.-Y., Su, C.-C., Liang, C.-C., Lin, C.-J., Huang, C.-T.: Dynamic security constrained multi-area unit commitment. IEEE Trans. Power Syst. **6**(3), 1049–1055 (1991)
63. Al-kalaani, Y., Villaseca, F.E., Renovich, F.: Storage and delivery constrained unit commitment. IEEE Trans. Power Syst. **11**(2), 1059–1066 (1996)
64. Fisher, M.L.: Optimal solution of scheduling problems using Lagrange multipliers: part I. Oper. Res. **21**, 1114–1127 (1973)
65. Muckstadt, J.A., Koenig, S.A.: An application of Lagrangian relaxation to scheduling in power-generation systems. Oper. Res. **25**(3), 387–403 (1977)

66. Bertsekas, D.P., Lauer, G.S., Sandell, N.R., Posbergh, T.A.: Optimal short-term scheduling of large-scale power systems. IEEE Trans. Autom. Control **Ac-28**(1), 1–11 (1983)
67. Merlin, A., Sandrin, P.: A new method for unit commitment at electricite de France. IEEE Trans. PAS **PAS-102**(5), 1218–1225 (1983)
68. Nieva, R., Inda, A., Guillen, I.: Lagrangian reduction of search-range for large-scale unit commitment. IEEE Trans. Power Syst. **PWRS-2**, 465–473 (1987)
69. Aoki, K., Satoh, T., Itoh, M.: Unit commitment in a large-scale power system including fuel constrained thermal and pumped-storage hydro. IEEE Trans. Power Syst. **PWRS-2**, 1077–1084 (1987)
70. Zhuang, F., Galiana, F.D.: Towards a more rigorous and practical unit commitment by Lagrangian relaxation. IEEE Trans. Power Syst. **3**(2), 763–773 (1988)
71. Aoki, K., Itoh, M., Satoh, T., Nara, K., Kanezashi, M.: Optimal long-term commitment in large-scale systems including fuel constraints thermal and pumped-storage hydro. IEEE Trans. Power Syst. **4**(3), 1065–1073 (1989)
72. Virmani, S., Adrian, E.C., Imhof, K., Mukherjee, S.: Implementation of a Lagrangian relaxation based unit commitment problem. IEEE Trans. Power Syst. **4**(4), 1373–1379 (1989)
73. Ruzic, S., Rajakovic, N.: A new approach for solving extended unit commitment problem. IEEE Trans. Power Syst. **6**(1), 269–275 (1991)
74. Wang, C., Shahidehpour, S.M.: Effects of ramp-rate limits on unit commitment and economic dispatch. IEEE Trans. Power Syst. **8**(3), 1341–1350 (1993)
75. Peterson, W.L., Brammer, S.R.: A capacity based Lagrangian relaxation unit commitment with ramp rate constraints. IEEE Trans. Power Syst. **10**(2), 1077–1084 (1995)
76. Wang, S.J., Shahidehpour, S.M., Kirschen, D.S., Mokhtari, S., Irisarri, G.D.: Short-term generation scheduling with transmission and environmental constraints using an augmented Lagrangian relaxation. IEEE Trans. Power Syst. **10**(3), 1294–1301 (1995)
77. Shaw, J.J.: A direct method for security-constrained unit commitment. IEEE Trans. Power Syst. **10**(3), 1329–11339 (1995)
78. Garver, L.L.: Power generation scheduling by integer programming – development of theory. American IEE Trans. **PAS-82**(2), 730–735 (1963)
79. Muckstadt, J.A., Wilson, R.C.: An application of mixed-integer programming duality to scheduling thermal generating systems. IEEE Trans. PAS-87 **12**, 1968–1977 (1968)
80. Dillon, T.S., Edwin, K.W., Kochs, H.D., Taud, R.J.: Integer programming approach to the problem of optimal unit commitment with probabilistic reserve. IEEE Trans. PAS **PAS-97**(6), 2154–2166 (1978)
81. Lauer, G.S., Bertsekas, D.P., Sandell, N.R., Posbergh, T.A.: Solution of large-scale optimal unit commitment problems. IEEE Trans. PAS **PAS-202**(1), 79–86 (1982)
82. Handschin, E., Slomski, H.: Unit commitment in thermal power systems with long-term energy constraints. IEEE Trans. Power Syst. **5**(4), 1470–1477 (1990)
83. Delson, J.K., Shaidehpour, S.M.: Linear programming applications to power system economics, planning and operations. IEEE Trans. Power Syst. **7**(3), 1155–1163 (1992)
84. Piekutowski, M., Rose, I.A.: A linear programming method for unit commitment incorporating generator configurations, reserve and flow constraints. IEEE Trans. Power Appar. Syst. **PAS-104**(12), 3510–3516 (1985)
85. Brannlund, H., Sjelvgren, D., Bubenko, J.A.: Short term generation scheduling with security constraints. IEEE Trans. Power Syst. **3**(1), 310–316 (1988)
86. Zhu, R., Fu, C., Rahamn, S.: Network programming technique for unit commitment. Electr. Power Energy Syst. **17**(2), 123–127 (1995)
87. Chowdhury, N., Billinton, R.: Unit commitment in interconnected generating systems using a probabilistic technique. IEEE Trans. Power Syst. **5**(5), 1231–1238 (1990)
88. Lee, F.L., Chen, Q.: Unit commitment risk with sequential rescheduling. IEEE Trans. Power Syst. **6**(3), 1017–1023 (1991)
89. Chowdhury, N.: Energy method of spinning reserve assessment in interconnected generation systems. IEEE Trans. Power Syst. **8**(3), 865–872 (1993)

90. Khan, M.E., Billinton, R.: Generating unit commitment in composite generation and transmission systems. IEE Proc. C **140**(5), 404–410 (1993)
91. Carpentiet, P., Cohen, G., Culioli, J.-C., Renaud, A.: Stochastic optimization of unit commitment: a new decomposition framework. IEEE Trans. Power Syst. **11**(2), 1067–1073 (1996)
92. Takriti, S., Birge, J.R., Long, E.: A stochastic model for unit commitment problem. A new decomposition framework. IEEE Trans. Power Syst. **11**(3), 1497–1508 (1996)
93. Shoults, R.R., Chang, S.K., Helmick, S., Grady, W.M.: A practical approach to unit commitment, economic dispatch and savings allocation for multiple-area pool operation with import/export constraints. IEEE Trans. Power Appar. Syst. **PAS-99**(2), 625–635 (1980)
94. Lee, F.N.: Short-term thermal unit commitment – a new method. IEEE Trans. Power Syst. **3**(2), 421–428 (1988)
95. Lee, F.N.: The application of Commitment Utilization Factor (CUF) to thermal unit commitment. IEEE Trans. Power Syst. **6**(2), 691–698 (1991)
96. Sheble, G.B.: Solution of the unit commitment problem by the method of unit periods. IEEE Trans. Power Syst. **5**(1), 257–260 (1990)
97. Mokhtari, S., Singh, J., Wollenberg, B.: A unit commitment expert system. IEEE Trans. Power Syst. **3**(1), 272–277 (1988)
98. Ouyang, Z., Shahidepour, S.M.: Short-term unit commitment expert system. Electr. Power Syst. Res. **20**, 1–13 (1990)
99. Tong, S.K., Shahidepour, S.M., Ouyang, Z.: A heuristic short-term unit commitment. IEEE Trans. Power Syst. **6**(3), 1210–1216 (1991)
100. Sasaki, F.H., Watanabe, M., Kubokkawa, J., Yorino, N.: A solution method using neural networks for the generator commitment problem. Electr. Eng. Japan **112**(7), 55–62 (1992)
101. Zhuang, F., Galiana, F.D.: Unit commitment by simulated annealing. IEEE Trans. Power Syst. **5**(1), 311–318 (1990)
102. Dasgupta, D., Mcgregor, D.R.: Thermal unit commitment using genetic algorithms. IEE Proc. Gen. Transm. Distrib. **141**(5), 459–465 (1994)
103. Van Meeteren, H.P.: Scheduling of generation and allocation of fuel, using dynamic and linear programming. IEEE Trans. Power Appar. Syst. **PAS-103**(7), 1562–1568 (1984)
104. Khodaverdian, E., Brameller, A., Dunnett, R.M.: Semi-rigorous thermal unit commitment for large scale electrical power systems. IEE Proc. **133**(4), 157–164 (1986). *Part-C*
105. Ouyang, Z., Shahidehpour, S.M.: An intelligent dynamic programming for unit commitment application. IEEE Trans. Power Syst. **6**(3), 1203–1209 (1991)
106. Chung-Ching, Su, Hsu, Y.-Y.: Fuzzy dynamic programming: an application to unit commitment. IEEE Trans. Power Syst. **6**(3), 1231–1237 (1991)
107. Ouyang, Z., Shahidehpour, S.M.: A hybrid artificial neural network-dynamic programming approach to unit commitment. IEEE Trans. Power Syst. **7**(1), 236–242 (1992)
108. Liang, R.-H., Hsu, Y.-Y.: A hybrid artificial neural network-differential dynamic programming approach for short-term hydro scheduling. Electr. Power Syst. Res. J. **33**, 77–86 (1995)
109. Girgis, A.A., Varadan, S.: Unit commitment using load forecasting based on artificial neural networks. Electr. Power Syst. Res. **32**, 213–217 (1995)
110. Ouyang, Z., Shahidehpour, S.M.: A multi-stage intelligent system for unit commitment. IEEE Trans. Power Syst. **7**(2), 639–646 (1992)
111. Wong, K.P., Cheung, H.N.: Thermal generator scheduling algorithm based on heuristic-guided depth-first search. IEE Proc. C **137**(1), 33–43 (1990)
112. Wong, K.P., Doan, K.P.: Artificial intelligence algorithm for daily scheduling of thermal generators. IEE Proc. C **138**(6), 518–534 (1991)
113. Doan, K., Wong, K.P.: Artificial intelligence-based machine-learning system for thermal generator scheduling. IEE Proc. Gen. Transm. Distrib. **142**(2), 195–201 (1995)
114. Sheble, G.B., Maifeld, T.T.: Unit commitment by genetic algorithm and expert system. Electr. Power Syst. Res. **30**, 115–121 (1994)

References

115. Orero, S. O., Irving, M. R.: Scheduling of generators with a hybrid genetic algorithm. In: Genetic Algorithms in Engineering Systems: innovation and Applications, IEE Conference, pp. 200–206. IEEE, Sheffield (1995)
116. Mantawy, H., Abdel-Magid, Y. L., Selim, S. Z.: A new hybrid algorithm for unit commitment. In: Accepted for presentation in the American Power Conference, APC'97, USA (1997)
117. Mantawy, H., Abdel-Magid Y. L., Selim, S. Z.: A new simulated annealing-based tabu search algorithm for unit commitment. In: Accepted for presentation in the IEEE International Conference on Systems Man and Cybernetics, SMC'97, USA (1997)
118. Aarts, E., Korst, J.: Simulated annealing and Boltzmann machines. In: A Stochastic Approach to Combinatorial Optimization and Neural Computing. John Wiley & Sons, Chichester (1989)
119. Selim, S.Z., Alsultan, K.: A simulated annealing algorithm for the clustering problem. Pattern Recognit. **24**(10), 1003–1008 (1991)
120. Metropolis, N., Rosenbluth, A., Rosenbluth, M., Teller, A., Teller, E.: Equations of state calculations by fast computing machines. J. Chem. Phys. **21**, 1087–1092 (1953)
121. Tado, M., Kubo, R., Saito, N.: Statistical Physics. Springer-Verlag, Berlin (1983)
122. Aarts, E.H.L., van Laarhoven, P.J.M.: Statistical cooling: a general approach to combinatorial optimization problems. Philips J. Res. **40**, 193–226 (1985)
123. Aarts, E. H. L., van Laarhoven, P. J. M.: A new polynomial time cooling schedule. In: Proceedings IEEE International Conference on Computer-Aided Design, Santa Clara, pp. 206–208 (1985)
124. Aarts, E.H.L., van Laarhoven, P.J.M.: Simulated annealing: a pedestrian review of the theory and some applications. In: Devijver, P.A., Kittler, J. (eds.) Pattern Recognition Theory and Applications. *NASI Series on Computer and Systems Sciences*, vol. 30, pp. 179–192. Springer-Verlag, Berlin (1987)
125. Mantawy, H., Abdel-Magid, Y. L., Selim, S. Z.: A simulated annealing algorithm for unit commitment. IEEE Tran. Power Syst. **13**(1), 197–204 (1998)

Chapter 5
Optimal Power Flow

Objectives The objectives of this chapter are
- Studying the load flow problem and representing the difference between the conventional load flow and the optimal load flow (OPF) problem
- Introducing the different states used in formulating the OPF
- Studying the multiobjective optimal power flow problem
- Introducing the particle swarm optimization algorithm as a tool to solve the optimal power flow problem

5.1 Introduction

The OPF problem has had a long history in its development for more than 25 years. A generalized formulation of the economic dispatch problem including voltage and other operating constraints was introduced and was later named the optimal power flow problem (OPF) [1].

The load flow problem in an electric power system is concerned with solving a set of static nonlinear equations describing the electric network performance. The problem is formulated on the basis of Kirchhoff's laws in terms of active and reactive power injections and voltages at each node in the system. Presently, the load flow program is a major tool for the electric power systems engineer in carrying out diverse functions in the planning and operation of the system.

In a load flow problem, the active power and voltage magnitudes are specified values for load buses; the active and reactive power demands are also given. One generator bus is taken as reference (slack bus) with specified voltage magnitude and phase angle. The problem is to find the reactive power generation and phase angles at the generating buses along with voltage magnitudes and their angles at the load buses. Available software uses the powerful Newton iterative method to obtain the solution.

For electric power systems with thermal (steam) generating resources, an important economic operation function is that of optimal load flow (OLF). In contrast to the load flow problem, where active power generations are specified, in an OLF, the optimal generations are sought to minimize the operating cost of the system.

Optimal load or power flow has been the subject of continuous intensive research and algorithmic improvements since its introduction in the early 1960s. The OPF problem seeks to find an optimal profile of active and reactive generations along with voltage magnitudes in such a manner as to minimize the total operating costs of an all-thermal electric power system while satisfying network security constraints. The problem is one of static optimization involving a large-scale system. Practical OPF algorithms for energy control center implementation were developed and reported in early 1984. OPF is a more realistic formulation than the conventional economic dispatch function, which does not account for network security constraints.

Optimal power flow is a software optimization tool for adjusting the power flows in a power network to achieve optimal value of a predefined objective such as production costs or losses. Although there are numerous OPF algorithms to find the optimal solution, there is one common feature that permeates all these OPF tools: all OPF tools incorporate a model of the power network.

By virtue of the OPF core feature of modeling the power network, it has become a common software tool made available to power engineers as they are called upon to plan and operate the power networks to accommodate higher and higher levels of power flows. An OPF program, when applied proficiently, may bring about great benefits to the operation and planning of a power system.

New users of OPF usually find it daunting to comprehend the complex OPF technology in order to harness its full power. In the following sections, the difficulties and burdens that an OPF places on a user are outlined. The pitfalls in using an OPF that a user needs to watch out for are discussed.

Difficulties: New users need to adapt to new algorithms when they start to make use of the OPF optimization tools. This section highlights some difficulties and burdens that an OPF tool imposes on a new user.

Global Optimization Control Versus Local Regulation: It is important to distinguish between global optimization control and local regulation control. Once the concept is grasped, it will end many hours of searching to comprehend why an OPF does not appear to be performing as one expects. For example, in an OPF of MVAR loss minimization, a particular transformer tap may be chosen as an optimization control among others, such as transformers, capacitors, reactors, and the like. Simultaneously, the same transformer tap may also be assigned to regulate the voltage of one of its terminal bus bars. The former control is deemed global because it contributes toward the overall objective of MVAR loss minimization. In contrast, the latter bus voltage regulation function is usually phrased as local control. This latter feature is also commonly available in a conventional power flow.

The same tap is adjusted to participate in both global optimization and local regulation, therefore it is conceivable that the two aims may sometimes conflict

5.1 Introduction

with each other. When such a conflict happens in a commercially available OPF, the program may disable one function, such as giving up local regulation for the sake of global optimization. The user needs to be aware of this situation. Otherwise, the OPF would appear to be not performing properly as far as local regulation of bus bar voltage is concerned.

Level of Security (Corrective or Preventive): One important network modeling issue of OPF is the strategy of dealing with network security; the requirement that there must be a feasible solution in the intact network as well as in postcontingency networks, that is, after a fault on one or more generating or transmission equipment items. This issue centers on the relationship between controls in the intact and postcontingency networks. Broadly, the strategy can be classified as either preventive or corrective security.

In a preventive strategy: there is no postcontingency rescheduling of control variables. As such, the solution for the intact network is determined with the enforcement of contingency constraints so that there are no violations of operating constraints following credible contingencies. This is achieved through constraining the intact network, thus requiring no corrective control actions. This approach is a conservative one.

In a corrective strategy: the control variables are rescheduled to rectify any violated operating constraints in postcontingency networks, Instead of being determined uniquely from the precontingency operating condition, the postcontingency controls vary within some specified control bands.

An OPF user needs to be consciously aware of which security strategy is going to be adopted for the OPF study at hand. The chosen strategy will not only have important bearing on the optimal solution, but also financial implications when it comes to implement the solution in the power system. In general, the solution found for a corrective strategy will cost less than that for a preventive strategy. It should be understood that the corrective strategy will expose the power system to certain undesirable effects of a contingency, albeit of short duration, until corrective control actions come into play.

Control Priority: It is important to understand in what order the control variables in an OPF are optimized. Common to many commercially available OPF, the users are normally given freedom to specify which controls are allowed to move first, second, and so on. This gives rise to a control priority list. For instance, when an OPF is used for generation rescheduling to alleviate circuit MW overloads, the priority list may look like the following.

- Phase-shifter taps
- Tie-line interchanges targets
- Generator MW set points
- Load shedding

This example assumes that adjusting the tie-line interchanges with neighboring utilities is preferred to the rescheduling of their own generations.

Although load shedding is the last resort to cure circuit overloads, phase-shifter taps will be the first controls to move because it costs literally nothing in monetary terms to move taps.

A user is completely free to make up the priority list, however, he or she must do so with due understanding of how the power system actually behaves. This ensures that the OPF will produce a plausible optimal solution.

Area of Control: One easily overlooked issue in the application OPF to power system studies is the proper treatment of external areas (e.g., a neighboring utility's networks). At one extreme, a user may completely ignore the external areas. But in practice external areas may have an important effect on the internal area. For instance, external areas may provide crucial voltage support in the bordering bus bars after a major contingency inside the internal area. At another extreme, external areas are extensively modeled in the OPF, but no due consideration is given to the control variables in the external areas. One example would be for a loss minimization OPF using control variables only in the internal area. The results would show disproportional losses in the external area and most likely are not reflecting reality.

Enforcement of Constraints: In a conventional power flow, equipment limits are normally supplied by the user for monitoring purposes, such as printing out of any violations of circuit flow limits. Only the power flow algorithm enforces a small set of limits, such as tap limits and generator MVAR limits. In contrast, an OPF enforces all equipment limits input by the user. This may easily lead to problem infeasibility if the limits are too restrictive or inconsistent. Careless input of limits should therefore be avoided.

A commercially available OPF normally offers a facility to relax limits in the case of infeasibility. Once a solution is obtained for the relaxed problem, the OPF will provide means to investigate how the original limits had caused convergence difficulties. Such a mechanism may provide valuable information concerning the power system being modeled. For instance, a region that requires relaxation of voltage limits may have implications of requiring new reactive compensation sources.

Some OPF programs require users to give them guidance as to which limits can be relaxed and in what sequence. This flexibility in fact places a great burden on the users who need to appreciate how an OPF algorithm performs before the preferred strategy for constraint relaxation can be formalized as input to the program.

Pitfalls: In the following, a number of pitfalls that a new OPF user may need to watch out for are discussed.

OPF Tends to Move Controls to Limits: There is a tendency for any optimization program to move control variables to extreme limits of the feasible range so as to achieve maximum or minimum objectives. Care should therefore be taken on two counts:

1. Are the control limits defined with confidence?
2. Will the optimal solution obtained be robust enough if the physical system is to deviate from the OPF model, due to say, a change in system load? In other words, a user must be careful not to overstretch the power system that is being optimized under the OPF model.

5.1 Introduction

One option to ensure solution robustness is to enforce in the optimal solution some reserve margin requirements, such as generation MW or MVAR reserve either regionally or systemwide. Another option is to rerun the OPF with slight variations of system demand so as to ensure that solutions are obtainable and reasonable for these variant conditions.

Peculiar and Unrealistic Solution: Users of OPF should always be on the lookout for any peculiar features of an optimal solution. Peculiarity is in the sense of lacking realism in the solution found. For instance, in the OPF studies of what reactive power compensation is needed for a network, it is very likely that the OPF will install more reactive compensation than is strictly required. One usually observes that certain generators available in the network are not committed online to provide reactive power support. As a result, more compensation is installed.

The crux of this difficulty arises from the fact that an OPF program is not a unit commitment program and hence has no scheduling capability to commit generators online. The moral of the lesson is then to ensure all available generators in the network are made online, perhaps by setting them at the minimum MW generation limit as input data, before the OPF is run.

Add Cost Signals to Reactive Minimization: Most experienced users of OPF will agree that what is mathematically feasible in an OPF model is not necessarily meaningful in the real world. One common complaint about using an OPF for "purely reactive power" optimization, such as minimization of MVAR losses, is that the active power variables in the optimal solution are not realistic at all (e.g., a cheap generator is giving out low MW output whereas an expensive one has high MW output). In fact, when one thinks about it, that is exactly what is expected because the OPF is asked to optimize the reactive power problem only. As such, the OPF has no guidance as to what is required in the active power problem, except perhaps all active power limits needed to be respected. To alleviate this difficulty, it is advisable to use a composite objective function of MVAR losses and active power generation costs with separate weighting factors. It has been shown that even when the ratio of the weighting of the MVAR losses to that of the active power cost is 100:1, the optimal solution thus produced is far more realistic than the one produced without active power consideration in the objective function.

OPF May Rely on a Converged Power Flow: Some commercially available OPF programs use linear programming (LP) optimization techniques. Their overall solution strategy may simply be summarized in several stages:

- Stage 1: nonlinear power flow
- Stage 2: linearization around a power flow solution
- Stage 3: LP optimization
- Iterate back to Stage 1

One major limitation of such a strategy is that the LP optimization process cannot proceed when Stage 1 produces no power flow solution.

Fortunately, such difficulty usually happens when the OPF first starts. Once the OPF gets into the iterative mode, Stage 1 power flow tends to pose no problem. It is therefore useful to keep in mind that some OPF programs do not necessarily guarantee a converged power flow solution. To get an OPF progressing on a difficult power flow problem, a user may need to employ the usual tricks of conventional power flow, such as relaxing the MVAR limits at all generation bus bars. This is to assist the OPF in producing a converged solution first before letting it get down working out the optimal solution.

Dependency on the Initial Solution: There are numerous vendors who supply OPF programs, therefore it is natural to find some programs that may not be as good as others. In the case where one's OPF does not converge or is giving obscure results, it is always worth trying to solve the problem again using another initial solution. Sometimes, a new initial point does the trick of helping the OPF to converge to a better and more realistic optimal solution. It may also be worth bringing the problem to the attention of the software vendor who has a duty and a commercial interest to ensure that the OPF package is free of defects of this obscure nature.

Any state-of-the-art OPF package is a powerful tool for performing a wide range of tasks for power system engineers. New users of OPF need to be aware of the new paradigm embedded into OPF, which is significantly different from conventional power flow.

OPF programs based on mathematical programming approaches are used daily to solve very large OPF problems. However, they are not guaranteed to converge to the global optimum of the general nonconvex OPF problem, although there is some empirical evidence on the uniqueness of the OPF solution within the domain of interest [2]. The existing OPF approaches have some problems, which include not only the robustness of optimization methodology used, but also the power system modeling.

The OPF problem is known as the twin subproblems of active power generation dispatch (economic dispatch problem, EDP) and reactive power generation dispatch. The main purpose of the EDP is to determine the generation schedule of the electrical energy system that minimizes the total generation and operation cost and does not violate any of the system operating constraints such as line overloading, bus voltage profiles, and deviations. On the other hand, the objective of reactive power dispatch is to minimize the active power transmission losses in an electrical system while satisfying all the system operating constraints. The objective function of the OPF can take different forms other than minimizing the generation cost and the losses in the transmission system.

The OPF problem is often infeasible, due to being either badly posed or under heavy operational stress. In the latter case, the online calculation of OPF problem is a critical function. Therefore, in cases where the original OPF does not have a feasible solution, it is desirable to be able to relax some "softer" constraints to produce the "best" engineering solution representing a solved power system operating state. If the feasibility is the result of an operator or system error in the definition of some constraint limits, it is also very useful to remove or "mark off" the offending constraints.

If both active and reactive powers are dispatchable in an electrical network, then the usual criterion for optimal operation is the minimization of generation cost. If only a reactive power is dispatchable, then active power loss minimization is frequently the desired objective This is also a convenient dummy objective if the main problem is to determine a feasible reactive power/voltage solution, or for the other purposes. Any other objective can be used based on the utility's interests and needs.

Performance and reliability of optimal power flow algorithms remain important problems in power system control and planning areas. A new application of OPF in the solution of some problems requires especially high performance of optimization algorithms.

Some of the control variables of OPF problems can be adjusted only in discrete steps, but present OPF solution methods treat all variables as continuous. The adjustments, if any, for discrete variables are made by arbitrary suboptimal procedures. These procedures may result in a significantly higher objective function cost than a solution in which the adjustment of the discrete variables is more nearly optimized. In conventional power flow solutions all of the control variables, including those that can be adjusted only in discrete steps, are treated as continuous until a first tentative solution has been reached. Then each discrete variable is rounded to its nearest step and a second and final solution is obtained using only the true continuous variables.

This procedure is valid for the conventional power flow problem because the only solution requirement is the feasibility, but it is not the case with the OPF problem where an objective function must also be minimized. Therefore, rounding to the nearest step does not minimize the objective function and it could make it impossible to obtain a feasible solution.

5.2 Power Flow Equations

Power flow studies are the backbone of power system analysis and design. They are necessary in planning and designing the future expansion of power systems as well as in determining the best operation of existing systems. In addition, power flow analysis is required for many other analyses such as transient stability and contingency studies. The principal information obtained from a power flow study is the magnitude and phase angle of the voltage at each bus and the real and reactive power flow in each line.

In any interconnected power system of N buses, the power injections at the buses are given by a set of $2N$ nonlinear simultaneous equations:

$$P_{Gk} - P_{Dk} = V_k \sum_{j=1}^{N} [V_j[G_{kj} \cos(\delta_k - \delta_j) + B_{kj} \sin(\delta_k - \delta_j)]] \quad (5.1)$$

$$Q_{Gk} - Q_{Dk} = V_k \sum_{j=1}^{N} [V_j[G_{kj}\cos(\delta_k - \delta_j) + B_{kj}\sin(\delta_k - \delta_j)]] \quad (5.2)$$

for $k = 1, 2, \ldots, N$

Where:

P_{Gk}, Q_{Gk} = active and reactive power generation at bus k
P_{Dk}, Q_{Dk} = active and reactive power demand at bus k
V_k, δ_k = voltage magnitude and angle at bus k
$G_{kj} + jB_{kj}$ = (k, j) element of the bus admittance matrix

The general practice in power flow studies is to identify the three types of buses in the network. At each bus two of four quantities δ_i, $|V_i|$, P_i, and Q_i are specified and the remaining two are calculated. Specified quantities are chosen in according to the following criteria.

5.2.1 Load Buses

At each nongenerator bus, called a load bus or a *PQ* bus, both P_{gi} and Q_{gi} are zero and the real power P_{di} and reactive power Q_{di} drawn from the system by the load are known from historical record, load forecast, or measurement. Therefore, the two unknown quantities those to be determined are δ_i and $|V_i|$.

5.2.2 Voltage Controlled Buses

Any bus of the system at which the voltage magnitude is kept at a constant is said to be a voltage-controlled bus. A generator bus is usually called a voltage-controlled or a *PV* bus. A prime mover of any generator can control the amount of generated megawatts, whereas generator voltage magnitude can be controlled by generator's excitation system. Therefore, at each generator bus both $|V_i|$, P_{gi} may properly be specified. The two unknown quantities that must be determined are δ_i and Q_i.

5.2.3 Slack Bus

The voltage angle of a slack bus serves as a reference for the angles of all other bus voltages. Thus, the usual practice is to set δ_1 to zero degree. Voltage magnitude of the slack bus $|V_1|$ is also specified. Therefore, the two unknown quantities (P_1 and Q_1) must be determined during the load flow analysis.

5.2 Power Flow Equations

Fig. 5.1 A generic bus

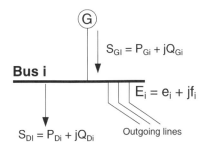

In the solution of the load flow Eqs. 5.1 and 5.2, one of the buses is chosen and named a reference bus or a slack bus and its voltage angle is set to zero. Power system variables and parameters can be classified into three groups:

(a) **Independent or control vector u:**
   This consists of the m operating variables that include active power of all generators except the slack generator and voltage magnitudes of all generators.

(b) **Dependent or state vector x:**
   This is the set of $n = 2N_L - N_G - 2$ state variables, which are obtained from the solution of n of the $2N_L - N_G - 2$ equations. Voltage angles of all buses excluding the slack bus and voltage magnitudes at load buses are classified as state variables. Once the state variables are calculated, the complete state of the system is known.

(c) **Constant vector p:**
   This contains the remaining system variables, which are specified.

The load flow equations may be written in a compact way as:

$$\underline{g}(\underline{x}, \underline{u}, \underline{p}) = \underline{0} \tag{5.3}$$

where

$\underline{g}$ is an n-dimensional vector

Figure 5.1 shows a standard bus in a power system. Complex or net power injected into a network at bus i can be stated as shown in Eq. 5.7. In addition, the bus power can be written as shown in Eq. 5.8. Sine, the power balance equates the real and imaginary parts of both the last two equations. This will result in standalone active and reactive power balance equations as indicated in Eqs. 5.9 and 5.10, respectively.

S_{Di} = assumed load demand at bus i
S_{Gi} = assumed generation at bus i

$$\text{Let } Y_{ij} = (Y_{bus})_{ij} = Y_{ij} \angle(\theta_{ij}) \tag{5.4}$$

$$\text{Then, from } I_{bus} = Y_{bus} V_{bus} \tag{5.5}$$

$$\text{Where } I_i = \sum Y_{ij} V_j \angle(\delta_j + \theta_{ij}) \tag{5.6}$$

$$S_i = V_i I_i^* = V_i \sum Y_{ij} V_j \angle (\delta_i - \delta_j - \theta_{ij}) \tag{5.7}$$

$$S_i = S_{Gi} - S_{Di} = (P_{Gi} - P_{Di}) + j(Q_{Gi} - Q_{Di}) \tag{5.8}$$

$$P_i = P_{Gi} - P_{Di} = V_i \sum_{j=1}^{n} Y_{ij} V_j \cos(\delta_i - \delta_j - \theta_{ij}) \tag{5.9}$$

$$Q_i = Q_{Gi} - Q_{Di} = V_i \sum_{j=1}^{n} Y_{ij} V_j \sin(\delta_i - \delta_j - \theta_{ij}) \tag{5.10}$$

where $i = 1, 2, \ldots, n$

During the course of solving a load flow problem, the calculated power does not match the schedule power, that is, the generation power to some extent is higher than the demand load. Therefore, mismatches can be defined as indicated in Eqs. 5.11 and 5.12, respectively. Based on the power balance, these mismatches must equal zero at the end of the problem. In other words, the load flow problem is to solve for all the buses Pi and Qi equations for values of voltages and angles, which cause the power mismatches ΔP_i and ΔQ_i to be equal to zero.

$$\Delta P_i = P_{Gi} - P_{Di} - P_i \tag{5.11}$$

$$\Delta Q_i = Q_{Gi} - Q_{Di} - Q_i \tag{5.12}$$

Throughout this research a fast decoupled load flow (FDLF) technique was used to solve the load flow problem and to determine all cost functions when all candidate control variables meet system constraints. The FDLF technique is formulated in Eq. 5.13.

$$\begin{bmatrix} \frac{\Delta \bar{P}}{V} \\ \frac{\Delta \bar{Q}}{V} \end{bmatrix} = \begin{bmatrix} \bar{B}' & 0 \\ 0 & \bar{B}'' \end{bmatrix} \begin{bmatrix} \Delta \bar{\delta} \\ \Delta \bar{V} \end{bmatrix} \tag{5.13}$$

This is conveniently expressed as follows.

$$\begin{bmatrix} \frac{\Delta \bar{P}}{V} \end{bmatrix} = [\bar{B}'][\Delta \bar{\delta}] \tag{5.14}$$

$$\begin{bmatrix} \frac{\Delta \bar{Q}}{V} \end{bmatrix} = [\bar{B}''][\Delta \bar{V}] \tag{5.15}$$

Where

$$\begin{bmatrix} \frac{\Delta \bar{P}}{V} \end{bmatrix} = \begin{bmatrix} \frac{\Delta P_2}{V_2} \\ \frac{\Delta P_3}{V_3} \\ \vdots \\ \frac{\Delta \dot{P}_n}{V_n} \end{bmatrix} \tag{5.16}$$

$$\left[\frac{\Delta \bar{Q}}{V}\right] = \begin{bmatrix} \frac{\Delta Q_{npv+1}}{V_{npv+1}} \\ \frac{\Delta Q_{npv+2}}{V_{npv+2}} \\ \vdots \\ \frac{\Delta \dot{Q}_n}{V_n} \end{bmatrix} \tag{5.17}$$

5.3 General OPF Problem Formulations

As described in Sect. 5.3, the mathematical formulation of the OPF problem can be stated as a nonlinearly constrained optimization problem (small letters represents column vectors unless otherwise stated):

$$\text{Minimize } F(x, u) \tag{5.18}$$

$$\text{Subject to}: g_E(x, u) = 0 \tag{5.19}$$

$$g_O(x, u) \le 0 \tag{5.20}$$

$$g_C(x, u) \le 0 \tag{5.21}$$

The OPF problem has many control variables to be adjusted, whereas the economic dispatch problem and reactive power generation dispatch have much less. The control variables $\underline{u}$ of the OPF problem can be stated in (5.22), and its state variables $\underline{x}$ are stated in (5.23).

$$\underline{u} = \begin{bmatrix} Q_C^T & TC^T & V_G^T & P_G^T \end{bmatrix} \tag{5.22}$$

Where

Q_C = reactive power supplied by all shunt reactors
TC = transformer load tap changer magnitudes
V_G = voltage magnitude at PV buses
P_G = active power generated at the PV buses

$$\underline{x} = \begin{bmatrix} V_L^T & \theta^T & P_{SG} & Q_G^T \end{bmatrix} \tag{5.23}$$

Where:

N_L = number of load buses
NG = number of PV buses, generator buses
V_L = voltage magnitude at PQ buses, load buses
θ = voltage angles of all buses, except the slack bus
P_{SG} = active generating power of the slack bus
Q_G = reactive power of all generator units (5.24)

As is any optimization problem, the OPF problem is formulated as a minimization or maximization to a certain objective function in which it is subjected to a variety of equality and inequality constraints. The proposed objective functions are presented in the coming section. The selection of these objective functions was made based on the literature review as well as the electrical system requirements. In addition, a set of constraints is then presented in detail.

5.3.1 The Objective Functions

Six objective functions have been considered in this chapter. These objective functions vary from fuel cost generation, active and/or reactive power transmission loss, reactive power reserve margin, security margin index, and emission, environmental, index. Some of these objective functions are conflicting in nature, which made the OPF problem so complicated. Then, a multiobjective concept was applied to solve OPF problem.

5.3.1.1 Minimization of Active Power Transmission Loss

The first objective function is the active power transmission loss. The OPF problem goal is to minimize the power loss, which can be stated as in Eq. 5.25. Another formulation of the real power loss can be represented as follows,

$$\text{Minimize } P_L = \sum_j \sum_k \left[A_{ji}(P_j P_k + Q_j Q_k) + B_{jk}(Q_j P_k - P_j Q_k) \right] \quad (5.25)$$

Where j = 1:N; k = 1:N

Or by minimizing $P_L = \Sigma P_i = \Sigma P_{gi} - \Sigma P_{di;}, i = 1, \ldots\ldots\ldots\ldots, N_b$ (5.26)

The term P_L in the above two equations represents the total I^2R loss in the transmission lines and transformers of the network. The individual current in the various transmission lines of the network cannot be calculated unless both the

5.3 General OPF Problem Formulations

voltage magnitude and angle at each bus in the electrical network are known. In practice, the P_{loss} is higher than zero.

5.3.1.2 Minimization of Generation Fuel Cost

The second objective function is the generation fuel cost of thermal units. Generally, the OPF generation fuel cost function can be expressed by a quadratic function as follows.

$$\text{Minimize (FT)} = \sum_{i=1}^{N_G} F_i(P_{Gi})$$

$$F_i(P_{Gi}) = a_i + b_i\, P_{Gi} + c_i\, P_{Gi}^2 \tag{5.27}$$

Where

- N_G is the number of generators including the slack generator in any electric network
- a_i is the basic cost coefficient of the ith generator
- b_i is the linear cost coefficient of the ith generator
- c_i is the quadratic cost coefficient of the ith generator
- P_{Gi} is the real power output of the ith generator. P_G is the vector of real power outputs of all generator units and it is defined as

$$Pg = [P_{G1}, P_{G2}, \ldots\ldots\ldots\ldots\ldots, P_{Gn}]^T \tag{5.28}$$

5.3.1.3 Maximization of Reactive Power Reserve Margin

The OPF of the third objective function is to maximize the reactive reserve margins and seek to distribute the reserve among the generators and SVCs in proportion to ratings. The objective can be obtained by minimizing the following function.

$$\text{Minimize } F = \sum_{i=1}^{N_G} \left[\frac{Q_i^2}{Q_{i\,\text{max}}}\right] \tag{5.29}$$

5.3.1.4 Minimization of Reactive Power Transmission Loss

Static network-related system voltage stability margin (VSM) depends on the availability of reactive power to support the transportation of real power from sources to sinks. Based on the premise, the total VAR loss is minimized as per Eq. 5.30. In practice, the Q_{loss} is not necessarily positive.

$$\text{Minimize } Q_{loss} = \Sigma Q_{Gi} - \Sigma Q_{Di}, \ i = 1, \ldots\ldots\ldots\ldots, N_b \qquad (5.30)$$

5.3.1.5 Minimization of Emission Index

Increased public awareness of environmental protection and the passage of the Clean Air Act Amendments of 1990 have forced utilities to modify their design or operational strategies to reduce pollution and atmospheric emissions of the thermal power plants. Several strategies to reduce atmospheric emissions have been proposed and discussed. These include installation of pollutant-cleaning equipment, switching to low emission fuels, replacement of aged fuel-burners with cleaner ones, and emission dispatching.

The first three options require installation of new equipment and/or modification of the existing ones that involve considerable capital outlay and, hence, they can be considered as long-term options. The emission dispatching option is attractive because both emission and fuel cost is minimized. In recent years, this option has received much attention inasmuch as it requires only a small modification of the basic economic dispatch to include emissions.

The emission index, or environmental index, is taken as the index from the viewpoint of environment conservation. The atmospheric pollutants such as sulphur oxides (SO_x) and nitrogen oxides (NO_x) caused by fossil-fueled thermal units can be modeled separately. However, the OPF problem seeks to minimize total (Ton/h) emission $E(P_G)$ of these pollutants, which can be stated as per Eq. 5.31. As indicated in this equation, the amount of emissions is given as a function of generator active power output, which is the sum of quadratic and exponential functions.

$$\text{Minimize (FT)} = \Sigma E_i (P_{Gi}); \text{ for } i = 1, \ldots, N_G$$

$$E_i(P_{Gi}) = \Sigma \, 10^{-2} \left(\alpha_i + \beta_i P_{Gi} + \gamma_i P_{Gi}^2 \right) + \zeta_i \exp\left(\lambda_i P_{Gi} \right) \qquad (5.31)$$

$\alpha_i, \beta_i, \gamma_i, \zeta_i,$ and λ_i are the coefficients of the ith generator emission characteristics.

5.3.1.6 Maximization of Security Margin Index

The last objective function is the security margin index (SMI). The OPF problem seeks to operate all the transmission lines connected in a network to their maximum capability. Therefore, the OPF problem will maximize the security margin by minimizing Eq. 5.32.

$$\text{Minimize SMI} = \sum_{i=1}^{N_l} \left(S_{li}^{\max} - S_{li} \right) \qquad (5.32)$$

5.3.2 The Constraints

The control variables for the OPF include: active power in all generator units, generator bus voltages, transformer tap positions, and switchable shunt reactors. The OPF constraints are divided into equality and inequality constraints. The equality constraints are power/reactive power equalities, the inequality constraints include bus voltage constraints, generator reactive power constraints, reactive source reactive power capacity constraints, the transformer tap position constraints, and so on. Therefore, all the above objective functions are subjected to the constraints below.

5.3.2.1 Equality Constraints

The equality constraints of the OPF reflect the physics of the power system. The physics of the power system are enforced through the power flow equations which require that the net injection of the real and reactive power at each bus to be zero as shown in Eq. 5.33–5.34.

- *Real Power Constraints:*

$$P_{Gk} - P_{Dk} = V_k \sum_{j=1}^{N} [V_j[G_{kj}\cos(\delta_k - \delta_j) + B_{kj}\sin(\delta_k - \delta_j)]] \quad (5.33)$$

- *Reactive Power Constraints:*

$$Q_{Gk} - Q_{Dk} = V_k \sum_{j=1}^{N} [V_j[G_{kj}\cos(\delta_k - \delta_j) + B_{kj}\sin(\delta_k - \delta_j)]] \quad (5.34)$$

5.3.2.2 Inequality Constraints

The inequality constraints of the OPF reflect the limits on physical devices in the power system as well as the limits created to ensure system security. This section delineates all the necessary inequality constraints needed for the OPF implementation in this thesis. These inequality constraints are as follows.

- *Bus Voltage Magnitude Constraints*:
  Both the load and generation voltage buses are restricted by lower and upper limits as follows.

$$V_{i-\min} \leq V_i \leq V_{i-\max} \quad (5.35)$$

$i \in N$: Set of total buses

- *Active Power Generation Constraints for all Units*:
  Active power generation constraints or generation capacity constraints have been incorporated for stable operation. This means that the active power output of each generator in any network is restricted by lower and upper limits as follows.

  $$P_{gi-min} \leq P_{gi} \leq P_{gi-max} \text{ for } i = 1 : N_G \quad (5.36)$$

  Where,

  P_{gi} = unit MW generated by ith generator
  P_{gi-max} = specified maximum MW generation by ith generator
  P_{gi-min} = specified minimum MW generation by ith generator

- *Reactive Power Generation Constraints for all Units*:
  Reactive power generated by each generator in an electrical system is restricted by lower and upper limits as shown in Eq. 5.37.

  $$Q_{Gi-min} \leq Q_{Gi} \leq Q_{Gi-max} \quad \text{for } i = 1 : N_G \quad (5.37)$$

- *Reactive Power Source Capacity Constraints*:
  All capacitors in a power system are used as reactive power suppliers. These capacitors are restricted by lower and upper reactive power limit as in Eq. 5.38. This limit will retain the amount of the exported reactive power into the power system as per the needs.

  $$q_{ci-min} \leq q_{ci} \leq q_{ci-max} \quad i \in N_c \quad (5.38)$$

  $$q_{ci} = q_{ci-min} + N_{ci} * \Delta q_{ci} \quad (5.39)$$

- *Transformer Tap Position Constraints*:
  Load tap changing transformers have a maximum and minimum tap ratio as shown in Eq. 5.40, which can be adjusted. The magnitude of the load tap changer is a discrete variable because the tap is changing with a certain increment. This increment depends on the size of the specified transformer.

  $$T_{i-min} \leq T_i \leq T_{i-max} \quad i \in N_T \quad (5.40)$$

  $$T_i = T_{i-min} + N_{Ti} * \Delta Ti \quad (5.41)$$

- *Line Thermal Limit Constraints for all Transmission Lines*:
  The power flow over a transmission line must not exceed a specified maximum limit because of the stability consideration.

  $$|S_i| \leq S_{i\,max} \quad \text{for } i = 1, \ldots\ldots, n_l \quad (5.42)$$

5.3 General OPF Problem Formulations

Where,

S_i : the complex power flow at line i
$S_{i\max}$: the maximum complex power flow at line i
n_1 : number of transmission lines in a system

5.3.3 Optimization Algorithms for OPF

A wide variety of classical optimization techniques have been applied in solving the OPF problems considering a single objective function such as nonlinear programming (NLP) [3–8], quadratic programming (QP) [9–14], linear programming (LP) [15–20], Newton-based techniques [21–30], sequential unconstrained minimization technique [31], interior point (IP) methods [32–38], and the parametric method [39]. The objective function employed for solving the optimal power flow problem regards power loss minimization or generation fuel cost minimization.

The nonlinear programming optimization algorithm deals with problems involving nonlinear objective and constraint functions. The constraints may consist of equality and/or inequality formulations. The inequality constraints can be specified as a variable that is restricted between predetermined values. Generally, nonlinear programming based procedures have many drawbacks such as insecure convergence properties and algorithmic complexity [35, 40].

The quadratic programming technique is a special form of nonlinear programming whose objective function is quadratic with linear constraints. Quadratic programming based techniques have some disadvantages associated with the piecewise quadratic cost approximation [35].

In the Newton method, the necessary conditions of optimality commonly referred to as the Kuhn–Tucker conditions are obtained. In general, these are nonlinear equations requiring iterative methods of solution. Newton-based techniques have a drawback of convergence characteristics that are sensitive to the initial conditions and they may even fail to converge due to the inappropriate initial conditions [41–47].

The sequential unconstrained minimization optimization techniques are known to exhibit numerical difficulties when the penalty factors become extremely large. Although the linear programming methods are fast and reliable, they have some disadvantages associated with piecewise linear cost approximation.

The interior point method converts the inequality constraints to equalities by the introduction of nonnegative slack variables. This method has been reported as computationally efficient; however, if the step-size is not chosen properly, the sublinear problem may have a solution that is infeasible in the original nonlinear domain. In addition, this method suffers from initial, termination, and optimality criteria and, in most cases, is unable to solve nonlinear quadratic objective functions.

Mixed integer programming (MIP) is a particular type of linear programming whose constraint equations involve variables restricted to being integers. Integer programming and mixed integer programming, like nonlinear programming, are extremely demanding of computer resources and the number of discrete variables is an important indicator of how difficult an MIP problem will be to solve.

There is some empirical evidence on the uniqueness of the OPF problem within the domain of interest. To avoid the prohibitive computational requirements of mixed integer programming, discrete control variables are initially treated as continuous and postprocessing discretization logic is subsequently applied [48, 49]. Whereas the effects of discretization on load tap changing transformers are small and usually negligible, the rounding of switch-able shunt devices may lead to voltage infeasibility, especially when the discrete VAR steps are large, and require special logic. The handling of nonconvex OPF objective functions and the unit-prohibited zones also present problems to mathematical programming OPF approaches.

The OPF problem in electrical power systems is as a static nonlinear, multi-objective optimization problem with both continuous and discrete control variables. Most of the published OPF algorithms seek to optimize only one objective function: either power loss minimization or fuel cost minimization. After scanning the literature about OPF only five papers contributed to the OPF multiobjective problem. Four papers used classical optimization techniques to solve the multiobjective optimal power flow (MOOPF) problem including: ε constraint technique [50]; weighting method [51, 52]; surrogate worth tradeoff technique [53] for solving the Multi-objective problem. In papers [50–52], linear programming was used to solve the MOOPF problem. In [50], three objective functions were considered and including generation fuel cost, environment, and security. In paper [51], two objective functions were taken into account and these include generation fuel cost and transmission loss. In [52], three objective functions were considered and these include generation fuel cost, reactive power reserve, and load voltage. The drawbacks of linear programming are insecure convergence properties and algorithmic complexity. In addition, in [53], the sequential unconstrained technique was used to solve the MOOPF problem with the same objective functions used in Reference [51].

Generally, most of the classical optimization techniques mentioned in the preceding section apply sensitivity analysis and gradient-based optimization algorithms by linearizing the objective function and the system constraints around an operating point. Unfortunately, the OPF problem is a highly nonlinear and a multimodal optimization problem; there exists more than one local optimum. Hence, local optimization techniques, which are well elaborated, are not suitable for such a problem. Moreover, there is no criterion to decide whether a local solution is also the global solution. Therefore, conventional optimization methods that make use of derivatives and gradients are not able to identify the global optimum. Conversely, many mathematical assumptions such as convex, analytical, and differential objective functions have to be given to simplify the problem. However, the OPF problem is an optimization problem with in general nonconvex, nonsmooth, and nondifferentiable objective functions. It becomes essential to develop optimization techniques that are efficient to overcome these drawbacks and handle such difficulties.

More recently, OPF has enjoyed renewed interest in a variety of formulations through the use of evolutionary optimization techniques to overcome the limitations of mathematical programming approaches. A wide variety of advanced optimization techniques have been applied in solving the OPF problems considering a single objective function including the genetic algorithm (GA) [1, 2, 54, 55], simulated annealing (SA) [56, 57], tabu search (TS) [58], and PSO algorithm [59]. The results reported were promising and encouraging for further research in this direction. Unfortunately, recent research has identified some deficiency in GA performance [60]. The degradation in efficiency is apparent in applications with highly *epistatic* objective functions, that is, where the parameters being optimized are highly correlated. In addition, the premature convergence of GA degrades its performance and reduces its search capability. The SA algorithm is a metaheuristic and many choices are required to turn it into an actual algorithm. There is a clear tradeoff between the quality of the solutions and the time required to compute them. The tailoring work required to account for different classes of constraints and to fine-tune the parameters of the algorithm can be rather delicate. The precision of the numbers used in implementation of SA can have a significant effect upon the quality of the outcome.

These are evolutionary programming (EP) methods, which use the mechanics of evolution to produce optimal solutions to a given problem. It works by evolving a population of candidate solutions toward the global optimum. The EP algorithms give better results than heuristic and classical algorithms.

Only one paper contributed to solving the multiobjective OPF problem using evolutionary programming [61]. A multiobjective hybrid evolutionary strategy (MOHES) is presented in paper [61] for solving the multiobjective OPF problem. The hybridization of GA and SA affects a beneficial synergism. This algorithm used Boltzmann probability selection criteria for solving the multiobjective OPF problem. The Boltzmann probability criterion is inspired by simulated annealing and was enhanced to include consideration of multiple objectives. Four objective functions were considered in this algorithm for optimization and these include generation fuel minimization, emission minimization, and power loss minimization and security margin maximization.

5.4 Optimal Power Flow Algorithms for Single Objective Cases [90–102]

The basic principle of the particle swarm optimization (PSO) algorithm was introduced earlier. In this section, implementation details for solving the single-objective OPF problem, using the PSO algorithm are presented. The objective functions presented earlier are individually applied as a single objective function in the optimization process. Some of these objective functions are by nature in conflict with each other. An IEEE-30 bus system is used as a test system to compare results obtained from different case studies. The detailed results are shown for all cases. In the next section, multiobjective optimization is introduced to solve the OPF problem considering multiobjective functions.

Fig. 5.2 Single objective OPF algorithm using PSO algorithm

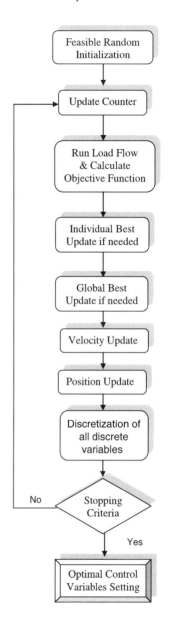

5.4.1 Particle Swarm Optimization (PSO) Algorithm for the OPF Problem

The PSO algorithm introduced previously has been used for solving the single-objective OPF problem. A modified PSO flowchart is presented in Fig. 5.2. Intensive simulations and analysis have been conducted to investigate the PSO algorithm parameters that obtain good results for the problem under study [62, 63].

5.4 Optimal Power Flow Algorithms for Single Objective Cases

Throughout this research, the following parameters are used in the PSO algorithm in order to enhance its performance.

- Both acceleration factors C_1 and C_2 are assumed to be 2.
- The number of particles was set to be 15 particles.
- The inertia factor was assumed to be decreasing linearly from about 0.9 to 0.4 during each run.
- The search will be terminated if one of the scenarios below is encountered:

  (a) $|g_{bestf}(i) - g_{bestf}(i-1)| < 0.0001$ for 50 iterations
  (b) Maximum number of iteration reached (500 iterations)

- Number of intervals N, which determine the maximum velocity v_k^{max} was selected to be 8.

It is observed that these values worked satisfactorily in all simulation results of this research. These parameters should be selected carefully for efficient performance of the PSO algorithm.

5.4.2 The IEEE-30 Bus Power System

The IEEE-30 bus system is used throughout this work to test the proposed algorithms. This system is part of the American Power Service Cooperation Network that is used as a standard test system to study different power problems and evaluate programs to analyze such problems [64]. This system consists of six generator units as well as 41 transmission lines. The one-line diagram of this system is shown in Fig. 5.3. The detailed system parameters are presented in Appendix Tables A.1 and A.2. The total active power load is 189.2 MW and the total reactive power is 126.2 MVAR. The values of fuel cost and emission coefficients of the six generators are given in Table A.3. The IEEE-30 control variables are also presented in detail in Table A.4. Throughout all cases, the IEEE-30 bus system base MVA has been assumed to be 100 MVA.

The basic power flow results of the IEEE-30 bus system without applying optimization techniques are shown in Table 5.1. The outcomes of the transmission line flows as well as the active power losses are tabulated in Table 5.2. Furthermore, values of the six objective functions, introduced in Sect. 3.3 of the previous chapter are calculated and tabulated in Table 5.3.

In the following sections, results of the individual single objective function of the OPF problem are presented.

5.4.3 Active Power Loss Minimization

As mentioned before the OPF objective function can be a minimization of the active power loss. This can be achieved by finding the optimal control parameter set,

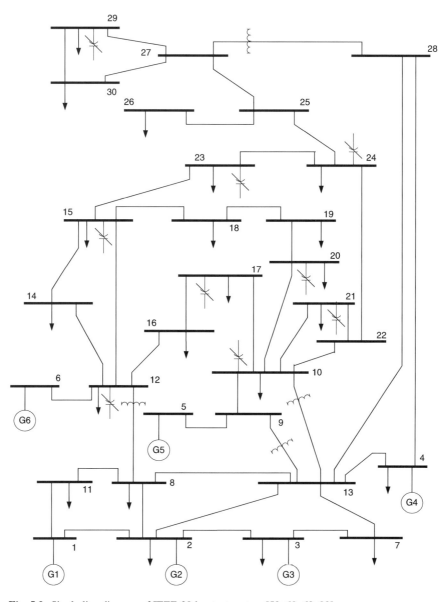

Fig. 5.3 Single-line diagram of IEEE-30 bus test system [59, 60, 62–83]

which gives the minimum objective function. In this section, the active power loss of the IEEE-30 bus test system has been minimized. The second scenario of the stopping mechanism has been encountered. The system parameter results of this case are shown in Table 5.4. In addition, the control variable parameters and the transmission line flow results of this case are tabulated in Tables 5.5 and 5.6, respectively.

5.4 Optimal Power Flow Algorithms for Single Objective Cases

Table 5.1 IEEE 30-bus basic power flow results before applying optimization, conventional power flow (bus results)

Bus No.	V (p.u.)	Delta (degree)	P_D (MW)	Q_D (MVAR)	P_G (MW)	Q_G (MVAR)
1	1.0500	0	0.00	0.00	0.9884	0.0333
2	1.0400	−1.7735	21.70	12.70	0.8000	0.2870
3	1.0100	−6.5518	94.2.0	19.00	0.5000	0.2243
4	1.0100	−5.6968	30.00	30.00	0.2000	0.4738
5	1.0500	−4.4931	0.00	0.00	0.2000	0.0246
6	1.0500	−6.2527	0.00	0.00	0.2000	0.1250
7	0.9990	−6.3455	22.80	10.90	0	0
8	1.0117	−4.5008	7.60	1.60	0	0
9	1.0459	−6.6640	0.00	0.00	0	0
10	1.0267	−8.5971	5.80	2.00	0	0
11	1.0192	−3.7736	2.40	1.20	0	0
12	1.0337	−7.7310	11.20	7.50	0	0
13	1.0049	−5.3001	0.00	0.00	0	0
14	1.0192	−8.6704	6.20	1.60	0	0
15	1.0151	−8.7888	8.20	2.50	0	0
16	1.0233	−8.3863	3.50	1.80	0	0
17	1.0202	−8.7505	9.00	5.80	0	0
18	1.0067	−9.4455	3.20	0.90	0	0
19	1.0050	−9.6358	9.50	3.40	0	0
20	1.0096	−9.4362	2.20	0.70	0	0
21	1.0138	−9.0650	17.50	11.20	0	0
22	1.0143	−9.0530	0.00	0.00	0	0
23	1.0055	−9.2329	3.20	1.60	0	0
24	1.0014	−9.4651	8.70	6.70	0	0
25	1.0120	−9.6332	0.00	0.00	0	0
26	0.9943	−10.0574	3.50	2.30	0	0
27	1.0274	−9.4559	0.00	0.00	0	0
28	1.0003	−5.7960	0.00	0.00	0	0
29	1.0077	−10.6757	2.40	0.90	0	0
30	0.9962	−11.5510	10.60	1.90	0	0

The active power loss minimization study has dramatically decreased to 3.1162 MW active power losses, which is considered as 42.75% lower than the basic case, that is, the case without optimizing the electrical system operation. The other system objective functions of this case are summarized in Table 5.7. It can be noticed that the reactive power reserve margin has improved because of minimizing the active power loss for the studied system whereas other objective functions became worse. Figure 5.4 shows the trend for minimizing the active-power transmission objective function using the PSO algorithm.

Table 5.2 IEEE 30-bus basic power flow results before applying optimization (line flow and losses)

Line No.	From bus	To bus	Line flow P_{line} (p.u.) + i Q_{line} (p.u.)	Line loss P_{loss} (p.u.) + i Q_{loss} (p.u.)
1	1	2	1.1728 − 0.1661i	0.0243 + 0.0152i
2	1	11	0.5920 − 0.0606i	0.0144 + 0.0147i
3	2	8	0.3400 − 0.0889i	0.0064 − 0.0201i
4	11	8	0.5535 − 0.0873i	0.0038 + 0.0021i
5	2	3	0.6305 + 0.0046i	0.0174 + 0.0294i
6	2	13	0.4495 − 0.0721i	0.0110 − 0.0063i
7	8	13	0.4935 + 0.0558i	0.0028 + 0.0001i
8	3	7	−0.1139 + 0.0333i	0.0007 − 0.0192i
9	13	7	0.3456 + 0.0485i	0.0031 − 0.0080i
10	13	4	0.1003 + 0.1019i	0.0002 − 0.0085i
11	13	9	0.2025 − 0.1383i	0.0000 + 0.0124i
12	13	10	0.1356 − 0.0382i	0 + 0.0112i
13	9	5	−0.1214 − 0.2926i	0 + 0.0196i
14	9	10	0.3239 + 0.1418i	−0.0000 + 0.0129i
15	8	12	0.3138 − 0.2300i	0 + 0.0418i
16	12	6	−0.1200 − 0.4694i	−0.0000 + 0.0311i
17	12	14	0.0782 + 0.0246i	0.0008 + 0.0016i
18	12	15	0.1770 + 0.0698i	0.0023 + 0.0045i
19	12	16	0.0666 + 0.0282i	0.0005 + 0.0010i
20	14	15	0.0154 + 0.0069i	0.0001 + 0.0001i
21	16	17	0.0312 + 0.0092i	0.0001 + 0.0002i
22	15	18	0.0562 + 0.0136i	0.0004 + 0.0007i
23	18	19	0.0239 + 0.0039i	0.0000 + 0.0001i
24	19	20	−0.0712 − 0.0302i	0.0002 + 0.0004i
25	10	20	0.0943 + 0.0397i	0.0009 + 0.0021i
26	10	17	0.0591 + 0.0495i	0.0002 + 0.0005i
27	10	21	0.1663 + 0.1087i	0.0013 + 0.0029i
28	10	22	0.0818 + 0.0515i	0.0007 + 0.0014i
29	21	22	−0.0100 − 0.0061i	0.0000 + 0.0000i
30	15	23	0.0518 + 0.0337i	0.0004 + 0.0008i
31	22	24	0.0711 + 0.0440i	0.0008 + 0.0012i
32	23	24	0.0195 + 0.0169i	0.0001 + 0.0002i
33	24	25	0.0027 + 0.0355i	0.0002 + 0.0004i
34	25	26	0.0355 + 0.0237i	0.0005 + 0.0007i
35	25	27	−0.0330 + 0.0114i	0.0001 + 0.0003i
36	28	27	0.1663 + 0.0345i	0.0000 + 0.0117i
37	27	29	0.0620 + 0.0169i	0.0010 + 0.0018i
38	27	30	0.0711 + 0.0169i	0.0018 + 0.0034i
39	29	30	0.0371 + 0.0061i	0.0004 + 0.0007i
40	4	28	0.0216 − 0.0384i	0.0000 − 0.0444i
41	13	28	0.1451 + 0.0161i	0.0003 − 0.0124i

5.4 Optimal Power Flow Algorithms for Single Objective Cases

Table 5.3 IEEE 30-bus system individual objective functions before applying optimization technique

Objective function	Objective value
Active power transmission loss (f1)	5.4434 MW
Over all Generation fuel costs (f2)	900.9102 $/h
Reactive power reserve margin (f3)	0.5472
Reactive power transmission loss (f4)	−9.4017 MVAR
Emission index (f5)	0.2697 t/h
Security margin (f6)	27.2555

Table 5.4 Objective function 1: active power loss minimization (bus result)

Bus No.	V (p.u.)	Delta (degree)	P_D (p.u.)	Q_D (p.u.)	P_G (p.u.)	Q_G (p.u.)
1	1.0593	0	0.0000	0.0000	0.5152	−0.0771
2	1.0558	−0.8349	0.2170	0.1270	0.8000	0.0627
3	1.0364	−4.8725	0.9420	0.1900	0.5000	0.2109
4	1.0442	−3.2124	0.3000	0.3000	0.3500	0.2809
5	1.0748	−0.3948	0.0000	0.0000	0.3000	0.1377
6	1.0583	−1.0556	0.0000	0.0000	0.4000	0.0760
7	1.0333	−4.4117	0.2280	0.1090	0	0
8	1.0472	−2.7076	0.0760	0.0160	0	0
9	1.0497	−3.5653	0.0000	0.0000	0	0
10	1.0410	−5.4702	0.0580	0.0200	0	0.0500
11	1.0500	−2.2895	0.0240	0.0120	0	0
12	1.0496	−3.9453	0.1120	0.0750	0	0.0500
13	1.0441	−3.2368	0.0000	0.0000	0	0
14	1.0384	−4.9998	0.0620	0.0160	0	0
15	1.0363	−5.3246	0.0820	0.0250	0	0.0100
16	1.0402	−4.9068	0.0350	0.0180	0	0
17	1.0381	−5.5710	0.0900	0.0580	0	0.0500
18	1.0293	−6.1595	0.0320	0.0090	0	0
19	1.0283	−6.4644	0.0950	0.0340	0	0
20	1.0332	−6.3385	0.0220	0.0070	0	0.0500
21	1.0325	−6.0262	0.1750	0.1120	0	0.0500
22	1.0332	−6.0181	0.0000	0.0000	0	0
23	1.0343	−6.1803	0.0320	0.0160	0	0.0500
24	1.0272	−6.6239	0.0870	0.0670	0	0.0450
25	1.0311	−6.9378	0.0000	0.0000	0	0
26	1.0136	−7.3462	0.0350	0.0230	0	0
27	1.0422	−6.8683	0.0000	0.0000	0	0
28	1.0403	−3.5933	0.0000	0.0000	0	0
29	1.0271	−8.1771	0.0240	0.0090	0	0.0150
30	1.0140	−8.9676	0.1060	0.0190	0	0

Table 5.5 Control variables of the first objective function

Control variable	Optimal value
P_{G2} (p.u.)	0.8000
P_{G3} (p.u.)	0.5000
P_{G4} (p.u.)	0.3500
P_{G5} (p.u.)	0.3000
P_{G6} (p.u.)	0.4000
V_{G1} (p.u.)	1.0593
V_{G2} (p.u.)	1.0558
V_{G3} (p.u.)	1.0364
V_{G4} (p.u.)	1.0442
V_{G5} (p.u.)	1.0748
V_{G6} (p.u.)	1.0583
TC_{L11} (p.u.)	1.0000
TC_{L12} (p.u.)	0.9875
TC_{L15} (p.u.)	1.0000
TC_{L36} (p.u.)	1.0250
Q_{C10} (p.u.)	0.0500
Q_{C12} (p.u.)	0.0500
Q_{C15} (p.u.)	0.0100
Q_{C17} (p.u.)	0.0500
Q_{C20} (p.u.)	0.0500
Q_{C21} (p.u.)	0.0500
Q_{C23} (p.u.)	0.0500
Q_{C24} (p.u.)	0.0450
Q_{C29} (p.u.)	0.0150

Table 5.6 Objective function 1: active power loss minimization (line flow and losses)

Line No.	From bus	To bus	Line flow P_{line} (p.u.) + i Q_{line} (p.u.)	Line loss P_{loss} (p.u.) + i Q_{loss} (p.u.)
1	1	2	0.2753 − 0.0538i	0.0013 − 0.0551i
2	1	11	0.2398 − 0.0233i	0.0023 − 0.0359i
3	2	8	0.2043 − 0.0319i	0.0021 − 0.0342i
4	11	8	0.2135 + 0.0006i	0.0005 − 0.0077i
5	2	3	0.3940 − 0.0005i	0.0066 − 0.0180i
6	2	13	0.2588 − 0.0306i	0.0035 − 0.0306i
7	8	13	0.2463 + 0.0036i	0.0007 − 0.0075i
8	3	7	−0.0546 + 0.0385i	0.0002 − 0.0213i
9	13	7	0.2849 + 0.0372i	0.0020 − 0.0121i
10	13	4	−0.0112 − 0.0056i	0−0.0098i
11	13	9	0.0302 − 0.0284i	0 + 0.0003i
12	13	10	0.0752 − 0.0170i	0 + 0.0031i
13	9	5	−0.3000 − 0.1181i	0 + 0.0196i
14	9	10	0.3302 + 0.0893i	0 + 0.0117i
15	8	12	0.0927 − 0.0090i	0 + 0.0020i
16	12	6	−0.4000 − 0.0553i	0 + 0.0207i
17	12	14	0.0819 + 0.0074i	0.0008 + 0.0016i
18	12	15	0.2040 + 0.0063i	0.0025 + 0.0049i

(continued)

Table 5.6 (continued)

Line No.	From bus	To bus	Line flow P_{line} (p.u.) + i Q_{line} (p.u.)	Line loss P_{loss} (p.u.) + i Q_{loss} (p.u.)
19	12	16	0.0948 + 0.0056i	0.0008 + 0.0016i
20	14	15	0.0192 − 0.0102i	0.0001 + 0.0001i
21	16	17	0.0591 − 0.0140i	0.0003 + 0.0007i
22	15	18	0.0707 − 0.0010i	0.0005 + 0.0010i
23	18	19	0.0382 − 0.0110i	0.0001 + 0.0002i
24	19	20	−0.0569 − 0.0452i	0.0002 + 0.0003i
25	10	20	0.0797 + 0.0037i	0.0005 + 0.0012i
26	10	17	0.0313 + 0.0228i	0 + 0.0001i
27	10	21	0.1596 + 0.0438i	0.0009 + 0.0019i
28	10	22	0.0769 + 0.0172i	0.0004 + 0.0009i
29	21	22	−0.0163 − 0.0201i	0.0001 + 0.0001i
30	15	23	0.0679 − 0.0230i	0.0005 + 0.0010i
31	22	24	0.0602 − 0.0038i	0.0004 + 0.0006i
32	23	24	0.0354 + 0.0101i	0.0002 + 0.0003i
33	24	25	0.0081 − 0.0167i	0.0001 + 0.0001i
34	25	26	0.0354 + 0.0236i	0.0004 + 0.0006i
35	25	27	−0.0274 − 0.0404i	0.0002 + 0.0005i
36	28	27	0.1603 + 0.0695i	0 + 0.0106i
37	27	29	0.0619 + 0.0057i	0.0008 + 0.0015i
38	27	30	0.0707 + 0.0123i	0.0015 + 0.0029i
39	29	30	0.0371 + 0.0102i	0.0003 + 0.0006i
40	4	28	0.0388 − 0.0149i	0.0001 − 0.0462i
41	13	28	0.1219 + 0.0250i	0.0002 − 0.0132i

Table 5.7 IEEE 30-bus system individual objective functions considering active power loss minimization

Objective function	Objective value
Active power loss minimization (f1)	3.1162 MW
Overall generation fuel cost (f2)	967.7009 $/h
Reactive power reserve margin (f3)	0.2415
Reactive power transmission loss (f4)	−20.0883 MVAR
Emission index (f5)	0.2728 t/h
Security margin (f6)	29.2449

5.4.4 Minimization of Generation Fuel Cost

The fuel cost of a thermal unit can be regarded as an essential criterion for economic feasibility. The generation fuel cost minimization was introduced in an earlier chapter. This can be stated again as

$$\text{Minimize (FT)} = \sum_{i=1}^{N_G} F_i(P_{Gi}) \tag{5.43}$$

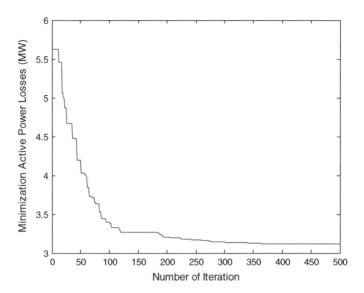

Fig. 5.4 P_{loss} minimization using PSO algorithm

$$F_i(P_{Gi}) = a_i + bi\, P_{Gi} + c_i P_{Gi}^2 \text{ for } i = 1 : N_G \qquad (5.44)$$

Because the IEEE-30 bus system has six generators P_G can be expressed as

$$P_G = [P_{G1} \quad P_{G2} \quad P_{G3} \quad P_{G4} \quad P_{G5} \quad P_{G6}]^T \qquad (5.45)$$

Therefore, in this section, the OPF problem is solved considering generation fuel cost minimization. This minimization can be achieved by finding the optimal control parameter set, which gives the minimum generation fuel cost objective function. The second scenario of the stopping mechanism has been encountered. The system parameter results of this case are shown in Table 5.8. In addition, the control variable parameters and the transmission line flow results are tabulated in Tables 5.9 and 5.10, respectively.

The generation fuel cost minimization study has resulted in 799.9859 $/h, which is considered 11.2% lower than the normal case, that is, the case without optimizing the electrical system operation. The other system objective functions of this case are summarized in Table 5.11. Figure 5.5 shows the trend for minimizing the generation fuel cost objective function using the PSO algorithm. Table 5.12 shows a comparison between the obtained minimized generation fuel cost results with the literature results. It was noticed that minimizing the generator fuel cost has directly resulted in maximizing the power loss inasmuch as they are in conflict by nature.

5.4 Optimal Power Flow Algorithms for Single Objective Cases

Table 5.8 Objective function 2: generation fuel cost minimization (bus result)

Bus No.	V (p.u.)	Delta (degree)	P_D (p.u.)	Q_D (p.u.)	P_G (p.u.)	Q_G (p.u.)
1	1.0830	0	0.0000	0.0000	1.7893	0.0350
2	1.0631	−3.3786	0.2170	0.1270	0.4898	0.1930
3	1.0321	−9.8602	0.9420	0.1900	0.2130	0.2702
4	1.0369	−7.7451	0.3000	0.3000	0.2119	0.3365
5	1.0301	−8.1929	0.0000	0.0000	0.1000	−0.0229
6	1.0530	−9.6210	0.0000	0.0000	0.1200	0.0240
7	1.0268	−8.9493	0.2280	0.1090	0	0
8	1.0420	−6.3773	0.0760	0.0160	0	0
9	1.0350	−9.5272	0.0000	0.0000	0	0
10	1.0449	−11.2947	0.0580	0.0200	0	0.0500
11	1.0500	−5.3115	0.0240	0.0120	0	0
12	1.0500	−10.4919	0.1120	0.0750	0	0.0500
13	1.0359	−7.4570	0.0000	0.0000	0	0
14	1.0401	−11.4577	0.0620	0.0160	0	0
15	1.0400	−11.7255	0.0820	0.0250	0	0.0400
16	1.0423	−11.1512	0.0350	0.0180	0	0
17	1.0416	−11.5212	0.0900	0.0580	0	0.0500
18	1.0332	−12.3519	0.0320	0.0090	0	0
19	1.0322	−12.5337	0.0950	0.0340	0	0
20	1.0371	−12.3453	0.0220	0.0070	0	0.0500
21	1.0364	−11.8398	0.1750	0.1120	0	0.0500
22	1.0370	−11.8295	0.0000	0.0000	0	0
23	1.0380	−12.3124	0.0320	0.0160	0	0.0500
24	1.0308	−12.3997	0.0870	0.0670	0	0.0500
25	1.0328	−12.1585	0.0000	0.0000	0	0
26	1.0154	−12.5658	0.0350	0.0230	0	0
27	1.0426	−11.7520	0.0000	0.0000	0	0
28	1.0343	−7.9781	0.0000	0.0000	0	0
29	1.0376	−13.3450	0.0240	0.0090	0	0.0500
30	1.0202	−14.0004	0.1060	0.0190	0	0

5.4.5 Reactive Power Reserve Maximization

The reactive reserve margin was introduced in the previous section. The ultimate goal of the reactive reserve margin maximization is to maximize the reactive reserve margins and seek to distribute the reserve among the generators in proportion to their ratings. This can be achieved by simply minimizing the following function.

$$Minimize\ F = \sum_{i=1}^{N_G} \left[\frac{Q_i^2}{Q_{i\max}}\right] \quad (5.46)$$

Therefore, in this section, the OPF problem is solved for the IEEE-30 bus system considering maximization of the reactive reserve margin of an objective function.

Table 5.9 Control variables of the second objective function

Control variable	Optimal value
P_{G2} (p.u.)	0.4898
P_{G3} (p.u.)	0.2130
P_{G4} (p.u.)	0.2119
P_{G5} (p.u.)	0.1000
P_{G6} (p.u.)	0.1200
V_{G1} (p.u.)	1.0830
V_{G2} (p.u.)	1.0631
V_{G3} (p.u.)	1.0321
V_{G4} (p.u.)	1.0369
V_{G5} (p.u.)	1.0301
V_{G6} (p.u.)	1.0530
TC_{L11} (p.u.)	0.9875
TC_{L12} (p.u.)	1.0875
TC_{L15} (p.u.)	1.0125
TC_{L36} (p.u.)	1.0125
Q_{C10} (p.u.)	0.0500
Q_{C12} (p.u.)	0.0500
Q_{C15} (p.u.)	0.0400
Q_{C17} (p.u.)	0.0500
Q_{C20} (p.u.)	0.0500
Q_{C21} (p.u.)	0.0500
Q_{C23} (p.u.)	0.0500
Q_{C24} (p.u.)	0.0500
Q_{C29} (p.u.)	0.0500

The maximization of reactive power reserve can be achieved by minimizing Eq. 5.4. The second scenario of the stopping mechanism has been encountered. System parameter results of this case are shown in Table 5.13. In addition, control variable parameters and transmission line flow results of this case are tabulated in Tables 5.14 and 5.15, respectively. Figure 5.6 shows the trend of maximization reactive power reserve by using the PSO algorithm.

5.4.6 Reactive Power Loss Minimization

As mentioned in the preceding section, the objective function of the OPF problem can be also a minimization of the reactive power loss. This can be achieved by finding the optimal control parameters set, which gives the minimum objective function. In this section, the reactive power loss of the IEEE-30 bus test system has been minimized (Table 5.16). The stopping mechanism scenario has been encountered. The system parameter results of this case are shown in Table 5.17. In addition, the control variable parameters and the transmission line flow results of this case are tabulated in Table 5.18 and 5.19, respectively. The reactive power loss

5.4 Optimal Power Flow Algorithms for Single Objective Cases

Table 5.10 Objective function 2: generation fuel cost minimization (line flow and losses)

Line No.	From bus	To bus	Line flow P_{line} (p.u.) + i Q_{line} (p.u.)	Line loss P_{loss} (p.u.) + i Q_{loss} (p.u.)
1	1	2	1.1847 − 0.0172i	0.0230 + 0.0080i
2	1	11	0.5870 + 0.0522i	0.0135 + 0.0089i
3	2	8	0.3421 + 0.0052i	0.0059 − 0.0227i
4	11	8	0.5495 + 0.0313i	0.0036 + 0.0012i
5	2	3	0.6366 + 0.0264i	0.0170 + 0.0256i
6	2	13	0.4543 + 0.0092i	0.0107 − 0.0089i
7	8	13	0.4956 + 0.0100i	0.0027 − 0.0003i
8	3	7	−0.1088 + 0.0809i	0.0009 − 0.0194i
9	13	7	0.3406 − 0.0006i	0.0029 − 0.0092i
10	13	4	0.1122 − 0.0623i	0.0002 − 0.0090i
11	13	9	0.1839 − 0.0557i	0 + 0.0073i
12	13	10	0.1417 + 0.1702i	0 + 0.0215i
13	9	5	−0.1194 + 0.0258i	0 + 0.0029i
14	9	10	0.3032 − 0.0888i	0 + 0.0103i
15	8	12	0.3105 + 0.0319i	0 + 0.0224i
16	12	6	−0.1200 − 0.0221i	0 + 0.0019i
17	12	14	0.0744 + 0.0051i	0.0006 + 0.0013i
18	12	15	0.1765 − 0.0074i	0.0019 + 0.0037i
19	12	16	0.0676 + 0.0088i	0.0004 + 0.0008i
20	14	15	0.0118 − 0.0122i	0.0001 + 0.0001i
21	16	17	0.0322 − 0.0100i	0.0001 + 0.0002i
22	15	18	0.0563 + 0.0052i	0.0003 + 0.0006i
23	18	19	0.0240 − 0.0044i	0 + 0.0001i
24	19	20	−0.0710 − 0.0385i	0.0002 + 0.0004i
25	10	20	0.0940 − 0.0024i	0.0008 + 0.0017i
26	10	17	0.0580 + 0.0185i	0.0001 + 0.0003i
27	10	21	0.1586 + 0.0453i	0.0009 + 0.0019i
28	10	22	0.0763 + 0.0182i	0.0004 + 0.0008i
29	21	22	−0.0173 − 0.0186i	0
30	15	23	0.0481 − 0.0135i	0.0002 + 0.0005i
31	22	24	0.0586 − 0.0013i	0.0004 + 0.0006i
32	23	24	0.0158 + 0.0201i	0.0001 + 0.0002i
33	24	25	−0.0130 + 0.0011i	0 + 0.0001i
34	25	26	0.0354 + 0.0236i	0.0004 + 0.0006i
35	25	27	−0.0485 − 0.0226i	0.0003 + 0.0006i
36	28	27	0.1815 + 0.0183i	0 + 0.0120i
37	27	29	0.0622 − 0.0193i	0.0009 + 0.0016i
38	27	30	0.0706 + 0.0025i	0.0015 + 0.0028i
39	29	30	0.0373 + 0.0200i	0.0004 + 0.0008i
40	4	28	0.0238 − 0.0168i	0 − 0.0458i
41	13	28	0.1581 − 0.0232i	0.0004 − 0.0125i

Table 5.11 IEEE 30-bus system individual objective functions considering generation fuel cost minimization

Objective function	Optimal value
Active power transmission loss (f1)	9.0706 MW
Generation fuel costs minimization (f2)	799.9859 $/h
Reactive power reserve margin (f3)	0.3198
Reactive power transmission loss (f4)	1.3704 MVAR
Emission index (f5)	0.4311 t/h
Security margin index (f6)	26.8580

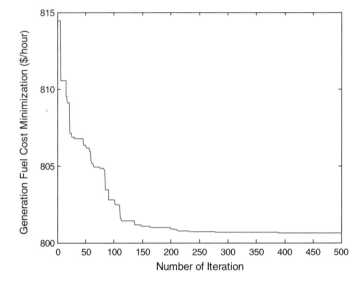

Fig. 5.5 Generation fuel cost minimization using PSO algorithm

Table 5.12 A comparison of fuel generation cost minimization algorithms

Literature	Method	Minimal generation fuel cost ($)
[84]	LP-programming algorithm	806.84
[54]	Genetic algorithm	800.805
[85]	Gradient-based algorithm	804.583
Following this section	PSO algorithm	799.98

minimization study has resulted in −20.4785 MVAR and other system objective functions of this case are summarized in Table 5.20. Figure 5.7 shows the trend for minimizing the active power transmission objective function using the PSO algorithm.

5.4.7 Emission Index Minimization

As mentioned in Sect. 3.3.5 of Chap. 3, the OPF problem can be solved with a minimization of the emission index. This can be achieved by dispatching the active

5.4 Optimal Power Flow Algorithms for Single Objective Cases

Table 5.13 Objective function 3: reactive power reserve margin maximization (bus result)

Bus No.	V (p.u.)	Delta (degree)	P_D (p.u.)	Q_D (p.u.)	P_G (p.u.)	Q_G (p.u.)
1	1.0586	0	0.0000	0.0000	0.5230	0.2331
2	1.0433	−0.6599	0.2170	0.1270	0.8000	0.1184
3	1.0030	−4.5996	0.9420	0.1900	0.5000	0.0965
4	1.0061	−2.7539	0.3000	0.3000	0.3500	0.0723
5	1.0303	0.1059	0.0000	0.0000	0.3000	0.0598
6	1.0177	−0.4973	0.0000	0.0000	0.3975	0.0709
7	1.0013	−4.1328	0.2280	0.1090	0	0
8	1.0257	−2.4530	0.0760	0.0160	0	0
9	1.0200	−3.2984	0.0000	0.0000	0	0
10	1.0129	−5.3432	0.0580	0.0200	0	0.0500
11	1.0321	−2.0895	0.0240	0.0120	0	0
12	1.0094	−3.6024	0.1120	0.0750	0	0.0500
13	1.0135	−2.8986	0.0000	0.0000	0	0
14	1.0012	−4.7650	0.0620	0.0160	0	0
15	1.0023	−5.2156	0.0820	0.0250	0	0.0500
16	1.0049	−4.6902	0.0350	0.0180	0	0
17	1.0078	−5.4378	0.0900	0.0580	0	0.0500
18	0.9971	−6.0954	0.0320	0.0090	0	0
19	0.9973	−6.4135	0.0950	0.0340	0	0
20	1.0029	−6.2765	0.0220	0.0070	0	0.0500
21	1.0046	−5.9346	0.1750	0.1120	0	0.0500
22	1.0053	−5.9275	0.0000	0.0000	0	0
23	1.0036	−6.1273	0.0320	0.0160	0	0.0500
24	1.0007	−6.5954	0.0870	0.0670	0	0.0500
25	1.0112	−6.9334	0.0000	0.0000	0	0
26	0.9934	−7.3563	0.0350	0.0230	0	0
27	1.0266	−6.8585	0.0000	0.0000	0	0
28	1.0082	−3.2847	0.0000	0.0000	0	0
29	1.0201	−8.4602	0.0240	0.0090	0	0.0450
30	1.0031	−9.1565	0.1060	0.0190	0	0

power of the generator units in a way that minimizes the emission index as stated in Eq. 5.47.

$$\text{Minimize E}(P_G) = \sum_{i=1}^{N_G} \left[10^{-2}(\alpha_i + \beta_i P_{Gi} + \gamma_i P_{Gi}^2) + \zeta_i \exp(\lambda_i P_{Gi})\right] \quad (5.47)$$

In this section, the emission index of the IEEE-30 bus test system has been minimized. The values of emission coefficients of the six generators are given in Table A.3. The first scenario of the stopping mechanism has been encountered at iteration no. 164. The system parameter results for this case are shown in Table 5.21. The control variable parameters and the transmission line flows results of this case are tabulated in Tables 5.22 and 5.23, respectively. The emission minimization study has resulted in 0.2592 t/h. The other system objective functions of this case are summarized in Table 5.24. Figure 5.8 shows the trend of minimizing the emission index objective function using the PSO algorithm.

Table 5.14 Control variables of the third objective function

Control variable	Optimal value
P_{G2} (p.u.)	0.8000
P_{G3} (p.u.)	0.5000
P_{G4} (p.u.)	0.3500
P_{G5} (p.u.)	0.3000
P_{G6} (p.u.)	0.3975
V_{G1} (p.u.)	1.0586
V_{G2} (p.u.)	1.0433
V_{G3} (p.u.)	1.0030
V_{G4} (p.u.)	1.0061
V_{G5} (p.u.)	1.0303
V_{G6} (p.u.)	1.0177
TC_{L11} (p.u.)	1.0125
TC_{L12} (p.u.)	1.0125
TC_{L15} (p.u.)	0.9625
TC_{L36} (p.u.)	1.0375
Q_{C10} (p.u.)	0.0500
Q_{C12} (p.u.)	0.0500
Q_{C15} (p.u.)	0.0500
Q_{C17} (p.u.)	0.0500
Q_{C20} (p.u.)	0.0500
Q_{C21} (p.u.)	0.0500
Q_{C23} (p.u.)	0.0500
Q_{C24} (p.u.)	0.0500
Q_{C29} (p.u.)	0.0450

Table 5.15 Objective function 3: reactive power reserve margin maximization (line flow and losses)

Line No.	From bus	To bus	Line flow P_{line} (p.u.) + i Q_{line} (p.u.)	Line loss P_{loss} (p.u.) + i Q_{loss} (p.u.)
1	1	2	0.2841 + 0.1587i	0.0020 − 0.0524i
2	1	11	0.2388 + 0.0743i	0.0027 − 0.0336i
3	2	8	0.2062 + 0.0209i	0.0023 − 0.0323i
4	11	8	0.2122 + 0.0959i	0.0007 − 0.0069i
5	2	3	0.3937 + 0.1081i	0.0075 − 0.0124i
6	2	13	0.2652 + 0.0735i	0.0042 − 0.0267i
7	8	13	0.2613 + 0.2248i	0.0014 − 0.0046i
8	3	7	−0.0557 + 0.0270i	0.0002 − 0.0200i
9	13	7	0.2862 + 0.0516i	0.0022 − 0.0104i
10	13	4	−0.0099 + 0.1753i	0.0004 − 0.0079i
11	13	9	0.0351 + 0.0306i	0 + 0.0004i
12	13	10	0.0797 + 0.0262i	0 + 0.0037i
13	9	5	−0.3000 − 0.0415i	0 + 0.0183i
14	9	10	0.3351 + 0.0716i	0 + 0.0124i
15	8	12	0.0781 − 0.0846i	0 + 0.0035i
16	12	6	−0.3975 − 0.0489i	0 + 0.0220i

(continued)

5.4 Optimal Power Flow Algorithms for Single Objective Cases

Table 5.15 (continued)

Line No.	From bus	To bus	Line flow P_{line} (p.u.) + i Q_{line} (p.u.)	Line loss P_{loss} (p.u.) + i Q_{loss} (p.u.)
17	12	14	0.0780 − 0.0044i	0.0007 + 0.0015i
18	12	15	0.1972 − 0.0419i	0.0026 + 0.0052i
19	12	16	0.0884 − 0.0179i	0.0008 + 0.0016i
20	14	15	0.0152 − 0.0220i	0.0002 + 0.0001i
21	16	17	0.0526 − 0.0375i	0.0003 + 0.0008i
22	15	18	0.0663 − 0.0081i	0.0005 + 0.0010i
23	18	19	0.0338 − 0.0181i	0.0001 + 0.0002i
24	19	20	−0.0613 − 0.0522i	0.0002 + 0.0004i
25	10	20	0.0842 + 0.0112i	0.0007 + 0.0015i
26	10	17	0.0378 + 0.0466i	0.0001 + 0.0003i
27	10	21	0.1586 + 0.0396i	0.0009 + 0.0020i
28	10	22	0.0763 + 0.0144i	0.0004 + 0.0009i
29	21	22	−0.0173 − 0.0244i	0.0001 + 0.0001i
30	15	23	0.0614 − 0.0361i	0.0005 + 0.0010i
31	22	24	0.0585 − 0.0109i	0.0004 + 0.0006i
32	23	24	0.0289 − 0.0031i	0.0001 + 0.0002i
33	24	25	−0.0001 − 0.0319i	0.0002 + 0.0003i
34	25	26	0.0354 + 0.0237i	0.0005 + 0.0007i
35	25	27	−0.0358 − 0.0559i	0.0005 + 0.0009i
36	28	27	0.1690 + 0.0566i	0 + 0.0115i
37	27	29	0.0622 − 0.0157i	0.0009 + 0.0016i
38	27	30	0.0706 + 0.0040i	0.0015 + 0.0029i
39	29	30	0.0373 + 0.0187i	0.0004 + 0.0008i
40	4	28	0.0397 − 0.0445i	0.0001 − 0.0430i
41	13	28	0.1298 + 0.0460i	0.0003 − 0.0121i

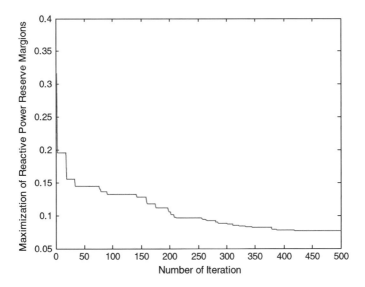

Fig. 5.6 Reactive power reserve margin maximization using PSO algorithm

Table 5.16 IEEE 30-bus system individual objective functions considering reactive power reserve margin maximization

Objective function	Objective value
Active power loss (f1)	3.6433 MW
Overall generation fuel cost (f2)	968.3029 $/h
Reactive power reserve margin Maximization (f3)	0.0771
Reactive power transmission loss (f4)	−16.5976 MVAR
Emission index (f5)	0.2739 t/h
Security margin index (f6)	28.1498

Table 5.17 Objective function 4: reactive power loss minimization (bus result)

Bus No.	V (p.u.)	Delta (degree)	P_D (p.u.)	Q_D (p.u.)	P_G (p.u.)	Q_G (p.u.)
1	1.0608	0	0.0000	0.0000	0.5162	−0.1029
2	1.0591	−0.8681	0.2170	0.1270	0.7995	0.1208
3	1.0425	−4.9193	0.9420	0.1900	0.5000	0.2527
4	1.0454	−3.1805	0.3000	0.3000	0.3500	0.2883
5	1.0428	−0.1597	0.0000	0.0000	0.3000	0.0555
6	1.0563	−1.1321	0.0000	0.0000	0.4000	0.0578
7	1.0362	−4.4030	0.2280	0.1090	0	0
8	1.0468	−2.6764	0.0760	0.0160	0	0
9	1.0334	−3.4788	0.0000	0.0000	0	0
10	1.0310	−5.4019	0.0580	0.0200	0	0.0500
11	1.0500	−2.2645	0.0240	0.0120	0	0
12	1.0500	−4.0263	0.1120	0.0750	0	0.0500
13	1.0447	−3.1984	0.0000	0.0000	0	0
14	1.0380	−5.0895	0.0620	0.0160	0	0
15	1.0352	−5.4080	0.0820	0.0250	0	0.0300
16	1.0362	−4.9177	0.0350	0.0180	0	0
17	1.0299	−5.5276	0.0900	0.0580	0	0.0500
18	1.0243	−6.1781	0.0320	0.0090	0	0
19	1.0211	−6.4443	0.0950	0.0340	0	0
20	1.0248	−6.2938	0.0220	0.0070	0	0.0400
21	1.0217	−5.9492	0.1750	0.1120	0	0.0250
22	1.0227	−5.9500	0.0000	0.0000	0	0
23	1.0302	−6.2357	0.0320	0.0160	0	0.0500
24	1.0189	−6.6422	0.0870	0.0670	0	0.0500
25	1.0203	−6.9598	0.0000	0.0000	0	0
26	1.0027	−7.3769	0.0350	0.0230	0	0
27	1.0300	−6.8906	0.0000	0.0000	0	0
28	1.0425	−3.5699	0.0000	0.0000	0	0
29	1.0220	−8.4404	0.0240	0.0090	0	0.0400
30	1.0057	−9.1515	0.1060	0.0190	0	0

5.4 Optimal Power Flow Algorithms for Single Objective Cases

Table 5.18 Control variables of the fourth objective function

Control variable	Optimal value
P_{G2} (p.u.)	0.7995
P_{G3} (p.u.)	0.5000
P_{G4} (p.u.)	0.3500
P_{G5} (p.u.)	0.3000
P_{G6} (p.u.)	0.4000
V_{G1} (p.u.)	1.0608
V_{G2} (p.u.)	1.0591
V_{G3} (p.u.)	1.0425
V_{G4} (p.u.)	1.0454
V_{G5} (p.u.)	1.0428
V_{G6} (p.u.)	1.0563
TC_{L11} (p.u.)	0.9875
TC_{L12} (p.u.)	1.0000
TC_{L15} (p.u.)	1.0125
TC_{L36} (p.u.)	1.0000
Q_{C10} (p.u.)	0.0500
Q_{C12} (p.u.)	0.0500
Q_{C15} (p.u.)	0.0300
Q_{C17} (p.u.)	0.0500
Q_{C20} (p.u.)	0.0400
Q_{C21} (p.u.)	0.0250
Q_{C23} (p.u.)	0.0500
Q_{C24} (p.u.)	0.0500
Q_{C29} (p.u.)	0.0400

5.4.8 Security Margin Maximization

Maximization of the security margin index is a goal of any electric power company in the world. The OPF problem can also incorporate this requirement by operating all the transmission lines in a network to their maximum capabilities. Therefore, the maximization of the security margin can be achieved by minimizing Eq. 5.48.

$$\text{Minimize SMI} = \sum_{i=1}^{N_l} (S_{li}^{\max} - S_{li}) \quad (5.48)$$

In this section, the security margin index has been maximized and the IEEE-30 bus system is considered as a case study. This can be achieved by finding the optimal control parameter set, which permits all the transmission lines to be as close to their maximum capability as possible. The first scenario of the stopping mechanism has been encountered at iteration no. 182. The system parameter results of this case are shown in Table 5.25. Also, the control variable parameters and the transmission line flow results are tabulated in Table 5.26 and 5.27, respectively.

Table 5.19 Objective function 4: reactive power loss minimization (line flow and losses)

Line No.	From bus	To bus	Line flow P_{line} (p.u.) + i Q_{line} (p.u.)	Line loss P_{loss} (p.u.) + i Q_{loss} (p.u.)
1	1	2	0.2765 − 0.0880i	0.0014 − 0.0552i
2	1	11	0.2396 − 0.0148i	0.0023 − 0.0360i
3	2	8	0.2049 − 0.0101i	0.0021 − 0.0343i
4	11	8	0.2133 + 0.0092i	0.0005 − 0.0077i
5	2	3	0.3954 − 0.0148i	0.0066 − 0.0185i
6	2	13	0.2574 − 0.0141i	0.0034 − 0.0310i
7	8	13	0.2371 − 0.0174i	0.0006 − 0.0077i
8	3	7	−0.0532 + 0.0664i	0.0004 − 0.0211i
9	13	7	0.2835 + 0.0092i	0.0020 − 0.0123i
10	13	4	−0.0120 − 0.0184i	0 − 0.0098i
11	13	9	0.0251 − 0.0089i	0 + 0.0001i
12	13	10	0.0745 + 0.0272i	0 + 0.0032i
13	9	5	−0.3000 − 0.0377i	0 + 0.0178i
14	9	10	0.3250 + 0.0287i	0 + 0.0110i
15	8	12	0.1024 + 0.0424i	0 + 0.0028i
16	12	6	−0.4000 − 0.0373i	0 + 0.0205i
17	12	14	0.0836 + 0.0095i	0.0008 + 0.0016i
18	12	15	0.2088 + 0.0154i	0.0026 + 0.0052i
19	12	16	0.0980 + 0.0270i	0.0009 + 0.0019i
20	14	15	0.0208 − 0.0082i	0.0001 + 0.0001i
21	16	17	0.0622 + 0.0072i	0.0003 + 0.0007i
22	15	18	0.0731 + 0.0161i	0.0006 + 0.0011i
23	18	19	0.0406 + 0.0060i	0.0001 + 0.0002i
24	19	20	−0.0545 − 0.0282i	0.0001 + 0.0002i
25	10	20	0.0772 − 0.0034i	0.0005 + 0.0012i
26	10	17	0.0282 + 0.0016i	0 + 0.0001i
27	10	21	0.1593 + 0.0538i	0.0009 + 0.0020i
28	10	22	0.0769 + 0.0197i	0.0004 + 0.0009i
29	21	22	−0.0166 − 0.0352i	0
30	15	23	0.0717 − 0.0092i	0.0005 + 0.0010i
31	22	24	0.0598 − 0.0164i	0.0004 + 0.0007i
32	23	24	0.0392 + 0.0238i	0.0003 + 0.0005i
33	24	25	0.0113 − 0.0108i	0 + 0.0001i
34	25	26	0.0354 + 0.0237i	0.0004 + 0.0007i
35	25	27	−0.0241 − 0.0345i	0.0002 + 0.0004i
36	28	27	0.1571 + 0.0375i	0 + 0.0095i
37	27	29	0.0621 − 0.0122i	0.0008 + 0.0016i
38	27	30	0.0706 + 0.0053i	0.0015 + 0.0029i
39	29	30	0.0373 + 0.0173i	0.0004 + 0.0007i
40	4	28	0.0380 − 0.0203i	0.0001 − 0.0464i
41	13	28	0.1193 − 0.0020i	0.0002 − 0.0134i

5.5 Comparisons of Different Single Objective Functions

Table 5.20 IEEE 30-bus system individual objective functions considering reactive power loss minimization

Objective function	Objective value
Active power transmission loss (f1)	3.1644 MW
Overall generation fuel costs (f2)	967.7024 $/h
Reactive power reserve margin (f3)	0.2500
Reactive power transmission loss minimization (f4)	−20.4785 MVAR
Emission index (f5)	0.2732 t/h
Security margin index (f6)	29.1831

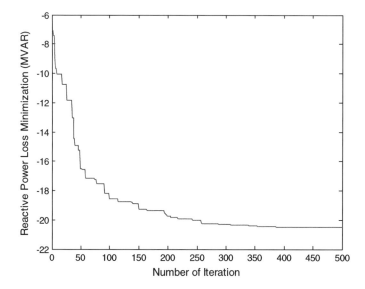

Fig. 5.7 Q_{loss} minimization using PSO algorithm

The maximization of the security margin of the IEEE-30 bus system has resulted in 21.6421 and other system objective functions of this case are summarized in Table 5.28. It was noticed that maximizing the security margin has negatively affected all other investigated objective functions. Figure 5.9 shows the trend of maximizing the security index using the PSO algorithm.

5.5 Comparisons of Different Single Objective Functions

In this section, a comparison between results of applying different single objective functions for solving the OPF problem is presented. Table 5.29 shows the values of the minimum objective function for each single objective case and other corresponding objective functions at the system parameters determined at this

Table 5.21 Objective function 5: emission index minimization

Bus No.	V (p.u.)	Delta (degree)	P_D (p.u.)	Q_D (p.u.)	P_G (p.u.)	Q_G (p.u.)
1	1.0266	0	0.0000	0.0000	0.5186	−0.1700
2	1.0253	−0.9329	0.2170	0.1270	0.8000	0.0667
3	1.0035	−5.1956	0.9420	0.1900	0.5000	0.1925
4	1.0152	−3.4888	0.3000	0.3000	0.3499	0.3125
5	1.0574	−0.7439	0.0000	0.0000	0.3000	0.0515
6	1.0729	−1.4652	0.0000	0.0000	0.4000	0.4496
7	1.0017	−4.7300	0.2280	0.1090	0	0
8	1.0256	−3.0614	0.0760	0.0160	0	0
9	1.0489	−3.9693	0.0000	0.0000	0	0
10	1.0109	−6.0282	0.0580	0.0200	0	0.0350
11	1.0265	−2.5790	0.0240	0.0120	0	0
12	1.0156	−4.4111	0.1120	0.0750	0	0.0500
13	1.0138	−3.4975	0.0000	0.0000	0	0
14	1.0044	−5.5350	0.0620	0.0160	0	0
15	1.0026	−5.8865	0.0820	0.0250	0	0.0500
16	1.0062	−5.4029	0.0350	0.0180	0	0
17	1.0045	−6.0777	0.0900	0.0580	0	0.0100
18	0.9966	−6.7719	0.0320	0.0090	0	0
19	0.9963	−7.0936	0.0950	0.0340	0	0
20	1.0017	−6.9579	0.0220	0.0070	0	0.0500
21	0.9999	−6.5828	0.1750	0.1120	0	0.0300
22	1.0004	−6.5747	0.0000	0.0000	0	0
23	0.9930	−6.5829	0.0320	0.0160	0	0
24	0.9891	−7.1471	0.0870	0.0670	0	0.0500
25	0.9865	−7.4429	0.0000	0.0000	0	0
26	0.9683	−7.8897	0.0350	0.0230	0	0
27	0.9940	−7.3478	0.0000	0.0000	0	0
28	1.0111	−3.8762	0.0000	0.0000	0	0
29	0.9797	−8.8341	0.0240	0.0090	0	0.0200
30	0.9653	−9.6854	0.1060	0.0190	0	0

Table 5.22 Control variables of the fifth objective function

Control variable	Optimal value
P_{G2} (p.u.)	0.8000
P_{G3} (p.u.)	0.5000
P_{G4} (p.u.)	0.3499
P_{G5} (p.u.)	0.3000
P_{G6} (p.u.)	0.4000
V_{G1} (p.u.)	1.0266
V_{G2} (p.u.)	1.0253
V_{G3} (p.u.)	1.0035
V_{G4} (p.u.)	1.0152
V_{G5} (p.u.)	1.0574
V_{G6} (p.u.)	1.0729

(continued)

5.5 Comparisons of Different Single Objective Functions

Table 5.22 (continued)

Control variable	Optimal value
TC_{L11} (p.u.)	1.1000
TC_{L12} (p.u.)	0.9000
TC_{L15} (p.u.)	0.9000
TC_{L36} (p.u.)	1.0000
Q_{C10} (p.u.)	0.0350
Q_{C12} (p.u.)	0.0500
Q_{C15} (p.u.)	0.0500
Q_{C17} (p.u.)	0.0100
Q_{C20} (p.u.)	0.0500
Q_{C21} (p.u.)	0.0300
Q_{C23} (p.u.)	0
Q_{C24} (p.u)	0.0500
Q_{C29} (p.u.)	0.0200

Table 5.23 Objective function 5: emission index minimization (line flow and losses)

Line No.	From bus	To bus	Line flow P_{line} (p.u.) + i Q_{line} (p.u.)	Line loss P_{loss} (p.u.) + i Q_{loss} (p.u.)
1	1	2	0.2755 − 0.0952i	0.0015 − 0.0512i
2	1	11	0.2430 − 0.0748i	0.0027 − 0.0321i
3	2	8	0.2037 − 0.0838i	0.0025 − 0.0312i
4	11	8	0.2164 − 0.0546i	0.0006 − 0.0071i
5	2	3	0.3937 + 0.0116i	0.0070 − 0.0136i
6	2	13	0.2597 − 0.0322i	0.0037 − 0.0275i
7	8	13	0.2547 + 0.2164i	0.0013 − 0.0049i
8	3	7	−0.0553 + 0.0277i	0.0002 − 0.0200i
9	13	7	0.2858 + 0.0508i	0.0022 − 0.0105i
10	13	4	−0.0124 − 0.0351i	0 − 0.0092i
11	13	9	0.0463 + 0.3555i	0 + 0.0215i
12	13	10	0.0732 − 0.1601i	0 + 0.0207i
13	9	5	−0.3000 − 0.0343i	0 + 0.0172i
14	9	10	0.3463 + 0.3682i	0 + 0.0255i
15	8	12	0.0863 − 0.3326i	0 + 0.0355i
16	12	6	−0.4000 − 0.4055i	0 + 0.0440i
17	12	14	0.0812 + 0.0062i	0.0008 + 0.0016i
18	12	15	0.2016 + 0.0011i	0.0026 + 0.0051i
19	12	16	0.0915 + 0.0051i	0.0008 + 0.0016i
20	14	15	0.0183 − 0.0114i	0.0001 + 0.0001i
21	16	17	0.0557 − 0.0145i	0.0003 + 0.0006i
22	15	18	0.0681 − 0.0052i	0.0005 + 0.0010i
23	18	19	0.0356 − 0.0152i	0.0001 + 0.0002i
24	19	20	−0.0595 − 0.0494i	0.0002 + 0.0004i
25	10	20	0.0823 + 0.0082i	0.0006 + 0.0014i
26	10	17	0.0347 + 0.0635i	0.0002 + 0.0004i
27	10	21	0.1643 + 0.0725i	0.0011 + 0.0024i
28	10	22	0.0802 + 0.0327i	0.0005 + 0.0011i

(continued)

Table 5.23 (continued)

Line No.	From bus	To bus	Line flow P_{line} (p.u.) + i Q_{line} (p.u.)	Line loss P_{loss} (p.u.) + i Q_{loss} (p.u.)
29	21	22	−0.0118 − 0.0119i	0
30	15	23	0.0672 + 0.0147i	0.0005 + 0.0010i
31	22	24	0.0679 + 0.0197i	0.0006 + 0.0009i
32	23	24	0.0347 − 0.0023i	0.0002 + 0.0003i
33	24	25	0.0148 − 0.0008i	0 + 0.0001i
34	25	26	0.0355 + 0.0237i	0.0005 + 0.0007i
35	25	27	−0.0207 − 0.0246i	0.0001 + 0.0002i
36	28	27	0.1537 + 0.0483i	0 + 0.0101i
37	27	29	0.0620 + 0.0023i	0.0009 + 0.0016i
38	27	30	0.0709 + 0.0111i	0.0017 + 0.0031i
39	29	30	0.0372 + 0.0117i	0.0004 + 0.0007i
40	4	28	0.0375 − 0.0133i	0.0001 − 0.0436i
41	13	28	0.1165 + 0.0055i	0.0002 − 0.0125i

Table 5.24 IEEE 30-bus system individual objective functions considering emission index minimization

Objective function	Objective value
Active power transmission loss (f1)	3.4507 MW
Overall generation fuel costs (f2)	968.4903 $/h
Reactive power reserve margin (f3)	0.5702
Reactive power transmission loss (f4)	−6.4128 MVAR
Emission index minimization (f5)	0.2592 t/h
Security margin index (f6)	27.4050

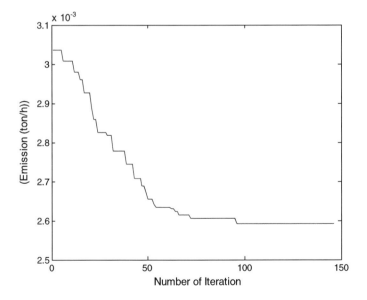

Fig. 5.8 Emission index minimization using PSO algorithm

5.5 Comparisons of Different Single Objective Functions

Table 5.25 Objective function 6: security margin index maximization (bus result)

Bus No.	V (p.u.)	Delta (degree)	P_D (p.u.)	Q_D (p.u.)	P_G (p.u.)	Q_G (p.u.)
1	1.1000	0	0.0000	0.0000	1.6225	0.7449
2	1.0520	−2.5480	0.2170	0.1270	0.5135	−0.0224
3	0.9724	−8.6139	0.9420	0.1900	0.1959	−0.1328
4	1.0013	−6.5932	0.3000	0.3000	0.1150	0.0476
5	0.9504	−4.6010	0.0000	0.0000	0.1928	0.0062
6	1.1000	−6.3508	0.0000	0.0000	0.3047	0.4135
7	0.9884	−7.7886	0.2280	0.1090	0	0
8	1.0299	−5.3239	0.0760	0.0160	0	0
9	0.9500	−7.1463	0.0000	0.0000	0	0
10	0.9711	−8.8979	0.0580	0.0200	0	0.0250
11	1.0430	−4.4483	0.0240	0.0120	0	0
12	1.0481	−8.4710	0.1120	0.0750	0	0.0500
13	1.0130	−6.2713	0.0000	0.0000	0	0
14	1.0304	−9.5029	0.0620	0.0160	0	0
15	1.0223	−9.6585	0.0820	0.0250	0	0.0500
16	1.0095	−8.9316	0.0350	0.0180	0	0
17	0.9794	−9.1798	0.0900	0.0580	0	0.0400
18	0.9935	−10.1894	0.0320	0.0090	0	0
19	0.9797	−10.3042	0.0950	0.0340	0	0
20	0.9780	−10.0467	0.0220	0.0070	0	0.0250
21	0.9665	−9.5212	0.1750	0.1120	0	0.0350
22	0.9689	−9.5201	0.0000	0.0000	0	0
23	1.0081	−10.2862	0.0320	0.0160	0	0.0500
24	0.9845	−10.4029	0.0870	0.0670	0	0.0150
25	1.0180	−10.7778	0.0000	0.0000	0	0
26	1.0004	−11.1969	0.0350	0.0230	0	0
27	1.0476	−10.6952	0.0000	0.0000	0	0
28	1.0017	−6.7706	0.0000	0.0000	0	0
29	1.0312	−11.9493	0.0240	0.0090	0	0.0100
30	1.0188	−12.7505	0.1060	0.0190	0	0

specific case. The basic case indicated in this table concerns the load flow without applying optimization techniques.

Case 1 is the active power loss minimization study. In this case, the active power loss of the studied network was considered for minimization. It is clear that minimizing this objective function has improved many objective functions except the generation fuel cost and emission index. It is concluded that minimizing the power loss is highly correlated with the generation fuel cost minimization, and emission index minimization.

Case 2 is the generation fuel cost minimization study. It was discovered that minimization of P_{loss} will worsen the generator fuel cost, emission index, and reactive power loss objective functions. Once again, it is concluded that minimizing the power loss is highly correlated with the generation fuel cost minimization and emission index minimization.

Table 5.26 Control variables of the sixth objective function

Control variable	Optimal value
P_{G2} (MW)	0.5135
P_{G3} (MW)	0.1959
P_{G4} (MW)	0.1150
P_{G5} (MW)	0.1928
P_{G6} (MW)	0.3047
V_{G1} (p.u.)	1.1000
V_{G2} (p.u.)	1.0520
V_{G3} (p.u.)	0.9724
V_{G4} (p.u.)	1.0013
V_{G5} (p.u.)	0.9504
V_{G6} (p.u.)	1.1000
TC_{L11} (p.u.)	0.9000
TC_{L12} (p.u.)	0.9625
TC_{L15} (p.u.)	1.0000
TC_{L36} (p.u.)	1.1000
Q_{C10} (p.u.)	0.0250
Q_{C12} (p.u.)	0.0500
Q_{C15} (p.u.)	0.0500
Q_{C17} (p.u.)	0.0400
Q_{C20} (p.u.)	0.0250
Q_{C21} (p.u.)	0.0350
Q_{C23} (p.u.)	0.0500
Q_{C24} (p.u.)	0.0150
Q_{C29} (p.u.)	0.0100

Table 5.27 Objective function 6: security margin index maximization (line flow and losses)

Line No.	From bus	To bus	Line flow P_{line} (p.u.) + i Q_{line} (p.u.)	Line loss P_{loss} (p.u.) + i Q_{loss} (p.u.)
1	1	2	1.0867 + 0.5430i	0.0240 + 0.0107i
2	1	11	0.5358 + 0.2019i	0.0126 + 0.0049i
3	2	8	0.3146 + 0.0177i	0.0052 − 0.0241i
4	11	8	0.4991 + 0.1850i	0.0035 + 0.0009i
5	2	3	0.6175 + 0.2810i	0.0202 + 0.0420i
6	2	13	0.4271 + 0.0843i	0.0102 − 0.0091i
7	8	13	0.4976 + 0.2768i	0.0037 + 0.0034i
8	3	7	−0.1488 − 0.0839i	0.0013 − 0.0162i
9	13	7	0.3828 + 0.1740i	0.0047 − 0.0026i
10	13	4	0.1997 + 0.2194i	0.0011 − 0.0054i
11	13	9	0.0636 − 0.1676i	0 + 0.0080i
12	13	10	0.0780 + 0.0086i	0 + 0.0036i
13	9	5	−0.1928 + 0.0024i	0 + 0.0086i

(continued)

5.5 Comparisons of Different Single Objective Functions

Table 5.27 (continued)

Line No.	From bus	To bus	Line flow P_{line} (p.u.) + i Q_{line} (p.u.)	Line loss P_{loss} (p.u.) + i Q_{loss} (p.u.)
14	9	10	0.2564 − 0.1780i	0 + 0.0119i
15	8	12	0.2315 − 0.0669i	0 + 0.0140i
16	12	6	0.3047 − 0.3829i	0 + 0.0305i
17	12	14	0.0903 + 0.0297i	0.0010 + 0.0021i
18	12	15	0.2199 + 0.0977i	0.0035 + 0.0069i
19	12	16	0.1140 + 0.1496i	0.0030 + 0.0064i
20	14	15	0.0273 + 0.0117i	0.0002 + 0.0002i
21	16	17	0.0759 + 0.1252i	0.0017 + 0.0040i
22	15	18	0.0880 + 0.0918i	0.0017 + 0.0034i
23	18	19	0.0544 + 0.0794i	0.0006 + 0.0012i
24	19	20	−0.0412 + 0.0442i	0.0001 + 0.0003i
25	10	20	0.0641 − 0.0602i	0.0008 + 0.0017i
26	10	17	0.0162 − 0.1022i	0.0004 + 0.0010i
27	10	21	0.1353 − 0.0023i	0.0007 + 0.0015i
28	10	22	0.0608 − 0.0151i	0.0003 + 0.0006i
29	21	22	−0.0404 − 0.0807i	0.0001 + 0.0002i
30	15	23	0.0735 + 0.0355i	0.0006 + 0.0013i
31	22	24	0.0200 − 0.0967i	0.0012 + 0.0019i
32	23	24	0.0408 + 0.0682i	0.0008 + 0.0017i
33	24	25	−0.0282 − 0.0840i	0.0015 + 0.0027i
34	25	26	0.0354 + 0.0237i	0.0004 + 0.0007i
35	25	27	−0.0651 − 0.1103i	0.0017 + 0.0033i
36	28	27	0.1995 + 0.1577i	0 + 0.0211i
37	27	29	0.0619 + 0.0093i	0.0008 + 0.0015i
38	27	30	0.0708 + 0.0136i	0.0015 + 0.0029i
39	29	30	0.0371 + 0.0088i	0.0003 + 0.0006i
40	4	28	0.0136 − 0.0275i	0 − 0.0429i
41	13	28	0.1868 + 0.1323i	0.0009 − 0.0100i

Table 5.28 IEEE 30-bus system individual objective functions considering maximization of security margin index

Objective function	Objective value
Active power transmission loss (f1)	11.0320 MW
Overall generation fuel costs (f2)	823.0009 $/h
Reactive power reserve margin (f3)	0.5888
Reactive power transmission loss (f4)	9.4976 MVAR
Emission index (f5)	0.3887 t/h
Security margin index maximization (f6)	21.6421

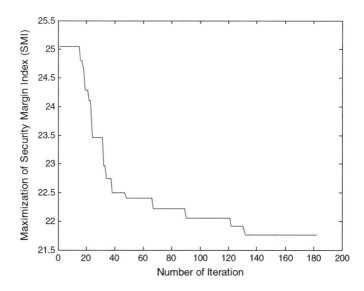

Fig. 5.9 Security margin maximization using PSO algorithm

Table 5.29 Comparisons of all single OPF objectives

	P_{LOSS} (MW)	GFC ($/h)	RPRM (p.u.)	Q_{LOSS} (MVAR)	EI (ton/h)	SMI (p.u.)
Basic Case	5.4434	900.9102	0.5472	−9.4017	0.2697	27.2555
Case 1	**3.1162**	967.7009	0.2415	−20.0883	0.2728	29.2449
Case 2	9.0706	**799.9859**	0.3198	1.3704	0.4311	26.8580
Case 3	3.6433	968.3029	**0.0771**	−16.5976	0.2739	28.1498
Case 4	3.1644	967.7024	0.2500	**−20.4785**	0.2732	29.1831
Case 5	3.8186	969.3794	0.6146	−1.3480	**0.2594**	27.3635
Case 6	11.0320	823.0009	0.5888	9.4976	0.3887	**21.6421**

Case 3 is the reactive reserve margin maximization study. In this case, it was observed that minimization of the reactive power reserve margin will improve the power loss whereas other objective functions become worse.

Case 4 is the reactive power loss minimization study. In this study, it was concluded that minimization of reactive power loss directly resulted in improving the power loss as well as the reactive power reserve margin.

Case 5 is the emission index minimization study. In this case, it was discovered that minimization of emission index has only resulted in improving the network power loss.

5.7 Basic Concept of Multiobjective Analysis

Case 6 is the security margin index maximization study. In this case, it was also found that minimizing the security margin index has only resulted in improving the generation fuel cost.

From the previous results and comments, conflict between some objectives is obvious. Therefore, the need for a multiobjective optimization algorithm for this problem is clear.

5.6 Multiobjective OPF Algorithm

Multiobjective programming problems consist of several objective functions that need to be achieved simultaneously. Such problems arise in many applications, where two or more competing and/or incommensurable objective functions have to be optimized concurrently. Due to the multicriteria nature of multiobjective problems, the optimality of a solution has to be redefined. A multiobjective programming problem is characterized by a p-dimensional vector of functions. This can be represented as

$$\bar{Z}(x) = [Z_1(x), \; Z_2(x), \; \ldots \ldots \; Z_p(x)] \tag{5.49}$$

Subject to:

$$x \in X$$

Where X is a feasible region:

$$\mathbf{X} = \{\mathbf{x} : \mathbf{x} \in \mathbf{R^n}, \; g_i(\mathbf{x}) \leq 0, \; x_j \geq 0 \; \forall i, j\} \tag{5.50}$$

where

$\mathbf{R}$ = set of real numbers
$g_i(\mathbf{x})$ = set of constraints
$\mathbf{x}$ = set of decision variables

The word optimization has been kept out of the definition of a multiobjective programming problem because one cannot in general optimize a priori a vector of objective functions [86]. The first step of the multiobjective problem consists of identifying the set of nondominated solutions within the feasible region X. Therefore, instead of seeking a single optimal solution, a set of "noninferior" solutions is sought.

5.7 Basic Concept of Multiobjective Analysis

The focus of multiobjective analysis in practice is to reform the mass of clearly dominated solutions, rather than determine the single best design. The result is the identification of a small subset of feasible solutions that are worthy of further

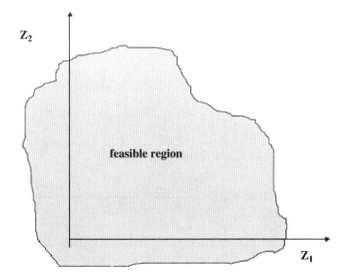

Fig. 5.10 Domain of the objective function in a two-objective problem

consideration. Formally, the result is known as nondominated solutions. The nondominated solutions are sometimes referred to by other names: noninferior, Pareto optimal, efficient, and so on.

The essential feature of the multiobjective problem is that the feasible region of generating solution is much more complex than for a single objective. In single optimization, any set of inputs x produces a set of results z that could be represented by a straight line going from bad (typically zero output) to best in the case of a maximization problem. In the multiobjective problem, any set of inputs x defines a multidimensional space of feasible solutions, as shown in Fig. 5.10. Then, there is no exact equivalent of a single optimal solution.

The nondominated solutions are the conceptual equivalents, in multiobjective problems, of a single optimal solution in a single objective problem. The main characteristic of the nondominated solution is that for each solution outside the set there is a nondominated solution for which all objective functions are unchanged or improved and at least one is strictly improved [87]. The preferred design of any problem should be one of the nondominated solutions. As long as all objectives worth taking into account have been considered, no design that is not among the nondominated solutions is worthwhile; it is dominated by some designs that are preferable on all accounts. This the main reason for determining the nondominated set for multiobjective analysis.

Given a set of feasible solutions X, the set of nondominated solutions is denoted as S and defined as in (6.3). It is obvious from the definition of S that as one moves from one nondominated solution to another nondominated solution and one objective function improves, then one or more of the other objective functions must decrease in value.

$$S = \{x : x \in X, \text{ there exists no other } x' \in X$$
$$\text{such that } z_q(x') > z_q(x) \text{ for some } q \in \{1, 2, \ldots, p\} \quad (5.51)$$
$$\text{and } z_k(x') \leq z_k(x) \text{ for all } k \neq q\}$$

5.8 The Proposed Multiobjective OPF Algorithm

5.8.1 Multiobjective OPF Formulation

In Chap. 4, the single objective OPF problem is solved for different. The PSO algorithm presented earlier is modified to suit the multiobjective OPF problem. A weighting method, introduced earlier, is used to generate the nondominated set. All the objective functions discussed in the previous chapters are used for solving the multiobjective OPF problem. These functions include active power loss minimization, generation fuel cost minimization, reactive power reserve margin maximization, reactive power transmission loss minimization, emission index minimization, and security margin index maximization. Therefore, the multiobjective OPF problem can be stated as follows.

Minimize $f(x)$ where

$$f(x) = w_1 f_1(x) + w_2 f_2(x) + w_3 f_3(x) + w_4 f_4(x) + w_5 f_5(x) + w_6 f_6(x) \quad (5.52)$$

$$\bar{f}(x) = \begin{bmatrix} f_1(x) \\ f_2(x) \\ f_3(x) \\ f_4(x) \\ f_5(x) \\ f_6(x) \end{bmatrix} = \begin{bmatrix} APLOSS \\ GFC \\ RPRM \\ RPLOSS \\ EI \\ SMI \end{bmatrix} \quad (5.53)$$

$$w_6 = 1 - \sum_{i=1}^{5} w_i$$

Subject to:

$$x \in X$$

Where X is a feasible region:

$$X = \{x : x \in \mathbf{R}^n, g_i(x) \leq 0, x_j \geq 0 \; \forall i, j\} \quad (5.54)$$

Where

R = set of real numbers
$g_i(\mathbf{x})$ = all constraints introduced earlier
x = 29 decision variables introduced earlier.

5.8.2 General Steps for Solving Multi-Objective OPF Problem

The major steps of the multiobjective OPF algorithm are given in Fig. 5.11. Below are the steps to solve the multiobjective OPF problem.

Step-1: Feasible initialization.
Step-2: Set the six weighting factors in order to get all the nondominated solutions.
Step-3: Perform the OPF algorithm as stated in Sect. 4.4 and in each run recalculate the load flow.
Step-4: If all nondominated solutions are found then go to Step 5 if not then go to Step 2.
Step-5: Use the hierarchical clustering technique to minimize the nondominated set to a smaller set, which will give some flexibility to the decision maker.
Step-6: Give the solutions to the decision maker.

5.9 Generating Nondominated Set

5.9.1 Generating techniques

A generating method considers a vector of objective functions and uses this vector to identify and generate the subset of nondominated solutions in the initial feasible region. In doing so, these methods deal strictly with the physical realities of the problem (i.e., the set of constraints) and make no attempt to consider the preferences of a decision maker (DM). The desired outcome, then, is the identification of the set of nondominated solutions to help the DM gain insight into the physical realities of the problem at hand.

There are several methods available to generate the set of nondominated solutions, and four of these methods are widely known. These methods are:

- Weighting method
- e-Constraint method
- Phillip's linear multiobjective method
- Zeleny's linear multiobjective method

The first two methods transform the multiobjective problem into a single objective programming format and then, by parametric variation of the parameters used

5.9 Generating Nondominated Set

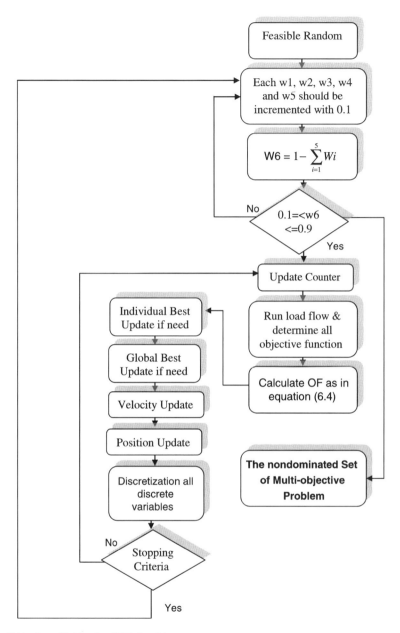

Fig. 5.11 A multiobjective OPF algorithm

to effect the transformation, the set of nondominated solutions can be generated. The weighting and constraint methods can be used to obtain nondominated solutions when the objective functions and/or constraints are nonlinear. The last two methods generate the nondominated set for linear models only. However, these two approaches do not require the transformation of the problem into a single objective programming format. These methods operate directly on the vector of objectives to obtain the nondominated solutions.

In this research, the weighting method has been used in order to generate the nondominated set of the multiobjective optimal power flow problem. Further insight into the weighting method is presented in the next section.

5.9.2 Weighting method

The weighting method belongs to the group of techniques for generating a nondominated set. It is based on the idea of assigning weights to the various objective functions, combining these into a single objective function, and parametrically varying the weights to generate the nondominated set. Mathematically, the weighting method can be stated as follows.

$$\text{Minimize } f(x) = w_1 f_1(x) + w_2 f_2(x) + \ldots\ldots\ldots + w_p f_p(x) \quad (5.55)$$

Subject to:

$$x \in X$$

Therefore, a multiobjective problem has been transformed into a single objective optimization problem for which the solution methods exist. The coefficient w_i operating on the ith objective function $f_i(x)$, is called a weight and can be interpreted as the relative weight or worth of that objective when compared to the other objectives. If the weights of the various objectives are interpreted as representing the relative preferences of some DM, then the solution to (5.55) is equivalent to the best-compromise solution, that is, the optimal solution relative to a particular preference structure. Moreover, the optimal solution to Eq. 5.55 is a nondominated solution, provided all the weights are positive. The weighting method will give an approximation of the nondominated set, because the nondominated set could have infinite solutions.

In this section, the lower value for the six weighting factors was 0.1 and the upper value was set to be 0.9 with a maximum increment of 0.1 in each run. The weighting factor method has generated a 126 nondominated solution. The nondominated solutions and their respective weighting factors are tabulated in Table A.5. The nondominated set is considered a big list that cannot be given to the decision maker. A cluster technique has been applied to minimize the nondominated set to a smaller set, which will give some flexibility to the decision maker. The clustering technique is presented in the next section.

5.10 Hierarchical Cluster Technique

The most commonly used term for techniques that seek to separate data into constituent groups is cluster analysis [88]. In a hierarchical classification, the data are not partitioned into a particular number of clusters at a single step. Instead the classification consists of a series of partitions, which may run from a single cluster containing all nondominated solutions, to n clusters each containing a single individual [88]. A hierarchical clustering technique has been implemented in order to reduce the nondominated set to a smaller set. This will definitely give the decision maker some flexibility in terms of selection. The hierarchical clustering methodology is a means to investigate grouping the data simultaneously over a variety of scales by creating a cluster tree. The tree is not a single cluster, but rather a multilevel hierarchy, where clusters at one level are joined as clusters at the next higher level. This allows deciding what level or scale of clustering is most appropriate.

To perform hierarchical cluster analysis in the nondominated solutions, the following steps should be followed [88, 89].

Step 1: Calculate the distance between the nondominated solutions using the formal Euclidean distance as in (5.56). As mentioned in the previous section, 126 nondominated solutions have been found. This set can be considered as a matrix of 126×6 and the number 6 is the number of the objective functions.

$$d^2 rs = (fr - fs)(fr - fs)' \qquad (5.56)$$

where $fr = [f_{r1}\ f_{r2}\ f_{r3}\ f_{r4}\ f_{r5}\ f_{r6}]$

Step 2: Group the objects into a binary hierarchical cluster tree by finding the average distance. This distance should be calculated between all pairs of objects in clusters r and s by using Eq. 5.57.

$$d(r,s) = \frac{1}{nr * ns} \sum_{i=1}^{nr} \sum_{j=1}^{ns} dist(xri, xsj) \qquad (5.57)$$

Step 3: Plot a dendrogram graph, which consists of U-shaped lines connecting objects in a hierarchical tree.

Step 4: Determine where to divide the hierarchical cluster tree. This depends on the decision maker.

Step 5: Find a candidate representative for each cluster by taking the member of the smallest distance, which is calculated between each cluster member and the average of all cluster members.

Figure 5.12 shows the steps required for performing hierarchical cluster analysis. The dendrogram graph of the nondominated set is shown in Fig. 5.13 after

Fig. 5.12 Hierarchical cluster analysis steps

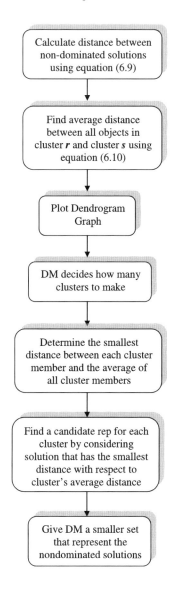

conducting the first three steps of the previous procedure. In Step 4, two, three, and four cluster divisions have been proposed. Table 5.30 shows two representative solutions in the case of two clusters and their control variables are shown in Table 5.31. The first solution (107) is better than the second solution (15) only in two objective functions that include generation fuel cost and security margin index, and solution (15) is better than solution (107) in four objective functions that include power loss, reactive power reserve margin, reactive loss, and emission.

5.10 Hierarchical Cluster Technique

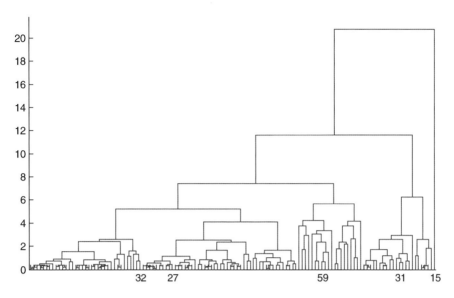

Fig. 5.13 MOOPF hierarchical cluster tree

Table 5.30 Considering two clusters

Objective Fun.	Sol. (107)	Sol. (15)
P_{loss} (MW)	7.783	6.0492
GFC ($/h)	805.4443	824.7188
RPRM (p.u.)	0.359	0.2484
Q_{loss} (MVAR)	−3.4097	−10.9965
Emission (ton/h)	0.3813	0.3320
SMI	26.6056	28.0342

Clustering:
C2 = {15}
C1 = {all the nondominated solutions} − C2

Table 5.32 shows three representative solutions in the case of three clusters, and their control variables are presented in Table 5.33. Table 5.34 gives four representative solutions in the case of four clusters, and their control variables are shown in Table 5.35. The proposed sets give the decision maker some flexibility in selecting the appropriate set that meets her needs. The candidate set depends mainly on the DM requirements. The two, three, and four clusters are proposed to provide the decision maker some flexibility in selecting the appropriate choice.

Table 5.31 Control variables of two representative solutions

Control variable	Sol. (107)	Sol. (15)
P_{G2} (p.u.)	0.4911	0.5154
P_{G3} (p.u.)	0.2404	0.3103
P_{G4} (p.u.)	0.2868	0.3252
P_{G5} (p.u.)	0.1599	0.1887
P_{G6} (p.u.)	0.171	0.2604
V_{G1} (p.u.)	1.0798	1.0773
V_{G2} (p.u.)	1.0608	1.0606
V_{G3} (p.u.)	1.0279	1.0379
V_{G4} (p.u.)	1.0393	1.035
V_{G5} (p.u.)	1.0371	1.055
V_{G6} (p.u.)	1.0146	1.0403
TC_{L11} (p.u.)	0.9875	1.025
TC_{L12} (p.u.)	1.075	1.0125
TC_{L15} (p.u.)	0.975	1
TC_{L36} (p.u.)	1.0375	1.0375
Q_{C10} (p.u.)	0.05	0.04
Q_{C12} (p.u.)	0.045	0.05
Q_{C15} (p.u.)	0.02	0.03
Q_{C17} (p.u.)	0.04	0.05
Q_{C20} (p.u.)	0.045	0.05
Q_{C21} (p.u.)	0.04	0.045
Q_{C23} (p.u.)	0.035	0.05
Q_{C24} (p.u.)	0.03	0.045
Q_{C29} (p.u.)	0.05	0.01

Table 5.32 Considering three clustering

Objective fun.	Sol. (32)	Sol. (31)	Sol. (15)
P_{loss} (MW)	8.3015	6.7370	6.0492
GFC ($/h)	804.0179	812.9552	824.7188
RPRM (p.u.)	0.2080	0.3189	0.2484
Q_{loss} (MVAR)	−1.8035	−8.3765	−10.9965
Emission (ton/h)	0.3846	0.3534	0.3320
SMI	26.2561	27.8074	28.0342

Clustering:
C2 = {10, 11, 12, 13, 14, 22, 23, 24, 25, 31, 78, 79, 80, 84, 85, 86, 89, 109, 110, 111, 114, 123}
C3 = {15}
C1 = {all the nondominated solutions} − C2 − C3

5.10 Hierarchical Cluster Technique

Table 5.33 Optimal control variables of three representative solutions

Control variable	Sol. (32)	Sol. (31)	Sol. (15)
P_{G2} (p.u.)	0.4895	0.4658	0.5154
P_{G3} (p.u.)	0.23	0.2625	0.3103
P_{G4} (p.u.)	0.2861	0.35	0.3252
P_{G5} (p.u.)	0.1415	0.1802	0.1887
P_{G6} (p.u.)	0.1515	0.2142	0.2604
V_{G1} (p.u.)	1.0857	1.077	1.0773
V_{G2} (p.u.)	1.0648	1.06	1.0606
V_{G3} (p.u.)	1.0311	1.0363	1.0379
V_{G4} (p.u.)	1.0215	1.039	1.035
V_{G5} (p.u.)	1.057	1.027	1.055
V_{G6} (p.u.)	1.013	1.0512	1.0403
TC_{L11} (p.u.)	1	1	1.025
TC_{L12} (p.u.)	1	1.025	1.0125
TC_{L15} (p.u.)	0.95	1.025	1
TC_{L36} (p.u.)	1.0625	1.0125	1.0375
Q_{C10} (p.u.)	0.05	0.05	0.04
Q_{C12} (p.u.)	0.05	0.04	0.05
Q_{C15} (p.u.)	0.04	0.04	0.03
Q_{C17} (p.u.)	0.05	0.03	0.05
Q_{C20} (p.u.)	0.045	0.03	0.05
Q_{C21} (p.u.)	0.035	0.035	0.045
Q_{C23} (p.u.)	0.03	0.035	0.05
Q_{C24} (p.u.)	0.05	0.05	0.045
Q_{C29} (p.u.)	0.015	0.04	0.01

Table 5.34 Considering four clustering

Objective Fun.	Sol. (59)	Sol. (27)	Sol. (31)	Sol. (15)
P_{loss} (MW)	9.1555	7.981	6.7370	6.0492
GFC ($/h)	804.8738	803.4925	812.9552	824.7188
RPRM (p.u.)	0.2136	0.2013	0.3189	0.2484
Q_{loss} (MVAR)	2.3297	−3.5686	−8.3765	−10.9965
Emission (ton/h)	0.3971	0.3838	0.3534	0.3320
SMI	25.8174	27.0139	27.8074	28.0342

Clustering:
C1 = {1, 2, 6, 16, 26, 36, 38, 46, 56, 59, 70, 71, 81, 82, 91, 97, 98, 106, 112, 119}
C3 = {10, 11, 12, 13, 14, 22, 23, 24, 25, 31, 78, 79, 80, 84, 85, 86, 89, 109, 110, 111, 114, 123}
C4 = {15}
C2 = {all the nondominated solutions} − C1 − C3 − C4

Table 5.35 Optimal control variables of the four representative solutions

Control variable	3.3.5 Sol. (59)	3.3.5 Sol. (27)	3.3.5 Sol. (31)	3.3.5 Sol. (15)
P_{G2} (p.u.)	0.4853	0.4857	0.4658	0.5154
P_{G3} (p.u.)	0.2301	0.2352	0.2625	0.3103
P_{G4} (p.u.)	0.2575	0.2929	0.35	0.3252
P_{G5} (p.u.)	0.1369	0.1503	0.1802	0.1887
P_{G6} (p.u.)	0.1465	0.1405	0.2142	0.2604
V_{G1} (p.u.)	1.0671	1.0841	1.077	1.0773
V_{G2} (p.u.)	1.045	1.0654	1.06	1.0606
V_{G3} (p.u.)	1.0087	1.0303	1.0363	1.0379
V_{G4} (p.u.)	0.9986	1.0285	1.039	1.035
V_{G5} (p.u.)	0.9874	1.053	1.027	1.055
V_{G6} (p.u.)	1.0132	1.0625	1.0512	1.0403
TC_{L11} (p.u.)	0.975	1	1	1.025
TC_{L12} (p.u.)	0.9625	0.9875	1.025	1.0125
TC_{L15} (p.u.)	0.9875	1.0125	1.025	1
TC_{L36} (p.u.)	1.1	1.025	1.0125	1.0375
Q_{C10} (p.u.)	0.05	0.025	0.05	0.04
Q_{C12} (p.u.)	0.05	0.05	0.04	0.05
Q_{C15} (p.u.)	0.05	0.03	0.04	0.03
Q_{C17} (p.u.)	0.025	0.035	0.03	0.05
Q_{C20} (p.u.)	0.04	0.03	0.03	0.05
Q_{C21} (p.u.)	0.05	0.05	0.035	0.045
Q_{C23} (p.u.)	0.05	0.04	0.035	0.05
Q_{C24} (p.u.)	0.05	0.04	0.05	0.045
Q_{C29} (p.u.)	0.02	0.05	0.04	0.01

5.11 Conclusions

This chapter has significantly accomplished many things in the area under discussion, which is the multiobjective optimal power flow. These achievements can be summarized as follows.

1. The single OPF objective function optimization algorithm based on the particle swarm optimization algorithm was developed and applied to a practical power system network. The developed OPF algorithm offers the following.

    (a) Provides a flexibility to add or delete any system constraints and objective functions. Having this flexibility will help electrical engineers analyzing other system scenarios and contingency plans.
    (b) Calculates the optimum generation pattern as well as all control variables in order to minimize the cost of generation together with meeting the transmission system limitations.
    (c) Finds the optimum setting for system control variables that achieve minimum objective functions. These control variables include: active power generation

except the slack bus, all PV-bus voltages, all transformer load tap changers, and the setting of all switched reactors or static VAR components.
(d) Provides a corrective dispatch action, which tells the system's operators how to modify the control variables in order to relieve the overload or voltage violation of any component in the system during any emergency case.
(e) Determines the maximum stress that a planned transmission system can withstand. This is especially useful for electrical system planning studies.

2. The developed OPF package has been applied to the IEEE-30 bus as a power system tool to optimize the following objective functions in different case studies. These include:

   (a) Active power loss minimization
   (b) Generation fuel cost minimization
   (c) Reactive power reserve maximization
   (d) Reactive power loss minimization
   (e) Emission index minimization
   (f) Security margin index maximization.

3. A new formulation for the multiobjective OPF problem has been proposed and this includes both the objective functions and the constraints.
4. A new multiobjective OPF algorithm has been developed using the PSO algorithm. The weighting factors concept has been successfully used to generate the nondominated set for the multiobjective OPF problem.
5. A hierarchical clustering technique was employed in order to provide the decision makers a candidate set that represents the nondominated solutions. The candidate set depends mainly on the DM requirements. In this work two, three, and four clustering has been proposed which will give the decision maker some flexibility in making the appropriate choice.

Appendix

Table A.1 IEEE 30-bus system line data

Line No.	From bus	To bus	R (p.u.)	X (p.u.)	Bover2 (p.u.)	Rating (MVA)	Tap ratio (p.u.)
1	1	2	0.0192	0.0575	0.0264	130.0	1.0
2	1	11	0.0452	0.1852	0.0204	130.0	1.0
3	2	8	0.0570	0.1737	0.0184	65.0	1.0
4	11	8	0.0132	0.0379	0.0042	130.0	1.0
5	2	3	0.0472	0.1983	0.0209	130.0	1.0
6	2	13	0.0581	0.1763	0.0187	65.0	1.0
7	8	13	0.0119	0.0414	0.0045	90.0	1.0
8	3	7	0.0460	0.1160	0.0102	70.0	1.0
9	13	7	0.0267	0.0820	0.0085	130.0	1.0
10	13	4	0.0120	0.0420	0.0045	32.0	1.0

(continued)

Table A.1 (continued)

Line No.	From bus	To bus	R (p.u.)	X (p.u.)	Bover2 (p.u.)	Rating (MVA)	Tap ratio (p.u.)
11	13	9	0.0000	0.2080	0.0000	65.0	0.978
12	13	10	0.0000	0.5560	0.0000	32.0	0.969
13	9	5	0.0000	0.2080	0.0000	65.0	1.0
14	9	10	0.0000	0.1100	0.0000	65.0	1.0
15	8	12	0.0000	0.2560	0.0000	65.0	0.932
16	12	6	0.0000	0.1400	0.0000	65.0	1.0
17	12	14	0.1231	0.2559	0.0000	32.0	1.0
18	12	15	0.0662	0.1304	0.0000	32.0	1.0
19	12	16	0.0945	0.1987	0.0000	32.0	1.0
20	14	15	0.2210	0.1997	0.0000	16.0	1.0
21	16	17	0.0824	0.1923	0.0000	16.0	1.0
22	15	18	0.1070	0.2185	0.0000	16.0	1.0
23	18	19	0.0639	0.1292	0.0000	16.0	1.0
24	19	20	0.0340	0.0680	0.0000	32.0	1.0
25	10	20	0.0936	0.2090	0.0000	32.0	1.0
26	10	17	0.0324	0.0845	0.0000	32.0	1.0
27	10	21	0.0348	0.0749	0.0000	32.0	1.0
28	10	22	0.0727	0.1499	0.0000	32.0	1.0
29	21	22	0.0116	0.0236	0.0000	32.0	1.0
30	15	23	0.1000	0.2020	0.0000	16.0	1.0
31	22	24	0.1150	0.1790	0.0000	16.0	1.0
32	23	24	0.1320	0.2700	0.0000	16.0	1.0
33	24	25	0.1885	0.3292	0.0000	16.0	1.0
34	25	26	0.2544	0.3800	0.0000	16.0	1.0
35	25	27	0.1093	0.2087	0.0000	16.0	1.0
36	28	27	0.0000	0.3960	0.0000	65.0	0.968
37	27	29	0.2198	0.4153	0.0000	16.0	1.0
38	27	30	0.3202	0.6027	0.0000	16.0	1.0
39	29	30	0.2399	0.4533	0.0000	16.0	1.0
40	4	28	0.0636	0.2000	0.0214	32.0	1.0
41	13	28	0.0169	0.0599	0.0065	32.0	1.0

Table A.2 IEEE 30-bus system bus data

Bus No.	V (pu)	P_G (pu)	P_D (pu)	Q_D (pu)	Q_{gmin} (pu)	Q_{gmax} (pu)	P_{gmin} (MW)	P_{gmax} (MW)
1	1.0500	0.00	0.0000	0.0000	−0.200	2.00	0.500	2.000
2	1.0382	48.84	0.2170	0.1270	−0.200	1.00	0.200	0.800
3	1.0114	21.51	0.9420	0.1900	−0.150	0.80	0.150	0.500
4	1.0194	22.15	0.3000	0.3000	−0.150	0.60	0.100	0.350
5	1.0912	12.14	0.0000	0.0000	−0.100	0.50	0.100	0.300
6	1.0913	12.00	0.0000	0.0000	−0.150	0.60	0.120	0.400
7	1.0000	0.00	0.2280	0.1090	0.0	0.0	0.0	0.0
8	1.0000	0.00	0.0760	0.0160	0.0	0.0	0.0	0.0
9	1.0000	0.00	0.0000	0.0000	0.0	0.0	0.0	0.0

(continued)

Appendix

Table A.2 (continued)

Bus No.	V (pu)	P_G (pu)	P_D (pu)	Q_D (pu)	Q_{gmin} (pu)	Q_{gmax} (pu)	P_{gmin} (MW)	P_{gmax} (MW)
10	1.0000	0.00	0.0580	0.0200	0.0	0.05	0.0	0.0
11	1.0000	0.00	0.0240	0.0120	0.0	0.0	0.0	0.0
12	1.0000	0.00	0.1120	0.0750	0.0	0.05	0.0	0.0
13	1.0000	0.00	0.0000	0.0000	0.0	0.0	0.0	0.0
14	1.0000	0.00	0.0620	0.0160	0.0	0.0	0.0	0.0
15	1.0000	0.00	0.0820	0.0250	0.0	0.05	0.0	0.0
16	1.0000	0.00	0.0350	0.0180	0.0	0.0	0.0	0.0
17	1.0000	0.00	0.0900	0.0580	0.0	0.05	0.0	0.0
18	1.0000	0.00	0.0320	0.0090	0.0	0.0	0.0	0.0
19	1.0000	0.00	0.0950	0.0340	0.0	0.0	0.0	0.0
20	1.0000	0.00	0.0220	0.0070	0.0	0.05	0.0	0.0
21	1.0000	0.00	0.1750	0.1120	0.0	0.05	0.0	0.0
22	1.0000	0.00	0.0000	0.0000	0.0	0.0	0.0	0.0
23	1.0000	0.00	0.0320	0.0160	0.0	0.05	0.0	0.0
24	1.0000	0.00	0.0870	0.0670	0.0	0.05	0.0	0.0
25	1.0000	0.00	0.0000	0.0000	0.0	0.0	0.0	0.0
26	1.0000	0.00	0.0350	0.0230	0.0	0.0	0.0	0.0
27	1.0000	0.00	0.0000	0.0000	0.0	0.0	0.0	0.0
28	1.0000	0.00	0.0000	0.0000	0.0	0.0	0.0	0.0
29	1.0000	0.00	0.0240	0.0090	0.0	0.05	0.0	0.0
30	1.0000	0.00	0.1060	0.0190	0.0	0.0	0.0	0.0

Table A.3 IEEE 30-bus system generation fuel cost and emission coefficients

		G1	G2	G3	G4	G5	G6
Cost	a	0	0	0	0	0	0
	b	200	175	100	325	300	300
	c	37.5	175	625	83.4	250	250
Emission	α	4.091	2.543	4.258	5.426	4.258	6.131
	β	−5.554	−6.047	−5.094	−3.550	−5.094	−5.555
	γ	6.460	5.638	4.586	3.380	4.586	5.151
	ζ	2.0E-4	5.0E-4	1.0E-6	2.0E-3	1.0E-6	1.0E-5
	λ	2.857	3.333	8.000	2.000	8.000	6.667

Table A.4 IEEE 30-bus system control variable constraints

	Optimized variable	Lower limit	Upper limit	Increment value
Active generated power (MW)	P_{G2}	0.20	0.80	N/A
	P_{G3}	0.15	0.50	
	P_{G4}	0.10	0.35	
	P_{G5}	0.10	0.30	
	P_{G6}	0.12	0.40	

(continued)

Table A.4 (continued)

	Optimized variable	Lower limit	Upper limit	Increment value
Generator bus voltage (pu)	V_{G1} V_{G2} V_{G3} V_{G4} V_{G5} V_{G6}	0.9	1.1	N/A
Tap position (pu)	TC_{L11} TC_{L12} TC_{L15} TC_{L36}	0.9	1.1	$0.9+16*1.25\%$
Capacitors (MVAR)	QC_{10}	0	0.05	$0.0 + 10*0.5\%$
	QC_{12}	0	0.05	
	QC_{15}	0	0.05	
	QC_{17}	0	0.05	
	QC_{20}	0	0.05	
	QC_{21}	0	0.05	
	QC_{23}	0	0.05	
	QC_{24}	0	0.05	
	QC_{29}	0	0.05	

References

1. Sun, D.I., Ashley, B., Brewer, B., Hughes, A., Tinney, W.F.: Optimal power flow by Newton approach. IEEE Trans. Power Appar. Syst. **103**(10), 2864–2880 (1984)
2. El-Hawary, M.E., Rao, R.S., Christensen, G.S.: Optimal hydro-thermal load flow: Formulation and a successive approximation solution for fixed head systems. J. Optimal Control Appl. Meth. **7**(4), 334–355 (1986)
3. Das, D.B., Patvardhan, C.: Useful multi-objective hybrid evolutionary approach to optimal power flow. IEE Proc-Gener. Transm. Distrib **150**(3), 275–282 (2003)
4. Kulworawanichpong, S.S.: Optimal power flow using Tabu search. In: IEEE Power Engineering Review, pp. 37–39. June 2002.
5. Prasad, N.P., Abdel-Moamen, M.A., Trivedi, P.K, Das, B.: A hybrid model for optimal power flow incorporating FACTS devices. In: Power Engineering Society Winter Meeting, 2001. IEEE, vol. 2, pp. 510–515. Feb 2001
6. Bakirtzis, A.G., Biskas, P.N., Zoumas, C.E., Petridis, V.: Optimal power flow by enhanced genetic algorithm. IEEE T Power Syst **17**(2), 229–236 (2002)
7. Aguado, J. A., Quintana, V.H.: Optimal power flows of interconnected power systems. In: IEEE Power Engineering Society Summer Meeting, vol. 2, pp. 814–819. Jul 1999
8. Kubokawa, J., Sasaki, H., Yorino, N.: A Fast solution method for multiobjective optimal power flow using an interactive approach. Electr Eng Japan **114**(2), 57–66 (1994)
9. Nangia, U., Jain, N.K., Wadhwa, C.L.: Optimal weight assessment based on a range of objectives in a multi-objective optimal load flow study. IEE Proc-Gener. Transm. Distrib **145**(1), 65–69 (1998)
10. Farag, A., Baiyat, S., Cheng, T.C.: Economic load dispatch multiobjective optimization using linear programming techniques. IEEE T Power Syst **10**(2), 731–738 (1995)

11. Zhiqiang, Y., Zhijian, H.: Economic dispatch and optimal flow based on chaotic optimization, Power system technology, 2002. In: Proceedings International Conference on PowerCon 2002, Kunming, China, vol. 4, pp. 2313–2317. 13–17 Oct 2002
12. Zhang, S., Irving, M.R.: Analytical algorithm for constraint relaxation in LP-based optimal power flow. IEE Proc **140**(4), 326–330 (1993)
13. Alsac, O., Stott, B.: Optimal load flow with steady state security. IEEE Trans Power Appar Syst **93**, 745–751 (1974)
14. Dommel, H., Tinny, W.: Optimal power flow solution. IEEE Trans Powr Appar Syst **PSA-87**(10), 1866–76 (1968)
15. Shoults, R., Sun, D.: Optimal power flow based on P-Q decomposition. IEEE Trans Power Appar Syst **PSA-101**(2), 397–405 (1982)
16. Happ, H.H.: Optimal power dispatch: A comprehensive survey. IEEE Trans Power Appar Syst **PSA-96**, 841–854 (1977)
17. Mamandur, K.R.C.: Optimal control of reactive power flow for improvements in voltage profiles and for real power loss minimization. IEEE Trans Power Appar Syst **PSA-100**(7), 3185–93 (1981)
18. Habiabollahzadeh, H., Luo, G.X., Semlyen, A.: Hydrothermal optimal power flow based on a combined linear and nonlinear programming methodology. IEEE Trans Power Appar Syst **PWRS-4**(2), 530–7 (1989)
19. Grudinin, N.: Combined quadratic-separable programming OPF algorithm for economic dispatch and security control. IEEE T Power Syst **12**(4), 1682–1688 (1997)
20. Momoh, J.A.: A generalized quadratic-based model for optimal power flow. In: IEEE International Conference on Conference Proceedings Systems, Man and Cybernetics, vol. 1, pp. 261–271. Nov 1989
21. Burchett, R.C., Happ, H.H., Vierath, D.R.: Quadratically convergent optimal power flow. IEEE Trans Power Appar Syst **PAS-103**, 3267–76 (1984)
22. AoKi, K., Nishikori, A., Yokoyama, R.T.: Constrained load flow using recursive quadratic programming. IEEE Trans Power Appar Syst **PAS-2**(1), 8–16 (1987)
23. Reid, G.F., Hasdorf, L.: Economic dispatch using quadratic programming. IEEE Trans Power Appar Syst **PAS-92**, 2015–2023 (1973)
24. Nanda, J.: New Optimal power-dispatch algorithm using Fletcher's quadraticc programming method. IEE Proc **136**(3), 153–161 (1989)
25. Almeida, K.C., Salgado, R.: Optimal power flow solutions under variable load conditions. IEEE Tran Power Appar Systems **15**(4), 1204–1211 (2000)
26. Torres, G.L., Quintana, V.H.: Optimal power flow by a nonlinear complementarily method. IEEE Trans Power Appar Syst **15**(3), 1028–1033 (2000)
27. Pudjianto, S.A., Strbac, G.: Allocation of Var support using LP and NLP based optimal power flows. IEE Proc.-Gener. Transm. Distrib. **149**(4), 377–383 (2002)
28. Stadlin, W., Fletcher, D.: Voltage verus reactive current model for dispatch and control. IEEE Trans Power Appar Syst **PAS-101**(10), 3751–8 (1982)
29. Mota-Palomino, R.: Sparse reactive power scheduling by a penalty-function linear programming technique. IEEE Trans Power Appar Syst **PAS-1**(3), 31–39 (1986)
30. Aoki, K., Kanezashi, M.: A modified Newton method for optimal power flow using quadratic approximation power flow. IEEE Trans Power Appar Syst **PAS-104**(8), 2119–2124 (1985)
31. Salgado, R., Brameller, A., Aitchison, P.: Optimal power flow solutions using the gradient projection method part 2: Modeling of the power system equations. Gener Trans Distrib, IEE Proc **137**(6), 429–435 (1990)
32. CIGRE.: Application of optimization techniques to study power system network performance, Task Force 38-04-02 Final Report, Chapter 2, Apr 1994
33. Frauendorfer, K., Glavitsch, H., Bacher, R.: Optimization in planning and operation of electrical power systems. Physica, Heidelberg (1992). (A Springer Company), ISBN-10: 3790807184
34. Saha, T.N., Maitra, A.: Optimal power flow using the reduced Newton approach in rectangular coordinates. Electr Power Energy Syst **20**(6), 383–389 (1998)

35. Hong, Y.Y., Liao, C.M., Lu, T.G.: Application of Newton optimal power flow to assessment of VAR control sequences on voltage security: Case studies for a practical power system. IEE Proc-C **140**(6), 539–543 (1993)
36. Baptista, E.C.: A new solution to the optimal power flow problem, 2001 IEEE Porto Power Tech Conference, Porto, Portugal, vol. 3, Sept. 10th–Sept. 13th, 2001
37. Talaq, J.H.: Minimum emissions power flow using Newton's method and its variants. Electr Power Syst Res J **39**, 233–239 (1996)
38. Zhang, S.: "Enhanced newton-raphson algorithm for normal control and optimal power flow solutions using column exchange techniques. IEE Proc Gener Trans Distrib **141**(6), 4647–657 (1994)
39. Sun, D.I., Ashley, B., Brewer, B., Hughes, A., Tinney, W.F.: Optimal power flow by Newton approach. IEEE Trans Power Appar Syst **PAS-103, No.10**, 2864–2880 (1984)
40. Santos, A.: Optimal power flow solution by Newton's method applied to AN augmented lagrangian function. IEE Proc Gener Transm Distrib **142**(1), 33–36 (1995)
41. Rahli, M.: Optimal power flow using sequential unconstrained minimization technique (SUMT) method under power transmission losses minimization. Electr Power Syst Res J **52**, 61–64 (1999)
42. Shengsong, L., Zhijian, H., Min, W.: A hybrid algorithm for optimal power flow using the chaos optimization and the linear interior point algorithm. In: Power System Technology, 2002. Proceedings International Conference on Power Con 2002, vol. 2, pp. 793–797. 13–17 Oct 2002
43. Momoh, J.A.: Improved interior point method for OPF problems. IEEE Trans Power Syst **14**(3), 1114–20 (1999)
44. Yan, X., Quintana, V.H.: Improved interior point based OPF by dynamic adjustment of step sizes and tolerances. IEEE Trans Power Syst **14**(2), 709–17 (1999)
45. Wu, Y.C., Debs, A.S.: Initialization, decoupling, Hot start, and warm start in direct nonlinear interior point algorithm for optimal power flows. IEE Proc-Gener. Transm. Distrib **148**(1), 67–75 (2001)
46. Bala, J.L.: An improved second order method for optimal load flow. IEEE Trans Power Appar Syst **PAS-97**(4), 1239–1244 (1978)
47. Almeida, K.C., Galiana, F.D., Soares, S.: A general parametric optimal power flow. IEEE T Power Syst **9**(1), 540–547 (1994)
48. Huneault, M., Galiana, F.D.: A survey of the optimal power flow literature. IEEE T Power Syst **6**(2), 762–770 (1991)
49. Momoh, J.A., El-Hawary, M.E., Adapa, R.: A review of selected optimal power flow literature to 1993 part-I: Nonlinear and quadratic programming approaches. IEEE T Power Syst **14**(1), 96–104 (1999)
50. Carpinter, J.: Contribution to the economic dispatch problem. Bulletin Society Francaise Electriciens **3**(8), 431–447 (1962)
51. Xie, K., Song, Y.H.: Dynamic optimal power flow by interior point methods. IEE Proc-Gener. Transm. Distrib **148**(1), 76–84 (2001)
52. Momoh, J.A., El-Hawary, M.E., Adapa, R.: A review of selected optimal power flow literature to 1993 part-II: Newton, linear programming & interior point methods. IEEE T Power Syst **14**(1), 105–111 (1999)
53. Stott, B., Marinho, J.L.: Linear programming for power system network security applications. IBID **PAS-98**, 837–848 (1979)
54. Stott, B., Hobson, E.: Power system security control calculation using linear programming. IEEE Trans **PAS-97**, 1713–1731 (1978)
55. Stott, B., Marinho, J., Alsac, O.: Review of linear programming applied to power system rescheduling. In: IEEE PICA Conference Proceedings, Cleveland, Ohio, pp. 142–154 (1979)
56. Alsac, O., Bright, J., Prais, M., Stott, B.: Further development in LP-based optimal power flow. In: IEEE/PES 1990 Winter Meeting, Atlantic, Georgia, Feb 1990

57. Cheng, D.T.: The challenges of using an optimal power flow. In: IEEE Power Engineering Review, pp. 62–63. Oct 1998
58. El-Hawary, M.E., Christensen, G.S.: Hydro-thermal load flow using functional analysis. J. Optimization Theory Appl. **12**, 576–587 (1973)
59. Vaahedi, E., Zein El-Din, H.: Considerations in applying optimal power flow to power system operation. IEEE T Power Syst **4**(2), 694–703 (1989)
60. Tinney, W.F., Bright, J.M., Demaree, K.D., Hughes, B.A.: Some deficiencies in optimal power flow. IEEE T Power Syst **3**, 676–683 (1988)
61. Burchett, R.C., Happ, H.H., Palmer, R.E., Vierath, D.R.: Quadratically convergent optimal power flow. IEEE Trans. Power Appar. Syst. **103**(11), 3264–3275 (1984)
62. Fogel, D.B.: Evolutionary computational toward a New philosophy of machine intelligence. IEEE Press, New York (1995)
63. Stott, B., Alsac, O.: Fast decoupled load flow. IEEE Trans PSA **PAS-93, No. 3**, 859–867 (1974)
64. Abido, M.A.: Optimal design of power-system stabilizers using particle swarm optimization. IEEE Trans Energy Conver **17**(3), 406–413 (2002)
65. Liu, E., Papalexopoulos, A.D., Tinney, W.F.: Discrete shunt controls in a Newton optimal power flow. IEEE T Power Syst **7**, 1519–1528 (1999)
66. Zhu, J.Z., Irving, M.R.: Combined active and reactive dispatch with multiple objectives using an analytic hierarchical process. IEE Proc-Gener. Transm. Distrib **143**(4), 344–352 (1996)
67. Nangia, U., Jain, N.K., Wadhwa, C.L.: Surrogate worth trade-off technique for multi-objective optimal power flows. IEE Proc-Gener, Transm Distrib **144**(6), 547–553 (1997)
68. Lai, L.L., MA, J.T., Yokohoma, R., Zhao, M.: Improved genetic algorithm for optimal power flow under both normal and contingent operation states. Electr Power Energy Syst **19**, 287–291 (1997)
69. Chen, L., Suzuki, H., Katou, K.: Mean field theory for optimal power flow. IEEE T Power Syst **12**, 1481–1486 (1997)
70. Miranda, V., Srinivasan, D., Proenca, L.M.: Evolutionary computation in power systems. Electr Power Energy Sys **20**, 89–98 (1998)
71. Abido, M.A.: Optimal power flow using particle swarm optimization. Electr Power Energy Syst **24**, 563–571 (2002)
72. Venkatesh, B., Rakesh Ranjan, Gooi, H.B.: Effect of minimizing var losses on voltage stability in a unified OPF framework. In: IEEE Power Engineering Review, pp. 45–47. Nov 2002
73. Abido, M.A.: Environmental/economic power dispatch using multi-objective evolutionary algorithms. IEEE T Power Syst **18**(4), 1529–1537 (2003)
74. El-Keib, A.A., Ma, H., Hart, J.L.: Economic dispatch in view of the clean Air Act of 1990. IEEE T Power Syst **9**, 972–978 (1994)
75. Talaq, J.H., El-Hawary, F., El-Hawary, M.E.: A summary of environmental/economic dispatch algorithms. IEEE T Power Syst **9**, 1508–1516 (1994)
76. Hu, X., Eberhart, R.C., Shi, Y.: Engineering optimization with particle swarm. In: IEEE International Conference on Evolutionary Computation. pp. 53–57 (2003)
77. Kennedy, J.: The particle swarm: Social adaptation of knowledge. In: Proceedings of 1997 IEEE International Conference Evolutionary Computation ICEC 97, Indianapolis, pp. 303–308 (1997)
78. Angeline, P.: Evolutionary optimization versus particle swarm optimization: Philosophy and performance differences. In: Proceedings of 7th Annual Conference Evolutionary Programming, San Diego, pp. 601–610 (1998)
79. Shi, Y., Eberhart, R.: Parameter selection in particle swarm optimization. In: Proceedings of 7th Annual Conference Evolutionary Programming, San Diego pp. 591–600 (1998)
80. Ozcan, E., Mohan, C.: Analysis of a simple particle swarm optimization system. Intell Eng Syst Artif Neural Net **8**, 253–8 (1998)

81. Kennedy, J., Eberhart, R.: Particle swarm optimization. IEEE int. Conf. Evol Comput **4**, 1942–1948 (1995)
82. Eberhart, R.C., Shi, Y.: Comparing inertia weights and constriction factors in particle swarm optimization. IEEE int. Conf. Evol Comput **1**, 84–88 (2000)
83. Gaing, Z.L.: Particle swarm optimization to solving the economic dispatch considering the generator constraints. IEEE T Power Syst **18**(3), 11871–195 (2003)
84. Papalexopoulos, A.D., Imparato, C.F., Wu, F.F.: Large-scale optimal power flow: Effects of initialization, decoupling & discretization. IEEE T Power Syst **4**(2), 748–759 (1989)
85. Hirotaka, Y., Kawata, K., Fukuyama, Y.: A particle swarm optimization for reactive power and voltage control considering voltage security assessment. IEEE T Power Syst **15**(4), 1232–1239 (2000)
86. Miranda, V., Fonseca, N.: EPSO-evolutionary particle swarm optimization, a New algorithm with applications in power systems. IEEE T Power Syst **2**, 745–750 (2000)
87. Shi, Y., Eberhart, R.C.: A modified particle swarm optimizer. In: Proceedings of IEEE International Conference on Evolutionary Computation, Anchorage, pp. 69–73. May 1998
88. Kennedy, J., Spears, W.: Matching algorithm to problems: An experimental test of the particle swarm optimization and some genetic algorithms on the multimodal problem generator. In: Proceedings of IEEE International Conference on Evolutionary Computation, Anchorage, pp. 78–83. May 1998
89. Angeline, P.: Using selection to improve particle swarm optimization, In: Proceedings of IEEE International Conference on Evolutionary Computation, Anchorage, pp. 84–89. May 1998
90. Huneault, M., Galliana, E.D.: A survey of the optimal power flow literature. IEEE T Power Syst **6**(2), 762–770 (1991)
91. Squires, R.B.: Economic dispatch of generation directly from power system voltages and admittances. AIEE Trans. Power Appar. Syst **PAS-79**(III), 1235–1244 (1961)
92. El-Hawary, M.E., Christensen, G.S.: Optimal economic operation of electric power systems. Academic, New York (1979)
93. El-Hawary, M.E., Tsang, D.H.: The hydro-thermal optimal load flow: A practical formulation and solution technique using Newton's approach. IEEE Trans. Power Syst. Eng **PWRS-1**(3), 154–167 (1986)
94. Vlachogiannis, J.G.: Fuzzy logic application in load flow studies. IEE Proc Gener Transm Distrib **148**(1), 34–40 (2001)
95. Xie, K., Song, Y.H.: Power market oriented optimal power flow via an interior point method. IEE Proc-Gener. Transm. Distrib **148**(6), 549–556 (2001)
96. Salgado, R., Brameller, A., Aitchison, P.: Optimal power flow solutions using the gradient projection method part 1: Theoretical basis. IEE Proc **137**(6), 424–428 (1990)
97. Zhenya, H., et al.: Extracting rules from fuzzy neural network by particle swarm optimization. In: Proceedings of IEEE International Conference on Evolutionary Computation, Anchorage, pp. 74–77. May 1998
98. Hartati, R.S.: Optimal active power flow solutions using a modified Hopfield neural network. In: IEEE T Power Syst, pp. 189–194 (2000)
99. Lee, K., Park, Y., Ortiz, J.: A united approach to optimal real and reactive power dispatch. IEEE Trans Power Appar Syst **104**(5), 1147–53 (1958)
100. Cohon, J.L., Marks, D.H.: A review and evaluation of multiobjective programming techniques. Water Res Res **12**, 845–851 (1975)
101. Cohon, J.L., Church, R.L., Sheer, D.P.: Generating multiobjective trade-offs: An algorithm for bicriterion problems. Water Res Res **15**, 1001–1010 (1979)
102. Wadhwa, C.L., Jain, N.K.: Multiple objective optimal load flow: A new perspective. IEE Proc-Gener, Trans Distri **137**(1), 13–18 (1990)

Chapter 6
Long-Term Operation of Hydroelectric Power Systems

Objectives The objectives of this chapter are

- Formulating the problem of long-term operation of a multireservoir power system connected in cascade (series)
- Implementing the minimum norm approach to solve the formulated problem
- Implementing the simulated annealing algorithm (SAA) to solve the long-term hydro scheduling problem (LTHSP)
- Introducing an algorithm enhancement for randomly generating feasible trial solutions
- Implementing an adaptive cooling schedule and a method for variable discretization to enhance the speed and convergence of the original SAA
- Using the short-term memory of the tabu search (TS) approach to solve the nonlinear optimization problem in continuous variables of the LTHSP

6.1 Introduction

In this chapter the long-term hydro scheduling problem (LTHSP) of a multireservoir hydro power plant connected in series on a river is considered [1–17]. The LTHSP optimization problem involves the use of a limited resource over a period of time. The resource is the water available for hydro generation. Most hydroelectric plants are used as multipurpose plants. In such cases, it is necessary to meet certain obligations other than power generation. These may include a maximum fore bay elevation, not to be exceeded because of danger of flooding, and a minimum plant discharge and spillage to meet irrigational and navigational commitments. Thus, the optimal operation of the hydro system depends on the conditions that exist over the optimization interval [9].

Other distinctions among power systems are the number of hydro stations, their location, and special unit operating characteristics. The problem is quite different if the hydro stations are located on the same stream or on different ones. An upstream station will highly influence the operation of the next downstream station. The latter, however, also influences the upstream plant by its effect on the tail water elevation and effective head. Close coupling of stations by such a phenomenon is a complicating factor [1–9].

The evaluation of the optimum monthly operating policy of a multireservoir hydroelectric power system is a stochastic nonlinear optimization problem with continuous variables. The problem of determining the optimal long-term operation of a multireservoir power system has been the subject of numerous publications over the past 50 years, and yet no completely satisfactory solution has been obtained because in every proposed method the problem is simplified.

Many classical methods [1–17] have been used to solve the LTHSP including dynamic programming, stochastic, aggregation linear programming, and decomposition methods. These methods need complex mathematical manipulation in addition to requiring excessive computing time and storage requirements.

Simulated annealing (SA) is a powerful technique to solve many optimization problems [18–23]. It has the ability to escape local minima by incorporating a probability function in accepting or rejecting new solutions. It was first proposed in the area of combinatorial optimization, that is, when the cost function is defined in a discrete domain. Therefore, many important problems are defined as functions of continuous variables as the problem under study. The application of the SA method to such problems requires an efficient strategy to generate feasible trial solutions. The other important point is the selection of an efficient cooling schedule that adapt itself according to statistics of the trial moves (acceptance/rejection) during the search.

In this chapter we present a simulated annealing algorithm (SAA) to solve the LTHSP of a multireservoir hydropower plant connected in series on a river. The proposed algorithm contains an adaptive strategy to control the step sizes of varying the variables between coarse and fine, according to the accepted and rejected trials. Moreover, a polynomial-time cooling schedule is used to control the temperature scheme based on the statistics of trial solutions generated during the search process [18, 23].

A major contribution of this work is introducing new rules for randomly generating feasible solutions satisfying all the system constraints including the well-known continuity equation of all reservoirs at all time periods.

An example system includes four series reservoirs and is taken from the literature [2, 9]. Comparing the results obtained with those of other classical optimization methods shows an improvement in the quality of solutions obtained with the proposed SAA.

6.2 Problem Formulation

The long-term optimization problem of the hydroelectric power system shown in Fig. 6.1 is to determine the discharge $u_{i,t}$, the storage $x_{i,t}$, and the spillage $s_{i,t}$ for each reservoir, $i = 1, \ldots, n$, at all scheduling time periods, $t = 1, \ldots, K$, under the following conditions [9].

The expected value of the water left in storage at the end of the last period studied is a maximum. The expected value of the MWh generated during the optimization interval is a maximum. The water conservation equation for each reservoir is adequately described by the following continuity difference equations.

$$x_{i,t} = x_{i,t-1} + I_{i,t} - u_{i,t} - s_{i,t} + s_{(i-1),t} \quad 1 \leq i \leq N, 1 \leq t \leq k \tag{6.1}$$

where $x_{i,t}$ is the storage of reservoir i at the end of period t, $I_{i,t}$ is the inflow to the reservoir i during the period t, and $s_{i,t}$ is the spillage from reservoir i during the period t. Spillage occurs when the discharge exceeds the maximum discharge value and the storage exceeds the maximum allowed storage of the reservoir.

To satisfy the multipurpose stream use requirements, the following operational constraints should be satisfied.

$$x_i^{\min} \leq x_{i,t} \leq x_i^{\max}, 1 \leq i \leq N; \; 1 \leq t \leq K, \tag{6.2}$$

$$u_{i,t}^{\min} \leq u_{i,t} \leq u_{i,t}^{\max}, 1 \leq i \leq N; \; 1 \leq t \leq K, \tag{6.3}$$

Where $x_{i,}^{\max}$ is the capacity of the reservoir, $x_{i,}^{\min}$ is the minimum storage, $u_{i,t}^{\min}$ is the minimum discharge through the turbine i during the period t, and $u_{i,t}^{\max}$ is the maximum discharge through the turbine i during the period t. If $u_{i,t} \geq u_{i,t}^{\max}$ and $x_{i,t}$ is equal to $x_i^{\max}$ then $u_{i,t} - u_{i,t}^{\max}$ is discharged through the spillways.

In mathematical terms, the long-term optimization problem for the power system shown in Fig. 6.1 is to determine the discharge $u_{i,t}$ and the storage $x_{i,t}$ (consequently the spillage $s_{i,t}$) that maximize

$$J = E\left[\sum_{i=1}^{N} V_i(x_{i,t}) + \sum_{i=1}^{N}\sum_{t=1}^{K} c_t G_i(x_{i,t-1}, u_{i,t})\right] \tag{6.4}$$

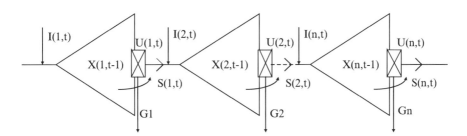

Fig. 6.1 The system under study

subject to satisfying the equality constraints given by Eq. 6.1, and the inequality constraints given by Eqs. 6.2 and 6.3. $V_i(x_{i,t})$ is the value of water left in storage in reservoir i at the end of the last period studied, $G_i(x_{i,t-1}, u_{i,t})$ is the generation of plant i during period t in MWh, and c_t is the value (in dollars) of the water on the river in month t.

Examining the above four equations, one can notice that the LTHSP is a nonlinear optimization problem with continuous variables. The problem size can be determined as the product of three times the number of reservoirs and the number of time periods in the required schedule.

In the LTHSP under consideration, one is interested in a solution that maximizes the total power generated during the scheduling time horizon, Eq. 6.4, and several constraints are satisfied.

6.3 Problem Solution: A Minimum Norm Approach

In this section we offer a formulation to the problem of a multireservoir power system connected in series on a river, which is based on the minimum norm approach. We repeat here the objective function for the reader's convenience.

The problem of the power system shown in Fig. 6.1 is to determine the discharge $u_{i,k}$, $i = 1, \ldots, n$, $k = 1, \ldots, K$ that maximizes

$$J = E\left[\sum_{i=1}^{n} V_i(x_{i,K}) + \sum_{i=1}^{n}\sum_{k=1}^{K} c_k G_i(u_{i,k}, x_{i,k-1})\right] \quad (6.5)$$

subject to satisfying the following constraints,

$$x_{i,k} = x_{i,k-1} + I_{i,k} + u_{(i-1),k} - u_{i,k} - s_{i,k} + s_{(i-1),k} \quad (6.6)$$

where $s_{i,k}$ denotes spill from reservoir i during month k:

$$x_i^m \leq x_{i,k} \leq x_i^M \quad (6.7)$$

$$u_{i,k}^m \leq u_{i,k} \leq u_{i,k}^M \quad (6.8)$$

E in Eq. 6.5 indicates the expected value m, M in the above two equations indicate the minimum and maximum values of the variables respective.

6.3.1 System Modeling

The conventional approach for obtaining the equivalent reservoir and hydroplant is based on the potential energy concept. Each reservoir on a river is mathematically represented by an equivalent potential energy balance equation. The potential energy balance equation is obtained by multiplying both sides of the reservoir balance-of-water equation by the water conversion factors of at-site and downstream hydroplants. We may choose the following for the function $V_i(x_{i,K})$.

$$V_i(x_{i,K}) = \sum_{j=1}^{n} h_j x_{i,K}, \quad i = 1, 2, ..., n \qquad (6.9)$$

where h_j is the average water conversion factor (MWh/Mm^3) at site j. In the above equation we assumed that the cost of this energy is one dollar/MWh (the average cost during the year).

The generation of a hydroelectric plant is a nonlinear function of the water discharge $u_{i,k}$ and the net head, which itself is a function of the storage. In this section, we assume a linear relation between the storage and the head (the storage-elevation curve is linear and the tailwater elevation is constant independent of the discharge). We may choose

$$G_i(u_{i,k}, x_{i,k-1}) = a_i u_{i,k} + b_i u_{i,k} x_{i,k-1} MWh \qquad (6.10)$$

where a_i and b_i are constants for the reservoir i. These are obtained by least error squares curve fitting to typical plant data available.

Now, the cost functional in Eq. 6.5 becomes

$$J = E\left[\sum_{i=1}^{n}\sum_{j=1}^{n} h_j x_{i,K} + \sum_{i=1}^{n}\sum_{k=1}^{K}\left(A_{i,k} u_{i,k} + u_{i,k} B_{i,k} x_{i,k-1}\right)\right] \qquad (6.11)$$

subject to satisfying the constraints given by Eqs. 6.6, 6.7, and 6.8 where

$$A_{i,k} = a_i c_k, i = 1, ..., n$$
$$B_{i,k} = b_i c_k, i = 1, ..., n$$

6.3.2 *Formulation*

The reservoir dynamic equation is added to the cost functional using the unknown Lagrange multiplier $\lambda_{i,k}$, and the inequality constraints (6.7) and (6.8) are added using the Kuhn–Tucker multipliers, so that a modified cost functional is obtained:

$$J_0(u_{i,k}, x_{i,k-1}) = E\left[\sum_{i=1}^{n}\sum_{j=1}^{n} h_j x_{i,K} + \sum_{i=1}^{n}\sum_{k=1}^{K}\{A_{i,k} u_{i,k} + u_{i,k} B_{i,k} x_{i,k-1}\right.$$
$$+ \lambda_{i,k}\left(-x_{i,k} + x_{i,k-1} + u_{(i-1),k} - u_{i,k}\right) + \left(e_{i,k}^M - e_{i,k}^m\right) x_{i,k}$$
$$\left.+ \left(f_{i,k}^M - f_{i,k}^m\right) u_{i,k}\}\right] \qquad (6.12)$$

Here terms explicitly independent of $u_{i,k}$ and $x_{i,k}$ are dropped. In the above equation $e_{i,k}^M, e_{i,k}^m, f_{i,k}^M,$ and $f_{i,k}^m$ are Kuhn–Tucker multipliers. These are equal to zero if the constraints are not violated and greater than zero if the constraints are violated.

Let R denote the set of n reservoirs, and define the $n \times 1$ column vectors

$$H = col[H_i, i \in R] \tag{6.13}$$

where

$$H_i = \sum_{j=1}^{n} h_j \tag{6.14}$$

$$x(k) = cp; [x_{i,k}, i \in R] \tag{6.15}$$

$$u(k) = col[u_{i,k}, i \in R] \tag{6.16}$$

$$\lambda(k) = col[\lambda_{i,k}, i \in R] \tag{6.17}$$

$$\mu(k) = col[\mu_{i,k}, i \in R] \tag{6.18}$$

$$\psi(k) = col[\psi_{i,k}, i \in R] \tag{6.19}$$

where

$$\mu_{i,k} = e_{i,k}^M - e_{i,k}^m, \quad i \in R \tag{6.20}$$

and

$$\psi_{i,k} = f_{i,k}^M - f_{i,k}^m, \quad i \in R \tag{6.21}$$

And the $n \times n$ diagonal matrix

$$B(k) = diag[B_{i,k}, \ i \in R] \tag{6.22}$$

Furthermore, define the $n \times n$ lower triangular matrix **M** by

$$m_{ii} = -1, \quad i \in R \tag{6.23}$$

6.3 Problem Solution: A Minimum Norm Approach

$$m_{(j+1)j} = 1, \qquad j = 1, \cdots, n-1 \qquad (6.24)$$

Then the modified cost functional in Eq. 6.12 becomes

$$J_0(u(k), x(k-1)) = E\left[H^T x(K) + \sum_{k=1}^{K} \{u^T(k)B(k)x(k-1)\right.$$

$$+ (\lambda(k) + \mu(k))^T x(k-1) - \lambda^T(k)x(k) + \left(A(k) + M^T \lambda(k)\right)$$

$$\left. + M^T \mu(k) + \psi(k))^T u(k)\}\right] \qquad (6.25)$$

We will need the identity:

$$\sum_{k=1}^{K} \lambda^T(k)x(k) = -\lambda^T(0)x(0) + \lambda^T(K)x(K) + \sum_{k=1}^{K} \lambda^T(k-1)x(k-1) \qquad (6.26)$$

Then we can write the cost functional (6.25) as

$$J_0[u(k), x(k-1)] = E\left[(H - \lambda(K))^T x(K) + \lambda^T(0)x(0) + \sum_{k=1}^{K} \{u^T(k)B(k)x(k-1)\right.$$

$$\left. + \left(A(k) + M^T \lambda(k) + M^T \mu(k) + \psi(k)\right)^T u(k)\}\right] \qquad (6.27)$$

Define the $2n \times 1$ column vectors

$$X(k) = col[x(k-1), u(k)] \qquad (6.28)$$

$$R(k) = col[(\lambda(k) - \lambda(k-1) + \mu(k)), \left(A(k)M^T \lambda(k) + M^T \mu(k) + \psi(k)\right)] \qquad (6.29)$$

and also the $2n \times 2n$ matrix $L(k)$ as

$$L(k) = \begin{bmatrix} 0 & \frac{1}{2}B(k) \\ \frac{1}{2}B(k) & 0 \end{bmatrix} \qquad (6.30)$$

Using these definitions, (6.27) becomes

$$J_0[x(K), X|k|] = E\left[(H - \lambda(K))^T x(K) + \lambda^T(0)x(0)\right.$$

$$\left. + \sum_{k=1}^{K} \{X^T(k)L(k)X(k) + R^T(k)X(k)\}\right] \qquad (6.31)$$

Equation 6.31 is composed of a boundary term and a discrete integral part, which are independent of each other. To maximize J_0 in Eq. 6.31, one maximizes each term separately:

$$\max J_0[x(K), X(k)] = \max_{x(K)} E\left[(H - \lambda(K))^T x(K) + \lambda^T(0)x(0)\right]$$

$$+ \max_{X(k)} E\left[\sum_{k=1}^{K} \{X^T(k)L(k)X(k) + R^T(k)X(k)\}\right] \quad (6.32)$$

6.3.3 Optimal Solution

There is exactly one optimal solution to the problem formulated in (6.32). The boundary part in (6.32) is optimized when

$$E[\lambda(K) - H] = [0] \quad (6.33)$$

because $\delta x(K)$ is arbitrary and $x(0)$ is constant.

We define the $2n \times 1$ vector $V(k)$ as

$$V(k) = L^{-1}(k)R(k) \quad (6.34)$$

Now, the discrete integral part of Eq. 6.32 can be written as

$$\max {}_2[X(k)] = \max_{X(k)} E\left[\sum_{k=1}^{K} \left\{\left(X(k) + \frac{1}{2}V(k)\right)^T L(Lk)\left(X(k) + \frac{1}{2}V(k)\right)\right.\right.$$

$$\left.\left. - \frac{1}{4}V^T(k)L(k)V(k)\right\}\right] \quad (6.35)$$

The last term in Eq. 6.35 does not depend explicitly on $X(k)$, so that it is necessary only to consider

$$\max J_2[X(k)] = \max_{X(k)} E\left[\sum_{k=1}^{K} \left\{\left(X(k) + \frac{1}{2}V(k)\right)^T L(k)\left(X(k) + \frac{1}{2}V(k)\right)\right\}\right] \quad (6.36)$$

Equation 6.36 defines a norm. This norm is considered to be an element of a Hilbert space because $X(k)$ is always positive. Equation 6.36 can be written as

$$\max J_2[X(k)] = \max_{X(k)} E\left\|X(k) + \frac{1}{2}V(k)\right\|_{L(k)} \quad (6.37)$$

6.3 Problem Solution: A Minimum Norm Approach

The maximization of $J_2 [X(k)]$ is mathematically equivalent to the minimization of the norm of Eq. 6.37. The minimum of the norm in Eq. 6.37 is clearly achieved when

$$E \left\| X(k) + \frac{1}{2} V(k) \right\| = [0] \tag{6.38}$$

Substituting from Eq. 6.34 into Eq. 6.38 for $V(k)$, one finds the optimal solution is given by

$$E[R(k) + 2L(k)X(k)] = [0] \tag{6.39}$$

Writing Eq. 6.39 explicitly and adding the reservoir dynamic equation, one obtains the long-term optimal equations as

$$E[-x(k) + x(k-1) + I(k) + Mu(k) + Ms(k)] = [0] \tag{6.40}$$

$$E[\lambda(k) - \lambda(k-1) + \mu(k) + B(k)u(k)] = [0] \tag{6.41}$$

$$E\left[A(k) + M^T \lambda(k) + M^T \mu(k) + \psi(k) + B(k)x(k-1)\right] = [0] \tag{6.42}$$

We can now state the optimal solution of Eqs. 6.40, 6.41, and 6.42 in component form as

$$E\left[-x_{i,k} + x_{i,k-1} + I_{i,k} + u_{(i-1),k} - u_{i,k} + s_{(i-1),k} - s_{i,k}\right] = 0; \quad i = 1, \cdots, n, \quad k = 1, \cdots, K \tag{6.43}$$

$$E\left[\lambda_{i,k} - \lambda_{i,k-1} + \mu_{i,k} + c_k b_i u_{i,k}\right] = 0, \quad i = 1, \cdots, n, \quad k = 1, \cdots, K \tag{6.44}$$

$$E\left[c_k a_i + \lambda_{(i-1),k} - \lambda_{i,k} + \mu_{(i+1),k} - \mu_{i,k} + \psi_{i,k} + c_k b_i x_{i,k-1}\right] = 0 \tag{6.45}$$

In addition to the above equations, one has the Kuhn–Tucker exclusion equations, which must be satisfied at the optimum as

$$e_{i,k}^m \left(x_i^m - x_{i,k}\right) = 0, \quad e_{i,k}^1 \left(x_{i,k} - x_i^M\right) = 0 \tag{6.46}$$

$$f_{i,k} \left(u_{i,k}^m - u_{i,k}\right) = 0, \quad f_{i,k}^1 \left(u_{i,k} - u_{i,k}^M\right) = 0 \tag{6.47}$$

Also we have the following limits on the variables:

if $x_{i,k} < x_i^m$, then we put $x_{i,k} = x_i^m$

Table 6.1 Characteristics of the installations [1]

Site	Minimum capacity x_i^m (Mm^3)	Maximum capacity x_i^M (Mm^3)	Maximum effective discharge (m^3/s)	Minimum effective discharge (m^3/s)	Average monthly productibility (MWh/Mm^3)	Reservoir's constants	
						a_i (MWh/Mm^3)	b_i MWh/(Mm^3)^2
1	0	9,628	400	0	18.31	11.8	1.3×10^{-3}
2	0	570	547	0	234.36	231.5	0.532×10^{-3}
3	0	50	594	0	216.14	215.82	12.667×10^{-3}
4	0	3,420	1,180	0	453.44	437.00	11.173×10^{-3}

$$\text{if } x_{i,k} < x_i^M, \quad \text{then we put } x_{i,k} = x_i^M$$

$$\text{if } u_{i,k} < u_i^m, \quad \text{then we put } u_{i,k} = u_{i,k}^m$$

$$\text{if } u_{i,k} < u_i^M, \quad \text{then we put } u_{i,k} = x_{i,k}^M \tag{6.48}$$

Equations 6.43, 6.44, 6.45, 6.46, 6.47, and 6.48 completely specify the optimal solution.

6.3.4 Practical Application

A computer program was written to solve Eqs. 6.43, 6.44, 6.45, 6.46, 6.47, and 6.48 iteratively using the steepest descent method for a system of four reservoirs connected in series. Table 6.1 gives the characteristics of the installations.

In Table 6.2 we give the optimal discharges and the profits realized during the year of high flow, year 1. Also, in Table 6.3 we report the results obtained for the same system during the year of low flow, year 2.

6.3.5 Comments

We presented in this section the application of the minimum norm theorem to the optimization of the total benefits from a multireservoir power system connected in series on a river. It has been found that this algorithm can deal with a large-scale power system with stochastic inflows. In this section new optimal equations are derived; if these equations are solved forward and backward in time, one can easily obtain the optimal long-term scheduling for maximum total benefits from any number of series reservoirs.

We compared the minimum norm approach with the dynamic programming and decomposition approaches, which are currently used by many utility companies. For the same system we obtained increased benefits using a smaller computing time as indicated above.

6.3 Problem Solution: A Minimum Norm Approach

Table 6.2 Optimal monthly releases from the reservoirs and the profits realized in year 1

Month k	$u_{1,k}$ (Mm³)	$u_{2,k}$ (Mm³)	$u_{3,k}$ (Mm³)	$u_{4,k}$ (Mm³)	Profits ($)
1	0	368	528	2,253	990,816
2	1,037	1,418	1,499	2,700	1,828,432
3	853	1,076	1,118	1,928	1,473,793
4	1,071	1,465	1,547	2,244	2,300,938
5	968	1,130	1,098	2,855	2,656,644
6	1,071	1,336	1,400	3,161	2,785,338
7	1,037	1,127	1,133	1,580	1,475,190
8	0	0	229	1,310	609,669
9	0	446	627	2,326	1,051,100
10	0	506	711	2,330	979,250
11	0	491	637	2,117	873,319
12	0	363	495	1,794	788,087
Value of water remaining in the reservoirs at the end of the year					10,414,598
Total profits					28,227,174

Table 6.3 Monthly releases from the reservoirs and the profits realized in year 2

Month k	$u_{1,k}$ (Mm³)	$u_{2,k}$ (Mm³)	$u_{3,k}$ (Mm³)	$u_{4,k}$ (Mm³)	Profits ($)
1	0	215	343	878	405,647
2	1,037	1,262	1,412	2,242	1,573,185
3	881	1,052	1,074	1,624	1,308,419
4	1,071	1,465	1,537	3,107	2,829,756
5	968	1,319	1,341	2,855	2,759,020
6	1,071	1,260	1,299	2,628	2,380,707
7	1,037	1,118	1,131	1,418	1,383,345
8	0	0	209	1,470	675,513
9	0	0	64	1,454	541,237
10	0	115	166	91	75,287
11	0	169	191	114	93,025
12	0	229	304	154	144,909
Value of water remaining in the reservoirs at the end of the year					7,165,203
Total profits					21,335,253

Table 6.4 Characteristics of the installations [18]

Site	x_i^M (Mm³)	Maximum effective discharge (m³/s)	Reservoir's constants		
			α_i	β_i	γ_i
1	9,628	400	11.41	0.15226×10^{-2}	-0.19131×10^{-7}
2	570	547	231.53	0.10282×10^{-1}	-0.13212×10^{-5}
3	50	594	215.82	0.12586×10^{-1}	0.2979×10^{-5}
4	3,420	1,180	432.20	-0.12972×10^{-1}	-0.12972×10^{-5}

6.3.6 A Nonlinear Model

In the previous two sections, a constant water conversion factor (MWh/Mm³) is used. For power systems in which the water heads vary by a small amount, this assumption is adequate, but for power systems in which this variation is large, using this assumption is not adequate. On the other hand, the linear storage-elevation curve used in modeling the reservoirs is adequate only for small-capacity reservoirs of rectangular cross-sections. In practice, most reservoirs are nearly trapezoidal in cross-section.

This section discusses the solution of the long-term optimal operating problem of multireservoir power systems having a variable water conversion factor and a nonlinear storage-elevation curve. The optimal solution is obtained using the minimum norm formulation in the framework of the functional analysis optimization technique.

6.3.6.1 Formulation

Again, the problem of the power system of Fig. 6.1 is to determine the optimal discharge $u_{i,k}$, $i = 1, \ldots, n$, $k = 1, \ldots, K$ subject to satisfying the following constraints.

1. The total expected benefits from the system (benefits from the energy generated by hydropower systems over the planning period plus the expected future benefits from the water left in storage at the end of that period) is a maximum.
2. The MWh generated per Mm^3 (1 Mm^3 = 10^6 m^3) discharge (the water conversion factor) as a function of the storage is adequately described by

$$h_{i,k} = \alpha_i + \beta_i x_{i,k} + \gamma_i (x_{i,k})^2, \qquad i = 1, \cdots, n, \quad k = 1, \cdots, K \quad (6.49)$$

where α_I, β_I, and γ_I are constants, These were obtained by least-squares curve fitting to typical plant data available. The reservoir dynamics is given by Eq. 6.6 and the operational constraints are given by Eqs. 6.7 and 6.8.

The MWh generated by a hydropower plant during a period k is given by

$$G_i\left[u_{i,k}, \tfrac{1}{2}(x_{i,k} + x_{i,k-1})\right] = \alpha_i u_{i,k} + \tfrac{1}{2}\beta_i u_{i,k}(x_{i,k} + x_{i,k-1})$$

$$+ \tfrac{1}{4}\gamma_{i,k}\left(x_{i,k} + x_{i,k-1}\right)^2, \quad i = 1, 2, \cdots, n \quad (6.50)$$

Substituting for $x_{i,k}$ from Eq. 6.6 into Eq. 6.50 yields

$$G_i\left[u_{i,k}, \tfrac{1}{2}(x_{i,k} + x_{i,k-1})\right] = b_{i,k} u_{i,k} + d_{i,k} u_{i,k} x_{i,k-1} + f_{i,k}(u_{i-1,k} - u_{i,k}) u_{i,k}$$

$$+ \gamma_i u_{i,k} y_{i,k-1} + \tfrac{1}{4}\gamma_i u_{i,k}(z_{i,k} + z_{i-1,k})$$

$$+ \gamma_i r_{i,k-1}(u_{i-1,k} - u_{i,k}) - \tfrac{1}{2}\gamma_i z_{i,k} u_{i-1,k} \quad (6.51)$$

6.3 Problem Solution: A Minimum Norm Approach

where

$$q_{i,k} = I_{i,k} + s_{i-1,k} - s_{i,k}, \quad i = 1, 2, \cdots, n \quad (6.52a)$$

$$b_{i,k} = \alpha_i + \tfrac{1}{2}\beta_i q_{i,k} + \tfrac{1}{4}\gamma_i (q_{i,k})^2, \quad i = 1, 2, \cdots, n \quad (6.52b)$$

$$d_{i,k} = \beta_i + \gamma_i q_{i,k}, \quad i = 1, 2, \cdots, n \quad (6.52c)$$

$$f_{i,k} = \tfrac{1}{2} d_{i,k}, \quad i = 1, 2, \cdots, n \quad (6.52d)$$

And the following are pseudo state variables.

$$y_{i,k} = (x_{i,k})^2, \quad i = 1, 2, \cdots, n, \quad k = 1, 2, \cdots, K \quad (6.53)$$

$$z_{i,k} = (u_{i,k})^2, \quad i = 1, 2, \cdots, n, \quad k = 1, 2, \cdots, K \quad (6.54)$$

$$r_{i,k-1} = u_{i,k} x_{i,k-1}, \quad i = 1, 2, \cdots, n, \quad k = 1, 2, \cdots, K \quad (6.55)$$

In mathematical terms, the object of the optimizing computation is to find the discharge $u_{i,k}$ that maximizes

$$J = E \left[\sum_{i=1}^{n} \sum_{j=1}^{n} x_{i,K} (\alpha_j + \beta_j x_{j,K} + \gamma_i u_{j,K}) \right.$$

$$+ \sum_{k=1}^{K} \{ c_k b_{i,k} + c_k d_{i,k} u_{i,k} x_{i,k-1} + c_k f_{i,k} u_{i,k} (u_{i-1,k} - u_{i,k})$$

$$+ c_k \gamma_i u_{i,k} y_{i,k-1} + \tfrac{1}{4} c_k \gamma_i u_{i,k} (z_{i,k} + z_{i-1,k})$$

$$\left. + c_k \gamma_i r_{i,k-1} (u_{i-1,k} - u_{i,k}) - \tfrac{1}{2} c_k \gamma_i z_{i,k} u_{i-1,k} \} \right] \quad in \ \$ \quad (6.56)$$

subject to satisfying the following constraints.
Equality constraints given by Eq. 6.6 and Eqs. 6.53, 6.54, and 6.55
Inequality constraints given by Eqs. 6.7 and 6.8
The cost functional in Eq. 6.56 can be written in vector form as

$$J = E \left[A^T x(K) + x^T(K) \beta x(K) + x^T(K) \gamma y(K) + \sum_{k=1}^{K} \{ B^T(k) u(k) \right.$$

$$+ u^T(k) d(k) x(k-1) + u^T(k) f(k) M u(k) + u^T(k) \gamma(k) y(k-1)$$

$$\left. + \tfrac{1}{4} u^T(k) \gamma(k) N z(k) + r^T(k-1) \gamma(k) - \tfrac{1}{2} z^T(k) \gamma(k) L u(k) \} \right] \quad (6.57)$$

and the equality and inequality constraints can be written in vector form asequality constraints given by

$$x(k) = x(k-1) + q(k) + Mu(k) \tag{6.58}$$

$$y(k) = x^T(k)Hx(k) \tag{6.59}$$

$$z(k) = u^T(k)Hu(k) \tag{6.60}$$

$$r(k-1) = u^T(k)Hx(k-1) \tag{6.61}$$

Inequality constraints are given by

$$x^m \leq x(k) \leq x^M \tag{6.62}$$

$$u^m(k) \leq u(k) \leq u^M(k) \tag{6.63}$$

where $x(k)$, $u(k)$, $y(k)$, $z(k)$, $r(k-1)$ are n-dimensional vectors at the end of period k; their components are $x_{i,k}$, $u_{i,k}$, $y_{i,k}$, $z_{i,k}$, and $r_{i,k-1}$, respectively; $d(k)$, $f(k)$, $\gamma(k)$ are $n \times n$ diagonal matrices whose elements are $d_{ii,k} = c_k d_{i,k}$, $f_{ii,k} = c_k f_{i,k}$, and $\gamma_{ii,k} = c_k \gamma_I$, $i = 1, \ldots, n$; M is a lower triangular matrix whose elements are $m_{ii} = -1$, $i = 1, \ldots, n$; $m_{(j+1)j} = 1, j = 1, \ldots, n-1$; N is a lower triangular matrix whose elements are $n_{ii} = 1, i = 1, \ldots, n$, $n_{(j+1)j} = 1, j = 1, \ldots, n-1$; and β is an $n \times n$ matrix whose elements are

$$\beta_{ii} = \beta_i, \quad i = 1, \cdots, n, \quad \beta_{(j+1)i} = \beta_{i(j+1)} = \tfrac{1}{2}\beta_{(j+1)}, \ j = 1, \cdots, n-1$$

γ is an $n \times n$ upper triangular matrix whose elements are $\gamma_{ii} = \gamma_I$; $\gamma_{i(j+1)} = \gamma_{(j+1)}$; $i = 1, \ldots, n, j = 1, \ldots, n-1$; H is a vector matrix in which the vector index varies from 1 to n and the matrix dimension of H is $n \times n$; A, $B(k)$ are n-dimensional vectors whose components are given by

$$A_i = \sum_{j=1}^{n} \alpha_j, \quad B_{i,k} = c_k b_{i,k}, \quad i = 1, \cdots, n$$

The augmented cost functional is obtained by adjoining to the cost function in Eq. 6.57, the equality constraints via Lagrange multipliers and the inequality constraints via Kuhn–Tucker multipliers as

$$\tilde{J} = E\big[A^T x(K) + x^T(K)\beta x(K) + x^T(K)\gamma y(K) + \sum_{k=1}^{K} \{B^T(k)u(k)$$

6.3 Problem Solution: A Minimum Norm Approach

$$+ u^T(k)d(k)x(k-1) + u^T(k)f(k)Mu(k) + u^T(k)\gamma(k)y(k-1)$$

$$+ \tfrac{1}{4}u^T(k)\gamma(k)Nz(k) + r^T(k-1)\gamma(k) - \tfrac{1}{2}z^T(k)\gamma(k)Lu(k)$$

$$+ \lambda^T(k)(-x(k) + x(k-1) + q(k) + Mu(k)) + \phi^T(k)(-y(k)$$

$$+ x^T(k)Hx(k)) + \mu^T(k)\bigl(-z(k) + u^T(k)Hu(k)\bigr)$$

$$+ \psi^T(k)\bigl(-r(k-1) + u^T(k)Hx(k-1)\bigr)$$

$$+ e^{mT}(k)(x^m - x(k)) + e^{MT}(k)\bigl(x(k) - x^M\bigr)$$

$$+ g^{mY}(k)(u^m(k) - u(k)) + g^{MT}(k)\bigl(u(k) - u^M(k)\bigr)\}] \tag{6.64}$$

where $\lambda(k)$, $\phi(k)$, $\mu(k)$, and $\psi(k)$ are Lagrange multipliers in \$/Mm³. These are obtained so that the corresponding equality constraints are satisfied, and $e^m(k)$, $e^M(k)$, $g^m(k)$, and $g^M(k)$ are Kuhn–Tucker multipliers; these are equal to zero if the constraints are not violated and greater than zero if the constraints are violated.

Employing the discrete version of integration by parts, and dropping constant terms we obtain

$$\tilde{J} = E\Big[x^T(K)(\beta + \phi^T(K)H)x(K) + (A - \lambda(K))^T x(K) - \phi^T(K)y(K)$$

$$+ x^T(K)\gamma y(K) + x^T(0)\phi^T(0)Hx(0) + \lambda^T(0)x(0) + \phi^T(0)y(0)$$

$$+ \sum_{k=1}^{K} \{x^T(k-1)\phi^T(k-1)Hx(k-1) + u^T(k)d(k)x(k-1)$$

$$+ u^T(k)\bigl(f(k)M + \mu^T(k)H\bigr)u^T(k)\gamma(k)y(k-1)$$

$$+ \tfrac{1}{4}u^T(k)\gamma(k)Nz(k) + r^T(k-1)\gamma(k)Mu(k) - \tfrac{1}{2}z^T(k)\gamma(k)luy(k)$$

$$+ u^T(k)\psi^T(k)Hx(k-1) + (\lambda(k) - \lambda(k-1) + v(k))^T x(k-1)$$

$$+ \bigl(M^T v(k) + B(K) + M^T \lambda(k) + \sigma(k)\bigr)^T u(k) - \phi^T(k-1)y(k-1)$$

$$\mu^T(k)z(k) - \psi^T(k)r(k-1)\}\Big] \tag{6.65}$$

where

$$v(k) = e^M(k) - e^m(k) \tag{6.66a}$$

$$\sigma(k) = g^M(k) - g^m(k) \tag{6.66b}$$

We define

$$Z^T(K) = [x^T(K), y^T(K)] \tag{6.67}$$

$$X^T(k) = [x^T(k-1), y^T(k-1), u^T(k), z^T(k), r^T(k-1)] \tag{6.68}$$

$$N(K) = \begin{bmatrix} \beta + \phi^T(K)H & \frac{1}{2}\gamma \\ \frac{1}{2}\gamma^T & 0 \end{bmatrix} \tag{6.69}$$

$$L(k) = \begin{bmatrix} \phi^T(k-1)H & 0 & \frac{1}{2}[\psi^T(k)H + d(k)] & 0 & 0 \\ 0 & 0 & \frac{1}{2}\gamma(k) & 0 & 0 \\ \frac{1}{2}[\psi^T(k)H + d(k)] & \frac{1}{2}\gamma(k) & f(k)M + \mu^T(k)H & \frac{1}{8}\gamma(k)N - \frac{1}{4}L^T\gamma(k) & \frac{1}{2}M^T\gamma(k) \\ 0 & 0 & \frac{1}{8}N^T\gamma(k) - \frac{1}{4}\gamma(k)L & 0 & 0 \\ 0 & 0 & \frac{1}{2}\gamma(k)M & 0 & 0 \end{bmatrix} \tag{6.70}$$

Furthermore, if we define the following vector such that

$$Q(K) = N^{-1}(K)W(K) \tag{6.71}$$

$$V|k| = L^{-1}(k)R(k) \tag{6.72}$$

then the augmented cost functional in Eq. 6.65 can be written as

$$\tilde{J} = E\Big[(Z(K) + \tfrac{1}{2}Q(K))^T N(K)(Z(K) + \tfrac{1}{2}Q(K)) - \tfrac{1}{4}Q^T(K)N(K)Q(K)$$
$$+ x^T(0)\phi^T(0)Hx(0) + \lambda^T(0) + \phi^T(0)y(0)\Big]$$

$$+ E\Bigg[\sum_{k=1}^{K}\Big\{(X(k) + \tfrac{1}{2}V(k))^T L(k)(X(k) + \tfrac{1}{2}V(K)) - \tfrac{1}{4}V^T(k)L(k)V(k)\Big\}\Bigg] \tag{6.73}$$

Because it is desired to maximize $\tilde{J}$ with respect to $Z(K)$ and $X(k)$, the problem is equivalent to

$$\max_{Z(K),X(k)} \tilde{J} = \max E\Big[(Z(K) + \tfrac{1}{2}Q(K))^T N(K)(Z(K) + \tfrac{1}{2}Q(K))\Big]$$

$$+ \max E\Bigg[\sum_{k=1}^{K}\Big\{(X(k) + \tfrac{1}{2}V(K))^T L(k)(X(k) + \tfrac{1}{2}V(k))\Big\}\Bigg] \tag{6.74}$$

because $Q(K)$ and $V(k)$ are independent of $Z(K)$ and $X(k)$, respectively, and $x(0)$ and $y(0)$ are constants.

It will be noticed that $\tilde{J}$ in Eq. 6.74 is composed of a boundary part and a discrete integral part, which are independent of each other. To maximize $\tilde{J}$ in Eq. 6.74, one maximizes each term separately.

The boundary part in Eq. 6.74 defines a norm. Hence we can write this part as

6.3 Problem Solution: A Minimum Norm Approach

$$\max_{Z(K)} J_1 = \max_{Z(K)} E \left\| Z(K) + \tfrac{1}{2} Q(K) \right\|_{N(K)} \tag{6.75}$$

Also, the discrete integral part in Eq. 6.74 defines a norm; we can write this part as

$$\max_{X(k)} J_2 = \max_{X(k)} E \left\| X(k) + \tfrac{1}{2} V(k) \right\|_{L(k)} \tag{6.76}$$

6.3.6.2 The Optimal Solution

There is only one optimal solution to the problems formulated in Eqs. 6.75 and 6.76. The maximization of J_1 is mathematically equivalent to the minimization of the norm in Eq. 6.75. The minimum of the norm in Eq. 6.75 is clearly achieved when

$$E\left[Z(K) + \tfrac{1}{2} Q(K) \right] = [0] \tag{6.77}$$

Substituting from Eqs. 6.67, 6.68, and 6.71 into Eq. 6.77, we obtain

$$E\left[A - \lambda(K) + 2\left(\beta + \phi^T(K) + \gamma y(K) \right) \right] = [0] \tag{6.78}$$

$$E\left[-\phi(K) + \gamma^T x(K) \right] = [0] \tag{6.79}$$

Equations 6.78 and 6.79 give the values of Lagrange multipliers at the last period studied. Also, the maximum of J_2 in Eq. 6.76 is achieved when

$$E\left[X(k) + \tfrac{1}{2} V(k) \right] = [0] \tag{6.80}$$

Substituting from Eqs. 6.68, 6.70, and 6.72 into Eq. 6.80, and adding the equality constraints (6.58, 6.59, 6.60, 6.61) we obtain

$$E\left[-x(k) + x(k-1) + q(k) + Mu(k) \right] = [0] \tag{6.81}$$

$$E\left[-y(k) + x^T(k) H(k) \right] = [0] \tag{6.82}$$

$$E\left[-z(k) + u^T(k) H u(k) \right] = [0] \tag{6.83}$$

$$E\left[-r(k-1) + u^T(k) H x(k-1) \right] = [0] \tag{6.84}$$

$$E\left[2\phi^T(k-1) H x(k-1) + \left(\psi^T(k) H + d(k) \right) u(k) + \lambda(k-1) + v(k) \right] = [0] \tag{6.85}$$

$$E\left[\gamma(k) u(k) - \phi(k-1) \right] = [0] \tag{6.86}$$

$$E\big[(\psi^T(k)H + d(k))x(k-1) + \gamma(k)y(k-1) + 2(f(k)M + \mu^T(k)H)u(k)\big] \quad (6.87)$$

$$E\big[(\tfrac{1}{4}N^T\gamma(k) - \tfrac{1}{2}\gamma(k)L)u(k)\big] = [0] \quad (6.88)$$

$$E[\gamma(k)Mu(k) - \psi(k)] = [0] \quad (6.89)$$

In addition to the above equations, we have Kuhn–Tucker exclusion equations that must be satisfied at the optimum

$$e_{i,k}^m(x_i^m - x_{i,k}) = 0, \qquad e_{i,k}^M(x_{i,k} - x_i^M) = 0 \quad (6.90)$$

$$e_{i,k}^m(u_i^m - u_{i,k}) = 0, \qquad e_{i,k}^M(u_{i,k} - u_{i,k}^M) = 0 \quad (6.91)$$

One also has the following limits on the variables.
If $x_{i,k} < x_i^m$, then we put $x_{i,k} = x_i^m$
If $x_{i,k} > x_i^M$, then we put $x_{i,k} = x_i^M$
If $u_{i,k} < u_{i,k}^m$, then we put $u_{i,k} = u_{i,k}^m$
If $u_{i,k} < u_{i,k}^M$, then we put $u_{i,k} = u_{i,k}^M$

Equations 6.81, 6.82, 6.83, 6.84, 6.85, 6.86, 6.87, 6.88, 6.89, 6.90, and 6.91 with Eqs. 6.78 and 6.79 completely specify the optimal long-term operation of a series multireservoir power system.

6.3.6.3 Practical Application

A computer program was written to solve these equations for the same system mentioned in the previous section. In Table 6.5 we give the inflows in Mm^3 to the first reservoir and the associated probabilities during each month. In Table 6.6, we give the monthly release from each reservoir and the profits realized during the optimization interval. In Table 6.7, we give the optimal monthly reservoir storage during the same year.

6.3.6.4 Comments

In this section, the problem of the long-term optimal operation of a series of reservoirs with a nonlinear model is discussed. The nonlinear model used was a quadratic function of the average storage to avoid underestimation in the hydroelectric production for rising water levels and overestimation for falling water levels.

The section dealt with a nonlinear cost functional and explained how to deal with such cost functionals; we introduced a set of pseudo state variables to cast

6.3 Problem Solution: A Minimum Norm Approach

Table 6.5 Monthly inflows and the associated probabilities for one of the reservoirs in the system

Month k	$I_{1,k}$ (Mm^3)	P	Month k	$I_{1,k}$ (Mm^3)	P
October	437.2	0.0668	April	46.8	0.0668
	667.8	0.2417		118.6	0.2417
	886.2	0.3830		186.6	0.3830
	1,104.6	0.2417		254.6	0.2717
	1,335.3	0.0668		326.4	0.0668
November	387.5	0.0668	May	120.2	0.0668
	550.3	0.2417		425.8	0.2417
	714.9	0.3830		715.2	0.3830
	879.4	0.2417		1,004.6	0.2417
	1,053.2	0.0668		1,310.2	0.3830
December	323.9	0.0668	June	791.5	0.0668
	414.1	0.2417		857.8	0.2417
	499.6	0.3830		1,204.6	0.3830
	585.0	0.2417		1,551.5	0.2417
	675.2	0.0668		1,917.8	0.0668
January	212.4	0.0668	July	448.1	0.0668
	269.3	0.2417		739.1	0.2417
	323.1	0.3830		1,014.7	0.3830
	377.0	0.2417		1,290.2	0.2417
	757.0	0.0668		1,581.3	0.0668
February	130.3	0.0668	August	255.1	0.0668
	168.8	0.2417		547.7	0.2417
	205.2	0.3830		824.7	0.3830
	241.6	0.2417		1,101.8	0.2417
	280.0	0.0668		1,394.4	0.0668
March	109.0	0.0668	September	263.4	0.0668
	147.5	0.2417		551.5	0.2417
	183.9	0.3830		824.8	0.3830
	220.3	0.2417		1,097.2	0.2417
	258.7	0.0668		1385.4	0.0668

Table 6.6 Optimal monthly releases from the four reservoirs and the profits realized

Month k	$u_{1,k}$ (Mm^3)	$u_{2,k}$ (Mm^3)	$u_{3,k}$ (Mm^3)	$u_{4,k}$ (Mm^3)	Profits ($)
1	0	315	474	1,867	817,745
2	1,025	1,289	1,378	2,270	1,566,862
3	1,071	1,256	1,305	2,029	1,596,784
4	1,071	1,367	1,520	1,894	1,997,250
5	953	1,035	1,058	2,459	2,151,664
6	1,071	1,035	1,058	2,459	2,151,664
7	1,024	999	1,022	1,039	1,137,618
8	0	234	273	950	533,420
9	0	483	660	2,194	1,042,055
10	337	675	773	1,370	703,741
11	0	305	407	1,060	457,890
12	0	305	407	1,695	723,348
Value of water remaining in the reservoirs at the end of the year					9,380,718
Total profits					74,537,888

Table 6.7 Optimal monthly reservoir storage

Month	$x_{1,k}$ (Mm^3)	$x_{2,k}$ (Mm^3)	$x_{3,k}$ (Mm^3)	$x_{4,k}$ (Mm^3)
1	7,575	570	0	3,254
2	7,265	570	0	3,409
3	6,693	570	13	3,417
4	5,954	393	0	3,417
5	5,196	342	0	2,235
6	4,308	446	0	1,104
7	3,470	540	0	1,360
8	4,185	570	50	1,731
9	5,390	533	22	1,962
10	6,067	570	50	2,853
11	6,892	570	50	3,408
12	7,717	570	50	3,328

the problem into a quadratic form that can be solved by the minimum norm formulation.

6.4 Simulated Annealing Algorithm (SAA) [28–32]

In applying the SAA to solve optimization problems, the basic idea is to choose a feasible solution at random and then get a neighbor to this solution. A move to this neighbor is performed if either it has a better (lower) objective value or, in the case where the neighbor has a higher objective function value, if $\exp(-\Delta E/Cp) \geq R(0, 1)$, where ΔE is the increase in objective value if we move to the neighbor and $R(0,1)$ is a random number between 0 and 1. The effect of decreasing Cpis that the probability of accepting an increase in the objective function value is decreased during the search. The major steps of the algorithm are summarized as follows.

Step (0): Set iteration counter $K = 0$.
Set the initial temperature of the cooling schedule that results in a high probability of accepting new solutions.
Initialize all step-size vectors ($U_{stepo}(i,t)$, $X_{stepo}(i,t)$, and $S_{stepo}(i,t)$) for the variables ($U(i,t), X(i,t)$, and $S(i,t)$), respectively, for all units ($i = 1, 2, \ldots N$) at all time periods ($t = 1, 2, \ldots, K$).
Step (1): Find, randomly, an initial feasible solution.
Step (2): Calculate the objective function at the initial solution (6.4).
Step (3): Generate an integer random number $J = IR(1,3)$
If $J = 1$ then generate the trial solution by randomly varying the reservoir discharge value.
If $J = 2$ then generate the trial solution by randomly varying the reservoir storage value.

6.4 Simulated Annealing Algorithm (SAA)

If $J = 3$ then generate the trial solution by randomly varying the reservoir spillage discharge value.

Step (4): Calculate the objective function at the generated trial solution.
Step (5): Perform the SA acceptance test to accept or reject the trial solution.
Step (6): Check for equilibrium at this temperature.
If equilibrium is reached go to Step (7). Else go to Step (3).
Step (7): If the prespecified maximum number of iterations is reached then stop. Else go to Step (8).
Step (8): If the step-size vector values for all variables (U_{step}, X_{step}, S_{step}) are less than a pre-specified value then stop. Else go to Step (9).
Step (9): Update the step-size vector values.
Decrease the temperature according to the polynomial time cooling schedule.
Go to Step (3)

6.4.1 Generating Trial Solution (Neighbor)

The most important part in the SAA is to have a good rule for finding a diversified and intensified neighborhood (trial solution) so that a large amount of the solution space can be explored.

Neighbors should be randomly generated, feasible, and span as much as possible the problem solution space. Because of the difficulty of satisfying the continuity equation (6.1) of all reservoirs at all time periods, generating a random trial feasible solution is not a simple matter. A major contribution of this work is the implementation of new rules to obtain randomly feasible solutions faster. Three different routines are implemented to generate trial solutions based on the random variation of one of the three variables (U, X, and S) at a time instant. The following steps describe the rules for generating trial solutions by randomly varying the reservoir discharge (U).

Step (1): Generate randomly a unit $i = IR(1,N)$, and a time instant $t = IR(1,K)$.
Step (2): Generate $R(0,1)$.
Let $\Delta U = R(0,1) * U_{step}(i,t)$, and
$U(i,t) = U(i,t) + $ (or $-$ randomly) ΔU.
Step (3): Check for the feasibility of $U(i,t)$:
If $U_{min}(i,t) \leq U(i,t) \leq U_{max}(i,t)$ then go to Step (4). Else go to Step (1).
Step (4): Use the continuity equation (6.1) to calculate the value storage of unit i at time t, $X(i,t)$.
Step (5): Check for the feasibility of $X(i,t)$:
If $X_{min}(i) \leq X(i,t) \leq X_{max}(i)$ then go to Step (6). Else go to Step (1).
Step (6): The variation of $U(i,t)$ will affect U, X, and S at units $i, i+1, \ldots, N$ at periods $t, t+1, \ldots, K$. The next step is then used to check the feasibility of all those values.

Use the continuity equation to calculate $X_{(i1,t1)}$ for $i_1 = i, i+1, \ldots, N$ and $t_1 = t, t+1, \ldots, K$.
At each value perform the following substeps.
a- If $X_{\min}(i1) \leq X(i1,t1) \leq X_{\max}(i1)$ then go to Step (7).
Else go to (6-b)
b- If $X(i1,t1) \geq X_{\max}(i1)$ then
Let $U_{\text{temp}} = U(i1,t1) + X(i1,t1) - X_{\max}(i1)$ and
$X(i1,t1) = X_{\max}(i1)$
Check if $U_{\text{temp}} \leq U_{\max}(i1,t1)$. Then
$U(i1,t1) = U_{\text{temp}}$. Else
$U(i1,t1) = U_{\max}(i1,t1)$ and
$S(i1,t1) = S(i1,t1) + U_{\text{temp}} - U_{\max}(i1,t1)$
c- If $X(i1,t1) \leq X_{\min}(i1)$ Then
Let $U_{\text{temp}} = U(i1,t1) + X(i1,t1) - X_{\min}(i1)$ and
$X(i1,t1) = X_{\min}(i1)$
Check If $U_{\text{temp}} \: U_{\min}(i1,t1)$ Then
$U(i1,t1) = U_{\text{temp}}$. Else
$U(i1,t1) = U_{\min}(i1,t1)$ and
$S(i1,t1) = S(i1,t1) + U_{\text{temp}} - U_{\min}(i1,t1)$.

Step (7): If all reservoir and time periods have been treated then go to Step (8); otherwise go to Step (6-a)

Step (8): Check for the continuity equation at all units and all time periods. If satisfied, then stop; otherwise go to Step (1).

6.4.2 Details of the SAA for the LTHSP

6.4.2.1 Generating an Initial Feasible Solution

The proposed SAA requires a starting feasible solution that satisfies all system constraints. This solution is randomly generated. The following steps are used in finding this starting solution.

Step (1): (a) Let $i = 1$ and $t = 1$.
(b) Let $U(i,t) = U_{\max}(i,t)$, $X(i,t) = X_{\max}(i)$ and $S(i,t) = 0$.

Step (2): Check if $S(i,t) \geq 0$; then go to Step (4). Else let $U1 = U_{\text{in}} * U_{\max}(i,t)$ and $X1 = X_{\text{in}} * X_{\max}(i)$, where U_{in} and X_{in} are prespecified parameters for finding initial feasible starting values for the variables (assumed 0.25).

Step (3): $U(i,t) = R(U1, U_{\max}(i,t))$.
$X(i,t) = R(X1, X_{\max}(i))$.
Go to Step (2).

Step (4): Check: If $i = N$ and $t = K$ then go to Step (5), Else increase i and t and go to Step (1-b).

Step (5): Check the continuity equation for all units at all time periods. If satisfied, then stop; else go to Step (1).

6.4.2.2 Testing the Simulated Annealing

The implementation steps of the SA test as applied to each iteration in the algorithm are described as follows.

Step (1): At the same calculated temperature c_p^k, apply the following acceptance test for the new trial solution.

Step (2): Acceptance test:
If $E_j \leq E_i$, or
if $\exp[(E_i - E_j)/Cp] \geq R(0,1)$, then accept the trial solution and set $X_i = X_j$ and $E_i = E_j$. Otherwise reject the trial solution, where X_i, X_j, E_i, E_j are the SA current solution, the trial solution, and their corresponding cost, respectively.

Step (3): Go to the next step in the algorithm.

6.4.2.3 Step Size Vector Adjustment

In this work the step vector is updated jointly with temperature, according to the acceptance rate of the attempted moves at the previous temperature stage [23]. All its components are updated simultaneously. It must be mentioned here that a good step vector adjustment scheme must not yield steps that are too large with too many uncorrelated and unaccepted moves, whereas steps that are too small result in an incomplete exploration of the variation domain, with small and frequent increasing of the objective function.

The following explains how the step vector is adjusted mathematically.

Step (1): Calculate $P(i) = \text{NTRACP}(i)/\text{NTR}(i)$, $i = 1, \ldots, N$,
where NTRACP (i) is the number of trials accepted when the variable i is changed.
NTR (i) number of trials attempted by changing variable i.

Step (2): If $P(i) > \text{PMAX}$, then $\text{STEP}(i) = \text{STEP}(i) * \text{STEP}_{\text{MAX}}$
If $P(i) < \text{PMIN}$, then $\text{STEP}(i) = \text{STEP}(i) * \text{STEP}_{\text{MIN}}$
where P_{MAX}, P_{MIN}, STEP_{MAX} and STEP_{MIN} are parameters taken in our implementation as 0.2, 0.8, 0.8, and 1.2, respectively [23].

6.4.2.4 Cooling Schedule

A finite-time implementation of the SA algorithm can be realized by generating homogeneous Markov chains of finite length for a finite sequence of descending values of the control parameter. To achieve this, one must specify a set of parameters that governs the convergence of the algorithm. These parameters form a cooling schedule. The parameters of the cooling schedules are: an initial value of the control parameter decrement function for decreasing the control parameter and

Table 6.8 Comparison with DEC-DP and M-NORM [33]

	Total profit ($)	Value of water in storage ($)
DEC-DP [2, 9]	28,145,330	10,403,170
M-NORM	28,227,174	10,414,598
SAA	28,799,568	10,307,287
% Increase of SAA	2.32	−2.02

a final value of the control parameter specified by the stopping criterion, and a finite length of each homogeneous Markov chain.

In this work a polynomial-time cooling schedule is used in which the temperature is decreased based on the statistics of the trial solution's acceptance or rejection during the search.

6.4.2.5 Equilibrium Test

The sequence of trial solutions generated in the SAA at a fixed temperature is stopped as soon as thermodynamic equilibrium, detected by some adequate condition, is reached. Then the temperature and step vectors are suitably adjusted [23].

The test of equilibrium is done as follows.

If the NTRACP $(T) < N1 * n$ and NTR $(T) < N2 * n$ then continue at the same temperature; otherwise end of temperature stage, where NTRACP (T), NTR (T) are the number of trials accepted and attempted at temperature T, respectively. n is the number of variables in the problem. $N1$ and $N2$ are end temperature stage parameters (taken as 12 and 100) [23].

6.4.3 Practical Applications

A computer model has been implemented based on the SAA described earlier and using the proposed rules for generating random trial solutions. To test the model, an example system with four reservoirs connected in series on a river is solved. The scheduling time horizon is 1 year divided into 12 periods. The full data of this example are given in [2, 9]. This example was solved by combining the decomposition–dynamic programming (DEC-DP) method [2, 9] and by the minimum norm (M-NORM) approach [16].

Table 6.8 shows a comparison of the results obtained by the DEC-DP approach, the M-NORM method, and our proposed SAA. As shown in the table, our SAA achieves a higher total profit with a percentage increase of 2.32% relative to the DEC-DP approach and a value of 2.02% relative to the M-NORM method. The value of water left in storage in the reservoirs at the end of the year (VWE) is also given in $ for the three methods.

Detailed results for the EXAMPLE test system are given in Tables 6.2 and 6.3. Table 6.2 shows the monthly discharge (releases) from the four reservoirs. Table 6.3

6.5 Tabu Search Algorithm

Table 6.9 The discharge (Mm^3)

Month	Reservoir number			
	1	2	3	4
1	43.87	411.97	572.7	2,298.13
2	1,037	1,418	1,500.39	2,704.04
3	1,071.08	1,290.33	1,330.8	2,143.63
4	1,068.41	1,464.95	1,546.35	2,450.56
5	968	1,322.77	1,338.41	2,854.51
6	1,064.36	1,160.65	1,178.01	3,160.28
7	1,037	1,105	1,109.42	1,363.13
8	0	127	405.17	711.3
9	0	510.31	695.25	1,695.27
10	0	314.82	470.73	3,030.41
11	0.13	491.05	636.48	2,220.36
12	0	363.09	495.24	1,948.48

gives monthly storage in the four reservoirs. The solution obtained requires zero spillage from all reservoirs during the 12 months except for the first month which requires 0.5, 0.6, 0.7, and 0.3 for the four reservoirs, respectively.

6.4.4 Conclusion

In this section we consider the long-term hydro scheduling problem. The problem is a hard nonlinear optimization problem in continuous variables. The SAA is used for solving the LTHSP. The proposed algorithm introduces three new innovations: new rules for randomly generating feasible trial solutions, an adaptive cooling schedule, and a method for variable discretization. This results in a significant reduction in the number of the objective function evaluations, and consequently fewer iterations are required to reach the optimal solution. The proposed SAA is applied successfully to solve a problem involving four series cascaded reservoir hydroelectric power systems. The results are compared with those obtained by DEC-DP [2, 9] and M-NORM [9]. Numerical results show a significant increase of the total profit per year for the solved test system compared to previously obtained results using classical techniques (Tables 6.9 and 6.10) [13].

6.5 Tabu Search Algorithm [34–41]

In this section a tabu search algorithm (TSA) is used to solve the LTHSP of a multireservoir hydro power plant connected in series on a river. In the proposed algorithm an approximated short-term memory tabu list is used to prevent cycling of solutions. Moreover, an adaptive approach to adjust the step-size vector between

Table 6.10 The storage (Mm^3)

Month	Reservoir number			
	1	2	3	4
1	7,473.13	569.9	49.27	3,419.58
2	7,265.13	519.9	48.88	3,416.92
3	6,772.05	524.65	49.4	3,414.09
4	6,097.64	274.11	0	2,996.52
5	5,394.64	14.35	2.36	1,782.42
6	4,563.28	0.06	0	58.82
7	3,719.28	0.06	2.37	0
8	3,938.28	0.05	4.06	1,179.7
9	5,039.28	103.75	0.12	3,419.68
10	6,926.28	569.93	49.21	3,420
11	8,076.15	570	49.79	3,419.12
12	8,900.15	569.91	49.63	3,419.89

coarse and fine, according to the accepted and rejected trials is implemented [18, 19]. Another important contribution of this work is introducing new rules for randomly generating feasible solutions satisfying all the system constraints including the well-known continuity equation of all reservoirs at all time periods.

The example system includes four series reservoirs and is taken from the literature [2, 9, 19]. Comparing the results obtained with those of other classical optimization methods and the SA method show an improvement in the quality of solutions obtained with the proposed TSA.

The algorithm introduces rules for generating feasible solutions with an adaptive step vector adjustment. Moreover an approximated tabu list for the continuous variables has been designed. The proposed implementation contributed to the enhancement of speed and convergence of the original tabu search algorithm. A significant reduction in the objective function over previous classical optimization methods and a simulated annealing algorithm has been achieved. Moreover, the proposed TS requires fewer iterations than simulated annealing to convergence. The algorithm has been applied successfully to solve a system with four series cascaded reservoirs. Numerical results show an improvement in the solutions compared to previously obtained results.

6.5.1 Problem Statement

The problem is stated above and we repeat this statement for the reader's convenience. The long-term optimization problem of the hydroelectric power system shown in Fig. 6.1 is to determine the discharge $u_{i,t}$, the storage $x_{i,t}$, and the spillage $s_{i,t}$ for each reservoir, $i = 1, \ldots n$, at all scheduling time periods $t = 1, \ldots, K$, given in Eqs. 6.1, 6.2, 6.3, 6.4, 6.5.

6.5.2 TS Method

TS is an iterative improvement procedure in that it starts from some initial feasible solution and attempts to determine a better solution in the manner of a greatest descent algorithm. However, TS is characterized by an ability to escape local optima (which usually cause simple descent algorithms to terminate) by using a short-term memory of recent solutions. TS has two main components: the tabu list (TL) and the aspiration level (AL) of the solutions associated with this TL [20–27].

TS is very successful in solving combinatorial optimization problems, that is, problems of a discrete nature [20–27]. To apply TS for a continuous variables problem it needs two major steps: a good method for variable discretization, and an efficient design for the TL to prevent cycling of the solution.

The proposed algorithm contains three major steps:

1. Generating randomly feasible trial solutions
2. Calculating the objective function of the given solution
3. Applying the TS procedures to accept or reject the solution at hand

The details of the TSA as applied to the LTHSP are given in the flowchart of Fig. 6.2.

6.5.3 Details of the TSA

6.5.3.1 Tabu List in the LTHSP

The problem deals with continuous variables, therefore repeating the solution exactly could never occur. For the significance of using a TL, some adaptation must be created. In the present work, values of variables stored in the TL are approximated to one-decimal digit numbers to facilitate and ease checking of TL contents.

For each of the three variables (U, X, and S), a TL of a three-dimensional array ($K \times N \times S$) is constructed, where K is the number if scheduling time periods, N is number of the reservoir, and Z is the TL size. Also, associated with each TL is an array for the AL.

6.5.3.2 TS Test

The implementation steps of the TS test as applied to each iteration in the algorithm are described as follows [20].

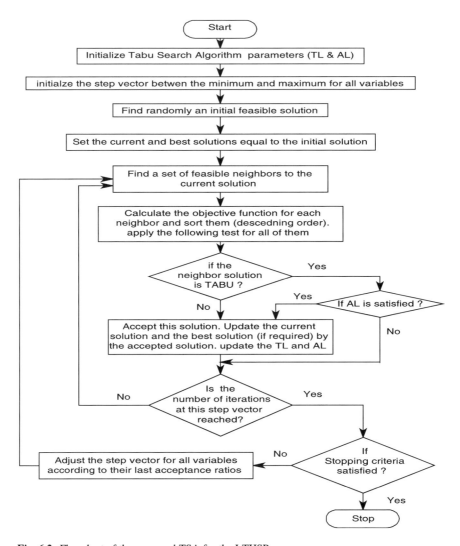

Fig. 6.2 Flowchart of the proposed TSA for the LTHSP

Step (1): If the new generated trial solution is NOT in the TL, then accept it as a current solution, update the TL and AL, and go to the next segment in the algorithm; else go to Step (2).

Step (2): Perform the AL test. If satisfied, then override the tabu state of this solution, update the AL and go to the next segment in the algorithm; else reject this solution and go to the next segment in the algorithm.

6.5.3.3 Generating Trial Solution (Neighbor)

The most important part in the TSA is to have a good rule for finding a diversified and intensified neighborhood (trial solution) so that a large amount of the solution space can be explored.

Neighbors should be randomly generated, feasible, and span the problem solution space as much as possible. Because of the difficulty of satisfying the continuity equation (6.1) of all reservoirs at all time periods, generating a random trial feasible solution is not a simple matter.

A major contribution of this work, along with previous work [19], is the implementation of new rules to obtain randomly feasible solutions faster.

Three different routines are implemented to generate trial solutions based on the random variation of one of the three variables (U, X, and S) at a time instant.

The following steps describe, as an example, the rules for generating trial solutions by randomly varying the reservoir discharge (X).

Step (1): Generate randomly a unit $i = IR(1,N)$, and a time instant $t = IR(1,K)$.
Step (2): Generate R(0,1).
　　Let $\Delta X = R(0,1)*X_{step}(i,t)$, and
　　$X(i,t) = X(i,t) + $ (or $-$ randomly) ΔX.
Step (3): Check for the feasibility of $X(i,t)$:
　　If $Xmin(i,t) \leq X(i,t) \leq Xmax(i,t)$ then go to Step (4). Else go to Step (1).
Step (4): Use the continuity Eq. 6.1 to calculate the value discharge of unit i at time t, $U(i,t)$.
Step (5): Check for the feasibility of $U(i,t)$:
　　If $U_{min}(i) \leq U(i,t) \leq U_{max}(i)$ then go to Step (6). Else go to Step (1).
Step (6): Because the variation of $X(i,t)$ will affect U, X, and S at units $i, i + 1, \ldots, N$ at periods $t, t + 1, \ldots, K$, the next step is then used to check the feasibility of all those values.
　　Use the continuity equation to calculate $X(i1,t1)$ for $i1 = i, i + 1, \ldots, N$ and $t1 = t, t + 1, \ldots, K$.
　　At each value perform the following substeps.
　　a- If $U_{min}(i1) \leq U(i1,t1) \leq U_{max}(i1)$, then go to Step (7). Else go to (6-b)
　　b- If $U(i1,t1) \geq U_{max}(i1)$, then
　　Let $X_{temp} = X(i1,t1) + U(i1,t1) - U_{max}(i1)$ and
　　$U(i1,t1) = U_{max}(i1)$
　　Check If $X_{temp} \leq X_{max}(i1,t1)$. Then
　　$X(i1,t1) = X_{temp}$. Else
　　$X(i1,t1) = X_{max}(i1,t1)$ and
　　$S(i1,t1) = S(i1,t1) + X_{temp} - X_{max}(i1,t1)$
　　c- If $U(i1,t1) \leq U_{min}(i1)$. Then
　　Let $X_{temp} = X(i1,t1) + U(i1,t1) - U_{min}(i1)$ and
　　$U(i1,t1) = U_{min}(i1)$
　　Check If $X_{temp} \geq X_{min}(i1,t1)$. Then

$X(i1,t1) = X_{temp}$, Else
$X(i1,t1) = X_{min}(i1,t1)$ and
$S(i1,t1) = S(i1,t1) + X_{temp} - X_{min}(i1,t1)$

Step (7): If all reservoir and time periods have been treated then go to Step (8), otherwise go to Step (6-a).

Step (8): Check for the continuity equation at all units and all time periods. If satisfied, then stop; otherwise go to Step (1).

6.5.4 Step-Size Vector Adjustment

In this work the step vector is updated jointly with temperature, according to the acceptance rate of the attempted moves at the previous temperature stage [18, 19]. All its components are updated simultaneously. It must be mentioned here that a good step vector adjustment scheme must not yield steps that are too large with too many uncorrelated and unaccepted moves, whereas steps that are too small result in an incomplete exploration of the variation domain, with small and frequent increasing of the objective function.

The following explains how the step vector is adjusted mathematically [18, 19].

Step (1) Calculate $P(i) = NTRACP(i)/NTR(i)$, $i = 1, \ldots, N$.
Where NTRACP (i) is the number of trials accepted when the variable i is changed.
NTR (i) is the number of trials attempted by changing variable i.

Step (2) If $P(i) > P_{MAX}$, then $STEP(i) = STEP(i) * STEP_{MAX}$
If $P(i) < P_{MIN}$, then $STEP(i) = STEP(i) * STEP_{MIN}$
where P_{MAX}, P_{MIN}, $STEP_{MAX}$, and $STEP_{MIN}$ are parameters taken in our implementation as 0.2, 0.8, 0.8, and 1.2, respectively [18].

6.5.5 Stopping Criteria

There may be several possible stopping conditions for the search. In our implementation we stop the search if either of the following two conditions is satisfied.

The number of iterations performed since the best solution last changed is greater than a prespecified maximum number of iterations.

The maximum allowable number of iterations is reached.

6.5.6 Numerical Examples

A computer model has been implemented based on the TSA described earlier and using the new proposed rules for generating random trial solutions.

Table 6.11 Comparison of DEC-DP and M-NORM

	Total profit ($)	Value of water left in storage ($)	% Increase of TSA
DEC-DP [2, 9]	28,145,330	10,403,170	5.52
M-NORM	28,227,174	10,414,598	5.21
SAA	28,799,568	10,307,287	3.12
TSA	29,699,340	10,978,910	–

Table 6.12 The discharge (Mm^3)

Month	Reservoir number			
	1	2	3	4
1	0.00	368.00	528.00	2,250.00
2	1,040.00	1,420.00	1,500.00	2,700.00
3	2.74	743.00	1,170.00	1,980.00
4	0.00	146.00	1,500.00	2,280.00
5	943.00	1,040.00	1,100.00	2,850.00
6	1,030.00	1,110.00	1,300.00	3,160.00
7	850.00	924.00	1,020.00	1,280.00
8	0.00	124.00	362.00	1,310.00
9	0.00	608.00	830.00	2,280.00
10	0.00	781.00	700.00	3,010.00
11	875.00	806.00	902.00	2,490.00
12	824.00	1,190.00	1,320.00	2,770.00

To test the model, an example system with four reservoirs connected in series on a river is solved. The scheduling time horizon is 1 year divided into 12 periods. The full data of this example are given in [2, 9]. This example was solved by a combined decomposition–dynamic programming method [2, 9, 19], by the minimum norm approach [9], and the simulated annealing algorithm [19].

Table 6.11 shows a comparison of the results obtained by the DEC-DP approach, the M-NORM method, the SAA, and the proposed TSA. As shown in the table, our SAA achieves a higher total profit with a percentage increase of 5.52% relative to the DEC-DP approach, a value of 5.21% relative to the M-NORM method, and 3.21% relative to the SAA. The value of water left in storage in the reservoirs at the end of the year is also given in $ for the three methods.

Detailed results for the example test system are given in Tables 6.12 and 6.13. Table 6.12 shows the monthly discharge (releases) from the four reservoirs. Table 6.13 gives monthly storage in the four reservoirs. The solution obtained requires zero spillage from all reservoirs during the 12 months except for two time periods. The spillage at the first month from the reservoir, respectively, is 0.5, 0.6, 0.7, and 0.3. The spillage at the eleventh month from unit 3 is $0.286E + 03$.

Table 6.13 The storage (Mm^3)

Month	Reservoir number			
	1	2	3	4
1	7,520.00	570.00	50.00	3,420.00
2	7,310.00	519.00	50.00	3,420.00
3	7,890.00	2.74	0.00	3,420.00
4	8,280.00	2.74	0.00	3,120.00
5	7,600.00	2.74	50.00	1,670.00
6	6,800.00	5.60	0.00	65.50
7	6,150.00	0.00	0.00	3.50
8	6,370.00	3.49	40.90	408.00
9	7,470.00	9.67	0.00	1,990.00
10	9,350.00	9.67	0.00	3,420.00
11	9,630.00	570.00	50.00	3,420.00
12	9,630.00	570.00	50.00	3,420.00

6.5.7 Conclusions

In this section we consider the long-term hydro scheduling problem. The problem is a hard nonlinear optimization problem in continuous variables. We present a new TSA for solving the LTHSP. The proposed algorithm introduces three new innovations: new rules for randomly generating feasible trial solutions, an efficient tabu list for such problems, and a method for variable discretization. This results in a significant reduction in the number of objective function evaluations, and consequently fewer iterations are required to reach the optimal solution. The proposed TSA is applied successfully to solve a problem involving four series cascaded reservoir hydroelectric power systems. The results are compared with those obtained by DEC-DP [2, 9], M-NORM, and SAA . Numerical results show a significant increase of the total profit per year for the solved test system compared to previously obtained results.

References

1. Arvanitidies, N.V., Rosing, J.: Composite respresentation of a multireservoir hydroelectric power system. IEEE Trans. Power App. Syst. **PAS-89**(2), 319–326 (1970)
2. Arvanitidies, N.V., Rosing, J.: Optimal operation of multireservoir systems using a composite representation. IEEE Trans. Power App. Syst. **PAS-89**(2), 327–335 (1970)
3. During, H., et al.: Optimal operation of multi-reservoir systems using an aggregation-decomposition approach. IEEE Trans. Power App. Syst. **PAS-104**(8), 2086–2092 (1985)
4. Turgeon, A.: A decomposition/projection method for the multireservoir operating. In: National TIMS/ORSA Meeting, Los Angeles, November 1978

5. Turgeon, A.: Optimal operation of multi-reservoir power system with stochastic inflows. Water Resour. Res. **16**(6), 275–283 (1980)
6. Turgeon, A.: A decomposition method for the long-term scheduling of reservoirs in series. Water Resour. Res. **17**(6), 1565–1570 (1981)
7. Olcer, S., Harsa, C., Roch, A.: Application of linear and dynamic programming to the optimization of the production of hydroelectric power. Optim. Contr. Appl. Methods **6**, 43–56 (1985)
8. Grygier, J.C., Stedinger, J.R.: Algorithms for optimizing hydropower system operation. Water Resour. Res. **21**(1), 1–10 (1985)
9. Halliburton, T.S., Sirisena, H.R.: Development of a stochastic optimization for multireservoir scheduling. IEEE Trans. Autom. Control **AC-29**, 82–84 (1984)
10. Marino, M.A., Loaiciga, H.A.: Quadratic model for reservoir management: applications to the central valley project. Water Resour. Res. **21**(5), 631–641 (1985)
11. Sage, A.P.: Optimal Systems Control. Prentice-Hall, Englewood Cliffs (1968)
12. El-Hawary, M.A., Christensen, G.S.: Optimal Economic Operation of Electric Power System. Academic, New York (1979)
13. Shamaly, A., Christensen, G.S., El-Hawary, M.A.: A transformation for necessary optimality conditions for systems with polynomial nonlinearities. IEEE Trans. Autom. Control **AC-24**(6), 983–985 (1979)
14. Shamaly, A., Christensen, G.S., El-Hawary, M.A.: Optimal control of large turboalternator. J. Optimiz. Theory Appl. **34**(1), 83–97 (1981)
15. Soliman, S.A., Christensen, G.S.: Discrete stochastic optimal long-term scheduling of series reservoir. In: Proceedings of 14th IASTED Internationjal Conference, Vancouver, 4–6 June 1986
16. Soliman, S.A., Christistensen, G.S., Abdel-Halim, M.A.: Optimal operation of multireservoir power system using functional analysis. J. Optimiz. Theory Appl. **49**(3), 449–461 (1986)
17. Marino, M.A., Loaiciga, H.A.: Dynamic model for multi reservoir operation. Water Resour. Res. **21**(5), 619–630 (1985)
18. Christensen, G.S., Soliman, S.A.: Optimal Long-Term Operation of Electric Power Systems. Plenum Press, New York (1988)
19. Maceira, M.E.P., Pereira, M.V.F.: Analytical modeling of chronological reservoir operation in probabilistic production costing. IEEE Trans. Power Syst. **11**(1), 171–180 (1996)
20. Christoforidis, M., et al.: Long-term/ mid-term resource optimization of a hydro-dominant power system using the interior point method. IEEE Trans. Power Syst. **11**(1), 287–294 (1996)
21. Escudero, L.F., et al.: Hydropower generation management under uncertainty via scenario analysis and parallel computation. IEEE Trans. Power Syst. **11**(2), 683–689 (1996)
22. da Cruz Jr., G., Soares, S.: Non-uniform composite representation of hydroelectric systems for long-term hydrothermal scheduling. IEEE Trans. Power Syst. **11**(2), 702–707 (1996)
23. Wong, K.P., Wong, S.Y.W.: Hybrid-genetic/ simulated annealing approach to short-term multi-fuel constrained generation scheduling. IEEE Trans. Power Syst. **12**(2), 776–784 (1997)
24. Guan, X., et al.: An optimization-based algorithm for scheduling hydrothermal power systems with cascaded reservoirs and discrete hydro constraints. IEEE Trans. Power Syst. **12**(4), 1775–1780 (1997)
25. Ponrajah, R.A., Galiana, F.D.: Systems to optimize conversion efficiencies at Ontario hydro's hydroelectric plants. IEEE Trans. Power Syst. **13**(3), 1044–1050 (1998)
26. Yu, Z., Sparrow, F.T., Bowen, B.H.: A New long-term hydro production scheduling method for maximizing the profit of hydroelectric systems. IEEE Trans. Power Syst. **13**(1), 66–71 (1998)
27. Mantawy, A.H., Abdel-Magid, Y.L., Selim, S.Z.: A simulated annealing algorithm for unit commitment. IEEE Trans. Power Syst. **13**(1), 197–204 (1997)
28. Aarts, E., Korst, J.: "Simulated Annealing and Boltzman Machines", A Stochastic Approach to Combinatorial Optimization and Neural Computing. John Willey & Sons, NY (1989)
29. Cerny, V.: Thermodynamical approach to the traveling salesman problem: an efficient simulation algorithm. J. Optimiz. Theory Appl. **45**(1), 41–51 (1985)

30. Selim, S.Z., Alsultan, K.: A simulated annealing algorithm for the clustering problem. Pattern Recogn. **24**(10), 1003–1008 (1991)
31. Metropolis, N., Rosenbluth, A., Rosenbluth, M., Teller, A., Teller, E.: Equations of state calculations by fast computing machines. J. Chem. Phys. **21**, 1087–1982 (1953)
32. Siarry, P., Berthiau, G., Haussy, J.: Enhanced simulated annealing for globally minimizing functions of many continuous variables. ACM Trans. Math. Software **23**(2), 209–228 (1997)
33. Mantawy, H., Soliman, S.A., El-Hawary, M.E.: An innovative simulated annealing approach to the long-term hydro scheduling problem. Int. Jr. of Elec. Power & Energy Syst. **25**(1), 41–46 (2002)
34. Mantawy, H., Abdel-Magid, Y.L., Selim, S.Z.: Unit commitment by Tabu search. IEE Proc-Gener. Transm. Distrib. **145**(1), 56–64 (1998)
35. Bland, J.A., Dawson, G.P.: Tabu search and design optimization. Comput. Aided Design **23**(3), 195–201 (1991)
36. Glover, F.: A user's guide to Tabu search. Ann. Oper. Res. **41**, 3–28 (1993)
37. Glover, F.: Artificial intelligence, heuristic frameworks and Tabu search. Managerial Dec. Econ. **11**, 365–375 (1990)
38. Glover, F., Greenberg, H.J.: New approach for heuristic search: a bilateral linkage with artificial intelligence. Europ. J. Oper. Res. **39**, 119–130 (1989)
39. Glover, F.: Future paths for integer programming and links to artificial intelligence. Comput. Oper. Res. **13**(5), 533–549 (1986)
40. Glover, F.: Tabu search-part I. ORSA J. Comput. **1**(3), 190–206 (Summer 1989)
41. Glover, F.: Tabu search-part II. ORSA J. Comput. **2**(1), 4–32 (Winter 1990)

Chapter 7
Electric Power Quality Analysis

Objectives The objectives of this chapter are

- Studying the applications to the simulated annealing (SA) optimization algorithm for measuring voltage flicker magnitude and frequency as well as the harmonic contents of the voltage signal, for power quality analysis
- Estimating voltage magnitude, frequency, and phase angle of the fundamental component
- Solving the nonlinear optimization problem, which minimizes the sum of the absolute value of the error in the estimated voltage signal in continuous variables
- Using the simulated annealing algorithm to estimate the parameters of a system steady power system, having a constant frequency within a data window and a variable frequency within a data window
- Solving the nonlinear optimization problem, which is the minimization of the sum of the squares of the errors, as a function of the signal amplitude, frequency, and phase angle
- Testing the algorithm using samples of the voltage or current signal of one phase of simulated and actual recorded data for noise-free and harmonic contaminated signals
- Studying the effects of critical parameters, such as sampling frequency and number of samples, on the estimated parameters

7.1 Introduction

Power quality is a new term that came to the power engineering community in the last two decades, and to power engineers it means how the voltage or current signal from the sinusoidal waveform is faring due to the widespread use of power electronic equipment in power systems, as well as the nonlinear loads, especially the arc furnace. Such usage introduces harmonics and flicker in power networks.

The definition of flicker in IEEE standards is the "impression of fluctuating brightness or color, when the frequency observed variation lies between a few Hertz and the fusion frequency of image" [1]. The flicker phenomenon may be divided into the two general categories of cyclic flicker and noncyclic flicker. Cyclic flicker is repetitive and is caused by periodic voltage fluctuations due to the operation of loads such as spot welders, compressors, or arc welders. Noncyclic flicker corresponds to occasional voltage fluctuations, such as the starting of large motors; some offloads will cause both cyclic and noncyclic flicker, such as an arc furnace, welder, and ac choppers.

Many algorithms have been proposed to measure the level of voltage flicker as well as the frequency at which it occurs. Reference [2] presents the application of the continuous wavelet transform for power quality analysis. The transform appears to be reliable for detecting and measuring voltage sags, flicker, and transients in power quality analysis.

Reference [3] pays attention to the fast Fourier transform (FFT) and its pitfalls. A lowpass digital filter is used, and the effects of system voltage deviation on the voltage-flicker measurements by direct FFT are discussed. The dc component leakage effect on the flicker components in the spectrum analysis of the effective value of the voltage and the windowing effect on the data acquisition of the voltage signal are discussed.

A digital flicker meter is presented in [4] based on the forward and inverse FFT and on filtering, in the frequency domain, for the implementation of the functional blocks of simulation of lamp–eye–brain response. References [4–6] propose a method based on Kalman filtering algorithms to measure the low frequency modulation of the 50/60 Hz signal. The method in these references allows for random and deterministic variation of the modulation. The approach utilizes a combination of linear and nonlinear Kalman filter modes.

Reference [7] presents a method for direct calculation of flicker level from digital measurements of voltage waveforms. The direct digital implementation uses a fast Fourier transform as the first step in computation. A pruned FFT, customized for the flicker level computation, is also proposed. Presented in [8] is a static-state estimation algorithm based on least absolute value error (LAV) for measurement of voltage flicker level. The waveform for the voltage signal is assumed to have, for simplicity, one flicker component. This algorithm accurately estimates the nominal voltage waveform and the voltage flicker component.

Simulated annealing (SA) is a powerful technique for solving optimization problems [9–15]. It has recently been applied successfully for solving many difficult optimization problems where the objective function is nonlinear in continuous variables. The application of the SA algorithm to such problems requires an efficient strategy to generate feasible trial solutions [9, 15]. The other important point is the selection of an efficient cooling schedule that adapts itself according to statistics of the trial moves (acceptance/rejection) during search.

Power quality varies because of the widespread use of power electronic equipment in power systems, as well as nonlinear loads, especially the arc furnace. This introduces harmonics and flicker in power networks.

7.1 Introduction

Many digital algorithms have been developed for measuring the harmonics and flicker in the network. Reference [16] applies a multirate digital signal-processing technique for harmonic analysis. An antialiasing filter is used with the FFT to estimate system harmonics. Reference [17] proposes a scheme based on Parseval's relation and energy concept, which defines a "group harmonics" identification algorithm for the estimation of the energy distribution in the harmonics of time-varying waveforms. By using this approach many of the drawbacks of FFT can be overcome.

A Newton solution for the harmonic phasor analysis of an ac/dc converter is proposed in [18]. The nonlinear equation for voltage contaminated with harmonics is solved, for the harmonics parameters, using Newton iterations. The solution includes the interaction of the converter with the dc system.

An experimental way to measure the contribution of city street gas discharge lamps is developed in [19]. The equipment used in this experiment is described. It is shown that an equivalent current harmonic spectrum exists which is independent of the type of gas and the lamp rated power.

Reference [20] applied a 12-state Kalman filter to the voltage or current samples to obtain the instantaneous values for a maximum of six harmonics as well as the existing harmonic distortion in real-time.

Reference [21] discusses problems associated with direct application of the FFT to compute harmonic levels on non-steady-state distorted waveforms, and various ways to describe recorded data in statistical terms.

The fast Fourier transform of a sequence that slides over a time-limited rectangular window, carried out in a nonrecursive manner to compute the harmonics for digital relaying, is applied in [22], where only certain isolated harmonics, rather than the full spectrum are needed. The proposed algorithm is compared with the discrete Fourier transform (DFT) and it is proven that this algorithm is less expensive than the DFT.

The concept of a switching function for harmonics measurement is applied in [23] for studying ac and dc harmonics generation by 24-pulse phase controlled converters under imbalanced operation. The characteristic and noncharacteristic ac and dc harmonics using symmetrical voltage components and rectifier switching-functions theory are presented.

Fuzzy linear regression for measurement of harmonics is presented in [24], where fuzzy parameters for each harmonic component are assumed, having a certain middle and a certain spread. The problem is converted to one of linear optimization, based on two inequality constraints to ensure that the optimal solution is within the assumed membership. A digital power analyzer for use in a power system for analysis of harmonics is described in [25]. The measurement algorithm is based on a recursive digital filter that processes input signal sample values.

Reference [26] presents a technique for harmonic extraction in a signal based on a modeling and identification method using a recursive least error squares algorithm. The parameters of the harmonics are estimated from the uniformly sampled signal.

A Hopfield neural network is used in [27] to determine simultaneously the supply-frequency variation and the fundamental-amplitude/phase variation, as well as the harmonics-amplitude/phase variation. The nonlinear least error squares algorithm based on Taylor-series expansion is used to compare with the neural network-based Hopfield algorithm. A set of data taken onsite was used.

An enhanced FFT-based (E-FFT) parametric algorithm suitable for online harmonic analysis of electrical power systems is presented in [28]. The E-FFT algorithm exploits its iteration loops in combination with the characteristic of steep-descent gradient search strategy. The E-FFT algorithm performs reasonably well with short data record length. Reference [29] offers the optimization of spectrum analysis for a harmonics signal, by means of scale fine-tuning to overcome the drawbacks of FFT. These include the picket fence and leakage effects.

Simulated annealing is a powerful technique for solving many optimization problems [11–14, 30, 31]. It has the ability of escaping local minima by incorporating a probability function in accepting or rejecting new solutions. It has been successfully used to solve combinatorial optimization problems, where the cost function is defined in a discrete domain. Recently, the SA algorithm has been successfully used to solve many important problems with functions of continuous variables, similar to the problem of this study. The application of the SA algorithm to such problems requires an efficient strategy to generate feasible trial solutions [30, 32]. The other important point is the selection of an efficient cooling schedule that adapts itself according to the statistics of the trial moves (acceptance/rejection) during the search.

7.2 Simulated Annealing Algorithm (SAA)

The algorithm aims to find the best estimate of voltage flicker amplitude and frequency as well as the power system voltage signal and frequency. The problem is formulated as an optimization problem, where the objective is to minimize the error between the estimated voltage signal and the actual available measurements. Implementation details of the SA algorithm are given in Chap. 2 of this book and we repeat the major steps in the following sections.

Step (0): Set iteration counter $ITR = 0$.
Set the initial temperature of the cooling schedule that results in a high probability of accepting new solutions.
Initialize step-size vector $G_{\text{stepo}}(i)$ for all values of variables $G(i)$, $i = 1,2,3,\ldots,N$.

Step (1): Find, randomly, initial values for the estimated parameters and set them as the current and best solution.

Step (2): Determine the error (objective function) for current estimated parameters.

Step (3): Generate randomly a new estimate (new trial solution) as a neighbor to the current solution.

Step (4): Calculate the objective function at the new estimate.

Step (5): Perform the SA acceptance test to accept or reject the trial solution.

Step (6): Check for equilibrium at this temperature. If equilibrium is reached go to Step (7). Else go to Step (3).
Step (7): If the prespecified maximum number of iterations is reached then stop. Else go to Step (8).
Step (8): If the step-size vector values for all variables (G_{step}) are less than a prespecified value then stop. Else go to Step (9).
Step (9): Update the step-size vector values. Decrease the temperature according to the polynomial time cooling schedule.
Go to Step (3).

7.2.1 Testing Simulated Annealing Algorithm

The implementation steps of the SA test as applied for each iteration in the algorithm are described as follows [9, 15].

Step (1): At the same calculated temperature $\mathbf{c_p}^k$ apply the following acceptance test for the new trial solution.
Step (2): Acceptance test:
If $E_j \leq E_i$, or
If $\exp[(E_i - E_j)/Cp] \geq R(0, 1)$, then accept the trial solution; set $X_i = X_j$ and $E_i = E_j$. Otherwise reject the trial solution, where X_i, X_j, E_i, E_j are the SA current solution, the trial solution, and their corresponding cost, respectively.
Step (3): Go to the next step in the algorithm.

7.2.2 Step-Size Vector Adjustment

In this work the step vector is updated jointly with the cooling schedule temperature, according to the acceptance rate of the attempted moves at the previous temperature stage [11, 13]. All its components are updated simultaneously. The following steps explain how the step vector is adjusted mathematically.

Step (1): Calculate P (i) = NTRACP (i)/NTR (i), i = 1, ..., N.
Where NTRACP (i) is the number of trials accepted, then variable i is changed.
NTR (i) is the number of trials attempted by changing the variable i.
Step (2): If P(i) > P_{MAX}, then STEP(i) = STEP(i)*STEP$_{MAX}$
If P(i) < P_{MIN}, then STEP(i) = STEP(i)*STEP$_{MIN}$
Where P_{MAX}, P_{MIN}, STEP$_{MAX}$, and STEP$_{MIN}$ are parameters taken in our implementation as 0.05, 0.5, 0.8, and 1.2, respectively [17].

7.2.3 Cooling Schedule

A finite-time implementation of the SAA can be realized by generating homogeneous Markov chains of finite length for a finite sequence of descending values of the control parameter. To achieve this, one must specify a set of parameters that govern the convergence of the algorithm. These parameters form a cooling schedule. The parameters of the cooling schedules are an initial value of the control parameter decrement function for decreasing the control parameter, a final value of the control parameter specified by the stopping criterion, and a finite length of each homogeneous Markov chain.

In this work a polynomial-time cooling schedule is used in which the temperature is decreased based on the statistics of the trial solution's acceptance or rejection during the search. Details of the implemented cooling schedule are described in [12, 13]

7.3 Flicker Voltage Simulation [2]

Generally speaking, the voltage during the time of flicker can be expressed as

$$v(t) = \left\{A_0 + \sum_{i=1}^{M} A_i \cos(\omega_{fi} t + \phi_{fi})\right\} \cos(\omega_0 t + \phi_0) \quad (7.1)$$

where A_O is the amplitude of the power system voltage, ω_o the power frequency, and ϕ_O the phase angle. Furthermore, A_i is the amplitude of the flicker voltage, ω_{fi} its frequency, and ϕ_{fi} the phase angle.

7.3.1 Problem Formulation

Assume that N samples of the voltage signal during the flicker period are available. The optimization problem is to estimate the above signal and voltage flicker parameters that minimize the sum of the absolute value of the error in the estimated signal. This can be expressed mathematically as

$$J = \sum_{j=1}^{N} \left| \left\{ v(t_j) - \left\{ A_O + \sum_{i=1}^{m} A_{fi} \cos(\omega_{fi} t_j + \phi_{fi}) \right\} \cos(\omega_O t_j + \phi_0) \right\} \right| \quad (7.2)$$

The above cost function is a nonlinear optimization problem. Many techniques are available to solve this problem. Each algorithm uses some kind of approximation to facilitate the estimation process. In this chapter the SA-based optimization algorithm is used to solve this problem. In the next section, we explain this algorithm.

7.3 Flicker Voltage Simulation

Table 7.1 Test signals

$v(t) = [1 + 0.01\cos(2\pi 1t)]\cos(2\pi 50t)$
$v(t) = [1 + 0.02\cos(2\pi 1t)]\cos(2\pi 50t)$
$v(t) = [1 + 0.05\cos(2\pi 7.5t)]\cos(2\pi 50t)$
$v(t) = [1 + 0.01\cos(2\pi 12.5t)]\cos(2\pi 50t)$
$v(t) = [1 + 0.02\cos(2\pi 17.5t)]\cos(2\pi 50t)$
$v(t) = [1 + 0.05\cos(2\pi 22.5.5t)]\cos(2\pi 50t)$
$v(t) = [1 + 0.01\cos(2\pi 27.5t)]\cos(2\pi 50t)$
$v(t) = [1 + 0.1\cos(2\pi 2t) + 0.5\cos(2\pi 5t)]\cos(2\pi 50t)$

$v(t) = \begin{bmatrix} 1 + 0.05\cos(2\pi 2..5t) + 0.333os(2\pi 7.5t) + 0.2\cos(2\pi 12.5t) + \\ 0.14\cos(2\pi 17..5t) + \\ + 0.11\cos(2\pi 22.5t) + 0.09\cos(2\pi 27.5t) \end{bmatrix} \cos(2\pi 50t)$

Table 7.2 Estimated parameters for each voltage signal

#	Signal parameters V_O	f_O	Voltage flicker parameters V_{f1}	f_{f1}	V_{f2}	f_{f2}	V_{f3}	f_{f3}	V_{f4}	f_{f4}	V_{f5}	f_{f5}	V_{f6}	f_{f6}
1	1.0	50	0.01	0.9993										
2	1.0	50	0.02	0.99984										
3	1.0	50	0.05	7.49999										
4	1.0	50	0.01	2.5										
5	1.0	50	0.02	7.5										
6	1.0	50	0.05	22.4999										
7	1.0	50	0.01	127.5										
8	1.0	50	0.5	5.0	0.1	2.0								
9	1.0	50	0.051	2.474	.333	7.5	0.20	12.5	0.14	17.5	0.11	22.5	0.09	27.5

7.3.2 Testing the Algorithm for Voltage Flicker

The algorithm is tested to estimate the parameters of a voltage signal contaminated with different voltage flicker components, having different amplitudes and different frequencies. These examples are given in [1] and are mentioned in Table 7.1.

All the above signals are sampled at 500 Hz (2 ms sampling time), and 200 samples are used (data window size = 20 cycles) to estimate the parameters of the signals according to the optimization criteria of Eq. 7.3. It has been found that the proposed algorithm explained in Sect. 7.2 is successful in estimating the parameters of each voltage signal, including the parameters of the flicker signal and the main voltage signal. Table 7.2 gives the results obtained.

Table 7.3 Effects of number of samples on the estimated parameter

M	V_O	f_O	V_{f1}	f_{f1}	V_{f2}	f_{f2}
50	1.021	50	.04882	4.988	.0804	2.32
100	0.9996	50.	0.05	4.997	0.100	1.994
150	0.9999	49.9999	0.04999	4.9998	0.0999	1.9997
200	1.000	49.9999	0.05002	5.0000	0.1000	1.99997
250	1.000	50.000	0.05000	5.000	0.1000	2.000

7.3.3 Effect of Number of Samples

The effects of the number of samples on the estimated parameters are studied in this section. The number of samples is changed from 50 samples to 250 samples (data window size is changed from 5 to 75 cycles), and the sampling frequency is kept constant at 500 Hz. Table 7.3 gives the results obtained for the waveform number.

Examining the above table reveals the following.

- The algorithm estimates the parameters of the voltage signal and the flicker voltage parameters accurately.
- The number of samples at the specified sampling frequency that produces the exact estimate is 250 samples.

7.3.4 Effects of Sampling Frequency

The effects of sampling frequency on the estimated parameters are studied in this section. With a number of samples equal to 250 samples for the same voltage signal of number 8, the sampling frequency is changed from 250 Hz to 1750 Hz. Table 7.4 gives the results obtained

Examining the table reveals the following.

- The algorithm produces accurate estimates for sampling frequencies up to 1,000 Hz, and produces bad estimates at frequencies 1,500 Hz and 1,750 Hz.
- It has been found that bad estimates are produced if the data window size is not an integer multiple of cycles. Thus we recommend using an integer number of cycles for the data window size.

7.4 Harmonics Problem Formulation [60–80]

Consider the Fourier expansion of a current or voltage wave with n harmonic terms:

$$y(t) = \sum_{n=1}^{N} A_n \sin(2\pi nft + \theta_n)$$

$$= \sum_{n=1}^{N} (a_n \sin 2\pi nft + b_n \cos(2\pi nft)) \quad (7.3)$$

7.5 Testing the Algorithm for Harmonics

Table 7.4 Effects of sampling frequency on the estimated parameters

F_S	V_O	f_O	V_{f1}	f_{f1}	V_{f2}	f_{f2}
250	1.000	50.0	0.100	2.000	0.050	5.000
500	1.000	50.0	0.100	1.9997	0.050	5.00
750	1.000	50.0	0.0999	2.000	0.050	5.00
1000	1.000	50.00	0.09998	2.0005	0.04999	5.00
1500	0.6713	50.00	0.41932	0.7922	0.0593	4.823
1750	0.62712	50.00	0.46421	0.7612	0.05864	4.847

Assume that m measuring samples are available (t_i, y_i), $i = 1, 2, \ldots, m$. The objective is to estimate the values of A_k, θ_k or to estimate a_k, b_k, and f so as to minimize the total squared error E given by:

$$E = \sum_{i=1}^{m} \left[y_i - \sum_{n=1}^{N} (A_k \sin(2\pi n f t + \theta_n)) \right]^2 \quad (7.4)$$

In this chapter two cases are studied. In the first case the signal fundamental frequency is known in advance and equals the nominal frequency. In the second case the signal frequency is unknown. The objective function in Eq. 7.2 is nonlinear in continuous variables. In the next section, implementation of the SA algorithm to find the best estimate for the parameters in (7.2) is described.

7.5 Testing the Algorithm for Harmonics

7.5.1 Signal with Known Frequency

This example is solved in [24] using fuzzy linear regression. The voltage signal waveform is contaminated with up to 19 harmonics and is given by

$$v(t) = 1.0 \sin(\omega t + 10°) + 0.1 \sin(3\omega t + 20°)$$
$$+ 0.08 \sin(5\omega t + 30°) + 0.08 \sin(9\omega t + 40°) + 0.06 \sin(11\omega t + 50°)$$
$$+ 0.05 \sin(13\omega t + 60°) + 0.03 \sin(19\omega t + 70°)$$

- The voltage signal is sampled at a sampling frequency of 2000 Hz and a 200-sample data window size is used. Table 7.5 gives the results obtained for the voltage harmonic amplitude as well as the phase angle. Examining this table reveals that the estimates of the proposed algorithm are accurate enough.

Example 1

Another test is performed where the proposed algorithm is tested on actual recorded data, sampled at 1500 Hz, and a 90-sample data window size is used. Figure 7.1

Table 7.5 Harmonics and phase angle estimate

Order of harmonic	Magnitude (p.u.)	φ (Degree)
1	0.99995	10.
3	0.099958	19.976
5	0.8	30.
9	0.8	40
11	0.06	50.0
13	0.05	60.0
19	0.03	70.00

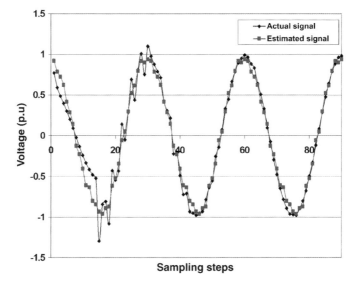

Fig. 7.1 Actual and estimated signals

gives the actual and estimated signals, and Table 7.6 gives the harmonic content of the signal. Examining this figure and Table 7.6 reveals the following.

- Although the voltage signal is highly distorted, the proposed algorithm succeeded in producing good estimates for the harmonic content of the signal.
- We recommend for the highly distorted data window an exponential decayed dc term, inasmuch as the voltage signal is produced for a faulted phase.
- We can consider harmonics of amplitude less than 3% of fundamental do not exist.

7.5.2 Signal with Unknown Frequency

In this section, we assume that the signal frequency is unknown. In this case Eq. 7.2 is a highly nonlinear cost function in the frequency and harmonics parameters. Two examples are solved in this section.

7.5 Testing the Algorithm for Harmonics

Table 7.6 Estimated harmonic content for the actual recorded data

Harmonic order	Amplitude (p.u.)	Phase angle (deg)
1	0.92937	88.92156
2	0	0.07784
3	0.00785	1.32428
4	0.00017	0.35137
5	0.00612	1.31303
6	0	0.02897
7	0.00121	0.76314
8	0.00027	0.26009
9	0	1.37055
10	0.02976	0
11	0.0224	12.25509
12	0.00139	0.18803
13	0.01191	2.8034
14	0.00022	0.06027
15	0	0.6211
16	0.00164	0.50068
17	0.00003	0.26156
18	0.02564	0
19	0.00035	0.44847
20	0.00012	0.06775
21	0.00055	0.30928
22	0	0.26512
23	0.0002	0
24	0.00462	0.38181
25	0.00025	0.18268
26	0.01513	0.18616
27	0	0.49983
28	0.04939	0.24213
29	0.00037	0.27831

Example 2

This example is the same as Example 1, but the signal frequency is assumed to be unknown. We use for simulation 50 Hz, nominal frequency, one time, and another time we use 49.8 Hz, near nominal frequency. Figure 7.2 shows the estimated and simulated waveforms. Examining this figure reveals that

- The proposed algorithm estimates the simulated results exactly, and gives harmonics parameters mentioned in the previous section, Sect. 7.1, Example 1.
- The two curves coincide, therefore the algorithm exactly estimates the signal frequency, nominal frequency, and near-nominal frequency (50 Hz or 49.8 Hz). It has been shown through extensive runs that the proposed algorithm estimates the signal frequency exactly regardless of the value of this frequency.

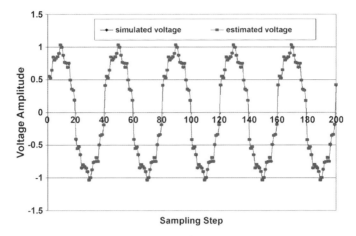

Fig. 7.2 Estimated and simulated signals

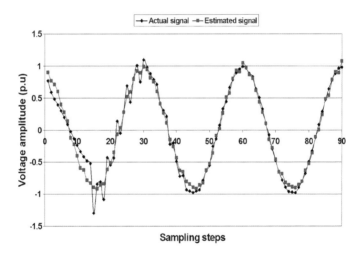

Fig. 7.3 Actual and estimated signals

Example 3

In this example, the proposed algorithm is tested on the same actual recorded data stated in the previous section, but we assume that the signal frequency is unknown and has to be estimated. The proposed algorithm accurately estimated the frequency. This frequency was found to be 50.15973 Hz. This value is very close to the value estimated by a frequency estimation algorithm of [14]. Figure 7.3 compares the actual signal with the estimated signal. Table 7.7 gives the harmonic content of the signal. Examining the figure and table reveals the following results.

7.6 Conclusions

Table 7.7 Estimated harmonics for the actual recorded dData

Harmonic order	Amplitude (p.u.)	Phase angle (deg)
1	0.89927	85.42126
2	0	1.03193
3	0.00738	0.15929
4	0	0.39935
5	0.00628	0.35735
6	0	0.56926
7	0.00027	0.08843
8	0.00067	0.23173
9	0.0005	0.28249
10	0.02979	0.10307
11	0.02426	11.10537
12	0	0.24713
13	0.01005	3.08373
14	0.00011	0.31609
15	0.00074	0.1531
16	0	0.48818
17	0.0001	0
18	0.0214	0.15004
19	0.00003	0.29182
20	0	0.40445
21	0.00148	1.13516
22	0.00024	0.37999
23	0.0001	0
24	0.00051	0
25	0.00029	0
26	0.00798	0.22898
27	0.00021	0.59784
28	0.04571	0
29	0.04999	0
30	0.00123	0.39088

- The proposed algorithm estimates the signal with good accuracy.
- The harmonic content is slightly different from that obtained in Table 7.2 because in this estimate we assume the frequency of the signal is unknown, which produces a different cost function.

7.6 Conclusions

We present in this section the applications of the simulated annealing algorithm for estimating the parameters of the power system voltage signal as well as the flicker voltage parameters. The proposed algorithm minimizes a nonlinear cost function of the error based on least absolute value. The proposed algorithm does not need any expansion to the function to become linear in the parameters, as has been done earlier in the literature. The proposed algorithm produces accurate estimates providing that the data window size is an integer number of cycles. Furthermore, the proposed algorithm is able to identify the parameters of the voltage flicker having any amplitude and frequency. The simulated annealing-based optimization algorithm is also implemented to estimate the harmonic content of a signal as well as the frequency of the fundamental. The algorithm does not need any model for the frequency or harmonics such as those available in the literature. The algorithm uses only samples of the voltage or current signal under study as well as the number of harmonics expected in the waveform. It has been shown that either the sampling frequency or the number of samples (provided that the sampling frequency satisfies the sampling theorem) does not affect the harmonic parameters. The proposed algorithm is suitable for offline and at steady-state estimation.

7.7 Steady-State Frequency Estimation [48–59]

Control and protection of power systems require accurate measurement of system frequency. A system operates at nominal frequency, 50/60 Hz. There is a balance in the active power; that is, the power generated equals the demand power plus losses when the frequency is steady. Imbalance in the active power causes the frequency to change. A frequency less than the nominal frequency means that the demand load plus losses is greater than the power generated, but a frequency greater than nominal frequency means that the system generation is greater than the load demand plus losses. As such, frequency can be used as a measure of system power balance.

Over the past three decades, many algorithms have been developed and tested to measure power system frequency and rate of change of frequency. A precise digital algorithm based on discrete Fourier transforms to estimate the frequency of a sinusoid with harmonics in real-time is proposed in [31]. This algorithm is called the smart discrete Fourier transform that avoids the errors due to frequency deviation and keeps all the advantages of the DFT.

Reference [32] presents an algorithm for frequency estimation of a power system from distorted signals. The proposed algorithm is based on the extended complex Kalman filter, which uses discrete values of a three-phase voltage, which are transformed into the well-known $\alpha\beta$-transform. Using this transform a nonlinear state-space formulation is obtained for the extended Kalman filter. The algorithm proposed in this reference is iterative and complex and needs much computing time and uses the three-phase voltage measurements to calculate the power system voltage frequency.

Reference [33] describes design, computational aspect, and implementation aspects of a digital signal-processing technique for measuring the operating frequency of a power system. It is suggested that this technique produces correct and noise-free estimates for near-nominal, nominal, and off-nominal frequencies in about 25 ms, and it requires modest computation. The proposed technique uses the per-phase digitized voltage samples and applies orthogonal FIR digital filters with the least error squares (LES) algorithm to extract the system frequency.

Reference [34] presents an iterative technique for measuring power system frequency to a resolution of 0.01–0.02 Hz for near-nominal, nominal, and off-nominal frequencies in about 20 ms. The algorithm in this reference uses per-phase digitized voltage samples together with a FIR filter and the LES algorithm to extract the signal frequency iteratively. This algorithm has beneficial features including fixed sampling rate, fixed data window size, and easy implementation.

References [35] and [36] present a new pair of orthogonal filters for phasor computation; the technique proposed in these references accurately extracts the fundamental component of fault voltage and current signal. Reference [37] describes an algorithm for power system frequency estimation. The algorithm, in this reference, applies an orthogonal signal component obtained with the use of two orthogonal FIR filters. The essential property of the algorithm proposed in this reference is outstanding immunity to both signals' orthogonal component magnitudes and FIR filter gain variations. Again this algorithm uses the per-phase digitized voltage samples.

Reference [38] presents a method of measurement of power system frequency, based on digital filtering and Prony's estimation. The discrete Fourier transform with a variable data window is used to filter out the noise and harmonics associated with the signal. The authors claimed that the results obtained using this algorithm are more accurate than when applying the method based on the measurement of angular velocity of the rotating voltage phasor. The response time of the proposed method is equal to three to four periods of the fundamental components. This method also uses the per-phase digitized voltage samples to compute the system frequency from a harmonics-polluted voltage signal. Reference [39] implements a digital technique for the evaluation of power system frequency. The algorithm is suitable for microprocessor implementation and uses only standard hardware. The algorithm works with any relative phase of the input signal and produces a new frequency estimate for every new input sample. This algorithm uses the orthogonal sine and cosine-filtering algorithm.

A frequency relay, which is capable of under/over frequency and rate of change of frequency measurements using an instantaneous frequency-measuring algorithm, is presented in [40]. It has been shown that filtering the relay input signal could adversely affect the dynamic frequency evaluation response. Misleading frequency behavior is observed in this method, and an algorithm has been developed to improve this behavior. The under/over frequency function of the relay will cause it to operate within 30 ms.

Digital state estimation is implemented to estimate the power system voltage amplitude and normal frequency and its rate of change. The techniques employed for static-state estimation are the least error squares technique [41–43] and least absolute value technique [44–47]. Linear and nonlinear Kalman filtering algorithms are implemented for tracking the system operating frequency, the rate of change of frequency and power system voltage magnitude from a harmonic-polluted environment of the system voltage at the relay location. Most of these techniques use the per-phase digitized voltage samples and assume that the three phase voltages are balanced and contain the same noise and harmonics, which is not the case in real-time, especially in the distribution systems, where different single-phase loads are supplied from different phases.

Artificial intelligence-based optimization methods, such as SA, have been successfully used in different applications.

Simulated annealing is a powerful technique for solving many optimization problems [10–15]. It has the ability to escape local minima by incorporating a probability function in accepting or rejecting new solutions. It has been successfully used to solve combinatorial optimization problems, where the cost function is defined in a discrete domain. Therefore, many important problems are defined as functions of continuous variables, including the problem under study. The application of the SA method to such problems requires an efficient strategy to generate feasible trial solutions [10–15]. The other important point is the selection of an efficient cooling schedule that adapts itself according to the statistics of the trial moves (acceptance/rejection) during search.

In this section the SA method is used to estimate the power system steady-state frequency. The proposed algorithm does not need prefiltering or modeling for the system frequency before and during the estimation process. The nonlinear optimization problem, which is a function of the signal amplitude, frequency, and phase angle, is solved using the SA method. This algorithm uses the samples of the voltage or current signal of one phase at the relay location. The proposed algorithm is tested using simulated and actual recorded data for noise-free and harmonic contamination signals. Effects of critical parameters, such as sampling frequency and number of samples, on the estimated parameters are tested.

7.7.1 A Constant Frequency Model, Problem Formulation [34–37]

Given m samples of the voltage magnitude signal, at the relay location, it is required to estimate the signal amplitude, frequency, and phase angle to perform the necessary control for the power system. The signal voltage can be written as

$$v(t) = \sqrt{2}V \sin(\omega t + \phi) + \xi(t) \qquad (7.5)$$

Where $v(t)$ is the voltage at time t, V is the rms value of the fundamental component, ω is the system frequency, and $\xi(t)$ is the error associated with the samples. This error may be due to noise in the a/d converter or due to harmonic contamination. Equation (7.5) can be written at any sample time Δt as

$$v(k\Delta T) = \sqrt{2}V \sin(2\pi f k \Delta T + \phi) + \xi(k\Delta T), \qquad (7.6)$$
$$k = 1, 2, ..., m; \ m \ total \ vailable \ voltage \ sample$$

7.7 Steady-State Frequency Estimation

where k is the order of samples and ΔT is the sampling time (1./sampling frequency). Equation (7.6) is a nonlinear equation in the signal parameters, V, f, and ϕ. Now, given the m voltage samples, the values of $V, f,$ and ϕ that minimize the error ξ based on least error squares can be obtained by minimizing the cost function given by

$$J = \frac{1}{2} \sum_{k=1}^{m} \left[v(k\,\Delta T) - \sqrt{2} V \, \sin(2\pi f k\,\Delta T + \phi) \right]^2 \tag{7.7}$$

In the next section, we propose an algorithm based on simulated annealing to solve this optimization problem.

7.7.2 Computer Simulation

The SA algorithm explained in Sect. 7.3 is implemented to solve the nonlinear problem formulated in Eq. 7.7; a number of simulated examples are used in the following sections.

7.7.2.1 Noise-free signal

In this test we assume that the signal is free of noise, and has a nominal frequency of 50 Hz, amplitude 1.0 p.u., and phase angle of 30°. The signal is sampled at 2,000 Hz, $\Delta T = 0.5$ ms, and a number of samples $= 40$ samples is used (data window size $= 1$ cycle). The proposed algorithm succeeded in accurately estimating the signal parameters in this short data window size for a signal frequency off-nominal, near-nominal, and at nominal, that is, at $f = 45$ Hz, 49.9, 50, and 55 Hz.

7.7.2.2 Effects of Number of Samples

In this section, the effects of the number of samples on the estimated parameters are studied, when the sampling frequency is kept constant at 2,000 Hz. Table 7.8 gives the results obtained for different number of samples.

Examining the Table 7.8 reveals the following.

- The proposed algorithm succeeded in estimating the signal parameters accurately for all sample amounts, and a cost function of zero value is almost obtained for all samples.
- The 40 samples are enough to estimate the signal parameters.

Table 7.8 Effect of number of samples on the estimated parameters, sampling frequency = 2,000 Hz

M	40	80	120	160
V (p.u.)	1.0011	1.0002	1.0002	1.00
f(Hz)	50.026	50.003	49.998	50.002
ϕ(deg)	29.909	29.980	30.02	29.975
$J \times 1E+05$	2.1261	0.19275	0.2306	0.25898
# of Iter.	130	545	915	320

Table 7.9 Effects of number of samples on the estimated parameters, number of samples = 80

F_s(Hz)	1,000	2,000	3,000	4,000	5,000
V	1.001	1.0002	1.0003	1.0008	1.003
f(Hz)	50.002	50.003	49.991	50.002	49.997
ϕ	29.945	29.98	30.052	29.993	30.011
J	00.00	0.000	0.000	0.000	0.000
Iter.	548	545	435	355	975

7.7.2.3 Effects of Sampling Frequency

Effects of sampling frequency in the estimated signal parameters are studied, where we change the sampling frequency from 1,000 Hz to 5,000 Hz, and 80 samples are used; the data window size is four cycles. Table 7.9 gives the results obtained.

Examining this table reveals that:

- Accurate estimates are obtained for the signal parameters at every sampling frequency.
- The cost function J is zero across the test although it is a nonlinear function, which means that the solutions obtained are optimal solutions.

7.7.3 Harmonic-contaminated Signal

Effects of harmonics in the estimated parameters are studied in this section, where we assume that the signal is contaminated with steady harmonics of order 3 and 5 harmonics of magnitudes 0.5 and 0.25 p.u., respectively.

7.7.3.1 Effects of Number of Samples

The effects of the number of samples on the estimated parameters when the voltage signal is contaminated with harmonics are studied. Table 7.10 gives the results obtained when the sampling frequency is chosen to be 2,000 Hz, and a number of samples is chosen from 40 to 160. The signal fundamental frequency equals 50 Hz.

Examining this table reveals that

- An inaccurate estimate is obtained for the signal parameters especially for the first two samples.

7.7 Steady-State Frequency Estimation

Table 7.10 Effect of number of samples when the signal is contaminated with harmonics, sampling frequency = 2,000 Hz

m	80	120	160	200	240
V	1.0013	1.009	1.0004	1.0569	1.05
f	48.631	49.313	49.613	49.899	49.94
ϕ(deg)	39.99	37.673	35.722	30.85	30.59
J	6.0083	9.2102	12.375	22.513	24.63
Iter.	525	395	610	467	260

Table 7.11 Effects of number of sampling frequency when the signal is contaminated with harmonics, number of samples = 80 samples

Fs(Hz)	1,000	2,000	3,000	4,000	5.000
V	1.000	1.001	1.012	1.0023	1.029
F	49.65	48.63	49.6	46.682	49.1
ϕ(deg)	35.194	39.99	37.37	39.99	34.99
J	6.196	6.01	5.882	5.526	5.646
Iter.	345	525	300	450	390

- Good estimates are obtained, for such a highly harmonic-contaminated signal, as the number of samples is increased to 120 and 160 samples.
- As the number of samples increases better estimates are obtained, as shown at $m = 240$ samples.

7.7.3.2 Effects of Sampling Frequency

Effects of sampling frequency on the estimated parameters when the signal is contaminated with harmonics are also studied in this test. Table 7.11 gives the results obtained when the number of samples equals 80 samples and the sampling frequency is changed from 1000 Hz to 5000 Hz.

Examining this table reveals that

- Accurate estimates are obtained for the frequency at sampling frequencies of 1,000, 3,000, and 5,000 Hz, and bad estimates are obtained at the other two sampling frequencies.
- The best sampling frequency in such harmonics contamination is 1,000 Hz.

The proposed algorithm together with the least error squares parameter estimation technique is tested when the voltage signal is contaminated with a dc component in the form of e^{-100t}. A number of 200 samples and a sampling frequency of Hz are used. The estimated signal parameters are

$$V = 1.11324 \text{ p.u.}, \quad f = 49.517 \text{ Hz, and } \phi = 30.7$$

The error in the voltage signal is about 11.324% and the signal frequency error is about -1%, which are acceptable in such a highly polluted signal.

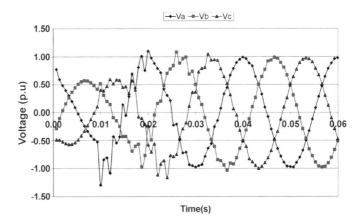

Fig. 7.4 Three-phase unbalanced system of voltages

Table 7.12 Estimated parameters for each phase signal, sampling frequency = 1,500 Hz, number of samples = 90

Phase	Phase A	Phase B	Phase C
V(p.u.)	0.9286	0.8667	0.8566
f(Hz)	50.06	50.495	50.00
ϕ	88.253	−39.20	−150.70
J_a	0.7507	0.924	0.954
# of iter.	330	220	270

7.7.4 Actual Recorded Data

The SA algorithm is tested using actual recorded data for a fault generated using the EMTP. The three phase signals for the voltage are given in Fig. 7.4. The signals are sampled at 1,500 Hz and 90 samples are used in the estimation process. The samples of voltage A are used first to estimate the parameters of the signals of this phase, the amplitude, the frequency f, and the phase angle, then the phase B samples and phase C samples. However, one-phase samples are enough for steady-state estimation. Table 7.12 gives the results obtained for each phase, at the same number of samples and same sampling frequency.

Examining this table reveals that:

- The three phases are unbalanced in magnitudes and phase angles.
- The frequency estimate is accurate enough for such types of fault and it is close to 50 Hz.
- The values of the cost function based on least error squares are slightly different, inasmuch as the voltage waveforms are different especially during the fault period.

The proposed technique succeeded in estimating the parameters of the voltage signal in a very bad environment, which is a fault on the system.

7.8 Conclusions

This section presents a new application of simulated annealing together with the least error squares algorithm for estimation of signal parameters. These include the signal amplitude, frequency, and phase angle. The SA algorithm shows high performance in estimating the signal parameters at off-nominal, near-nominal, and nominal. Effects of critical parameters such as sampling frequency, number of samples, and harmonic contamination are studied. It has been shown that in most cases the algorithm succeeded in estimating the signal parameters.

7.8.1 A Variable Frequency Model

In the following, we assume that the frequency of the voltage signals has a linear variation as

$$f = f_O + bt \qquad (7.8)$$

where f_O is the nominal frequency 50 or 60 Hz, and b is the rate of change of frequency measured by Hz/s. Then,

$$\omega(t) = 2\pi f_O t = 2\pi f_O + 2\pi bt \qquad (7.9)$$

The angle of the voltage signal in this case is given by

$$\theta(t) = \int \omega(t)\, dt = (2\pi f_O + \pi bt)t + \phi \qquad (7.10)$$

The voltage signal can be written as

$$v(t) = \sqrt{2}V \sin \theta(t) = \sqrt{2}V \sin[(2\pi f_O t + \pi bt)t + \phi] \qquad (7.11)$$

The above voltage can be written at any sample k; $k = 1, 2, \ldots, m$; m is the total number of samples in the data window size as

$$v(t) = \sqrt{2}V \sin\left[(2\pi f_O k\Delta T + \pi b k^2 (\Delta T)^2) + \phi\right] + \xi(k) \qquad (7.12)$$

Where

V	Is the rms. of the signal amplitude
ΔT	Is the sampling time $= 1/F_s$; F_s is the sampling frequency
k	Is the sampling step; $k = 1, \ldots, m$
ϕ	Is the voltage phase angle
$\xi(k)$	Are the noise terms, which may contain harmonics

Equation (7.12) describes the voltage signal for a time-variant frequency. If $b = 0$, Eq. (4) becomes a signal voltage with a constant frequency.

7.8.1.1 Problem Formulation

Given m samples of the voltage signal at the relay location, these samples may or may not be contaminated with harmonics and/or noise. It is required to estimate the signal parameters, voltage amplitude, nominal frequency f_0, rate of change of frequency b, and the phase angle ϕ, so that the sum of the absolute value of the error is a minimum. This can be expressed mathematically:

$$J = \sum_{k=1}^{m} \left| v(k\Delta T) - \sqrt{2}V \sin\left[(2\pi f_0 k\Delta T + \pi b k^2 (\Delta T)^2) + \phi \right] \right| \qquad (7.13)$$

The above cost function is a highly nonlinear objective function. The techniques used earlier tried to do some sort of approximations for this cost function, such as Taylor series expansion, to make this cost function linear in the parameters and the linear programming-based simplex method is used to solve the resulting problem. However, this is true and may produce accurate estimates if the power system frequency variation is small and close to the initial guessed value. But if the frequency variation is too large, the estimates will be poor.

7.8.1.2 The Algorithm of Solution

The proposed algorithm aimed to find the best estimate for signal amplitude, frequency, and phase angle of a power system having constant frequency during data windows. To obtain the optimal estimated parameters, the problem is formulated as a nonlinear optimization problem. An efficient simulated annealing algorithm with an adaptive cooling schedule and a new method for variable discretization are implemented. The proposed algorithm minimizes the sum of absolute errors between the sampling signal and the estimated one at all sampling time periods (6). Implementation details of the SA algorithm are given Sect. 7.3.

7.8.2 Simulated Example

In this section the proposed algorithm is tested using simulated examples. Two tests are performed, in the first test the signal is assumed to be a noise-free signal, and the effects of the number of samples and sampling frequency on the estimated parameters are studied. The voltage signal waveform is given as

$$v(t) = \sqrt{2} \sin(2\pi 50 t + 0.2\pi t^2 + 30^o)$$

7.8 Conclusions

Table 7.13 Effects of number of samples, sampling frequency = 1,000 Hz

m	V(p.u.)	$f_O = a$, (Hz)	b(Hz/s)	ϕ
50	1.0	50.0	0.06818	29.996
100	1.0	50.0	0.099771	30.0
150	1.0	50.0	0.10132	30.0
200	1.0	50.0	0.10037	30.0
250	1.0	50.0	0.099781	29.999

This signal is sampled using a sampling frequency of 1,000 Hz and the number of samples equal to 200 is used to estimate the signal parameters. It has been found that the proposed algorithm estimates the signal parameters accurately. These estimates are

$$V = 1.0 p.u \quad f_O = 50.0, \quad b = 0.10, \quad \phi = 30.0^O$$

7.8.2.1 Effects of Number of Samples

Effects of the number of samples on the estimated parameters are studied in this section, where the sampling frequency is kept constant at 1,000 Hz and the number of samples changes from 50 to 250. Table 7.13 gives the results obtained for the test.

Examining Table 7.13 reveals the following remarks.

- For a number of samples greater than 50 the SAA produces an accurate estimate for the signal parameters.
- At a number of samples equal to 50 an inaccurate estimate for the rate of change of the frequency and the phase angle is obtained, and an accurate estimate for the voltage amplitude and nominal frequency is produced.

7.8.2.2 Effects of Sampling Frequency

Effects of sampling frequency on the estimated signal parameters have been studied in this section, where the number of samples is kept constant at 200 samples, and the sampling frequency changes from 250 Hz and 1500 Hz. Table 7.14 gives the results obtained for this test.

Examining this table, we note that

- The proposed algorithm, at the specified number of samples and sampling frequency, produces very accurate estimates for the signal parameters.
- We recommend, for this test, a number of samples equal to 200 samples and a sampling frequency of 750 Hz to produce accurate estimates.

Table 7.14 Effects of number of sampling frequency, $m = 200$ samples

F	V(p.u.)	$f_O = a$, (Hz)	b(Hz/s)	ϕ
250	1.0	50.0	0.09996	30.0
500	1.0	50.0	0.09996	30.0
750	1.0	50.0	0.100	30.0
1,000	1.0	50.0	0.10037	30.0
1,250	1.0	50.0	0.10016	30.0
1,500	1.0	50.0	0.10019	30.0

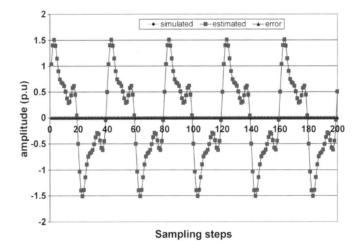

Fig. 7.5 Simulated and estimated signals with the error

7.8.2.3 Effects of Harmonics

Due to the widespread use of power electronic devices today in power system operation and control, the voltage waveforms are polluted with all kinds of harmonics. In this test, we assume that the signal is contaminated with the third and fifth harmonics. A number of samples equal to 200 with a sampling frequency equal to 2,000 Hz are used in this test. The following results are obtained.

$$V = 1.00(p.u), f_o = 50.0 Hz, \ b = 0.10198, \ \phi = 30.0^o$$

Here, we assume that the harmonics frequencies are an integral number of the nominal frequency, which is assumed to be 50 Hz. Examining these results, one notices that the SAA produces very accurate estimates for the signal parameters from a harmonics-polluted signal. Figure 7.5 shows the simulated waveform and the estimated one together with the error.

It can be noticed from the figure that the SAA produces the same signal exactly because the error in all samples is zero.

7.8.3 Exponential Decaying Frequency

Another test is conducted in this section, where we assume that the frequency of the voltage signal has the form of

$$f = f_o + be^{-ct}$$

where $f_o, b,$ and c are the parameters to be estimated. The voltage signal equation in this case becomes

$$v(t) = \sqrt{2}V \sin(2\pi f_o t - \frac{2\pi b}{c} e^{-ct} + \phi_o) \qquad (7.14)$$

This type of variable frequency could be obtained at transient operation in power systems. The cost function to be minimized based on least absolute error in this case is given by

$$J = \sum_{i=1}^{m} \left| v(t_i) - \sqrt{2}V \sin(2\pi f_o t_i - \frac{2\pi b}{c} e^{-ct_i} + \phi_o) \right| \qquad (7.15)$$

In this test, we assume that $f_o = 50.0$, $b = 0.1$, $c = -10$, and $\phi = 30°$ The signal is sampled at 1000 Hz and 200 samples are used. The results obtained for this simulation are:

$$V = 1.0 \; f_o = 50.0, \; b = 0.0996, \; c = 10.18 \; and \; \phi = 29.92$$

The error in the estimated value of b equals 0.4% whereas the estimated value of c equals 1.8%. These are acceptable errors for such a highly nonlinear estimation.

7.8.3.1 Actual Recorded Data

The proposed algorithm is tested on actual recorded data generated from EMTP due to a short circuit. The frequency is first assumed to be a linear time-variant and second with exponential decaying. The results for the linear time-variant are

$$V = 0.976(p.u), f_o = 49.8Hz, b = 0.370 \; Hz/s \; and \; \phi = 90.94°$$

Figure 7.6 compares the actual and the estimated signal waveforms. Examining this curve carefully reveals that

- A large error in the estimated wave is produced in the first quarter of the cycle, because the frequency is constant to the nominal value 50 Hz, and the model assumes a linear variation in the third part of the data window size.

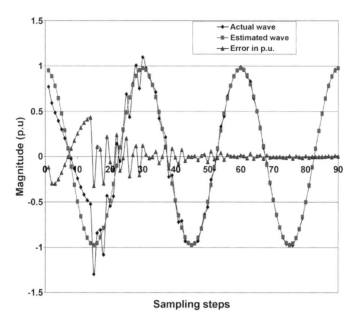

Fig. 7.6 Actual and estimated signals with error

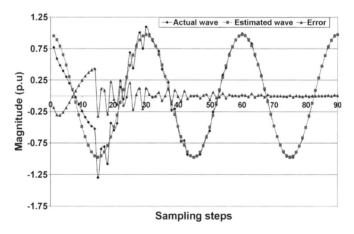

Fig. 7.7 Actual and estimated signals with error

- During the fault, the error goes to a small value until the end of the data window size. This means that the model for the frequency is adequate for this part of the data window size.

In the second test we assume an exponential decay model for the frequency. The following results are obtained (Fig. 7.7).

$$V = 0.976(p.u), f_o = 49.93 Hz,\ b = 0.03,\ c = 14.26\ and\ \phi = 90.73$$

Examining this figure we reach the same remarks as for Figure 7.7, except in the decaying model where the decaying term goes to zero very quickly because the coefficient c is relatively large. However, for the two frequency models used, the frequency is a time-variant, thus it needs a dynamic estimation algorithm to track the frequency variation at each instant.

7.9 Conclusions

In this section the simulated annealing algorithm is used to estimate the frequency of the power system, where we assume a time-variant frequency model for the voltage signal. The proposed algorithm uses the digitized sample of the voltage waveform at the relay location and is tested using simulated and actual data. The algorithm is able to predict the frequency and rate of change of frequency from a highly nonlinear function, and does not need any approximations.

It has been shown that the proposed algorithm is not sensitive to the number of samples used in the estimation, but the sampling frequency should satisfy the sampling theory, and the data window size must be an integral number of cycles.

References

1. Fallon, C.M., McDermott, B.A.: Development and testing of a real-time voltage flicker meter, Proceeding of IEEE Proceeding of Transmission and Distribution Conference, pp. 31–36. Los Anglos, California, USA (1996)
2. Poisson, O., Rioual, P., Meunier, M.: New signal processing tools applied to power quality analysis. IEEE Trans. Power Del. **14**(2), 561–566 (1999)
3. Chang, W.N., Wu, C.J.: A flexible voltage flicker teaching facility for electric power quality education. IEEE Trans. Power Syst. **13**(1), 27–33 (1998)
4. Girgis, A. A., Makram, E.B., Malkinson, T.J.: Measurement of voltage flicker magnitude and frequency using a kalman filtering based approach. Can. Conf. Elect. Comp. Eng. **2**, 659–662 (1996)
5. S. Nuccio.: A digital instrument for measurement of voltage flicker, Proceedings of IMTC/97, Instrumentation and Measurement Technology Conference, **1**, 281–284. Ottawa, Ontario, Canada (1997)
6. Girgis, A.A., Stephens, J.W., Makram, E.B.: Measurement and predication of voltage flicker magnitude and frequency. IEEE Trans. Power Del. **10**(3), 1600–1605 (1995)
7. Srinivasan, K.: Digital measurement of voltage flicker. IEEE Trans. Power Del. **6**(4), 1593–1598 (1991)
8. Soliman, S.A., El-Hawary, M.E.: Measurement of voltage flicker magnitude and frequency in a power system for power quality analysis. Electr. Mach. Power Syst. J. **27**, 1289–1297 (1999)
9. Mantawy, A.H., Soliman, S.A., El-Hawary, M.E.: An innovative simulated approach to long-term Hydro scheduling problem. Int. J. Elec. Power **25**(1), 41–46 (2002)
10. Mantawy, A.H., Abdel-Magid, Y.L., Selim, S.Z.: A simulated annealing algorithm for unit commitment. IEEE Trans. Power Syst. **13**(1), 197–204 (1997)
11. Aarts, E., Jan, K.: "Simulated annealing and boltzman machines", A stochastic approach to combinatorial optimization and neural computing. John Wiley & Sons Ltd, New York (1989)
12. Cerny, V.: Thermodynamical approach to the traveling salesman problem: an efficient simulation algorithm. J. Optimiz. Theory App. **45**(1), 41–51 (1985)

13. Selim, S.Z., Alsultan, K.: A simulated annealing algorithm for the clustering problem. Pattern Recogn. **24**(10), 1003–1008 (1991)
14. Metropolis, N., Rosenbluth, A., Rosenbluth, M., Teller, A., Teller, E.: Equations of state calculations by fast computing machines. J. Chem. Phys. **21**, 1087–1982 (1953)
15. Patrick, S., Gerard, B., Jacques, H.: Enhanced simulated annealing for globally minimizing functions of many continuous variables. ACM Trans. Math. Softw. **23**(2), 209–22 (1997)
16. Miller, A.J.V., Dewe, M.B.: The application of multi-rate digital signal processing techniques to the measurement of power system harmonic levels. IEEE Trans. Power Del. **8**(2), 531–539 (1993)
17. Moo, C.S., Chang, Y.N., Mok, P.P.: A digital measurement scheme for time-varying transient harmonics. IEEE Trans. Power Del. **10**(2), 588–594 (1995)
18. Smith, B.C., Waston, N.R., Wood, A.R., Arrillaga, J.: A Newton solution for the harmonic Phasor analysis of AC/DC. IEEE Trans. Power Del. **11**(2), 965–971 (1996)
19. Rios, S., Castaneda, R., Veas, D.: Harmonic distortion and power factor assessment in city street gas discharge lamps. IEEE Trans. Power Del. **11**(2), 1013–1020 (1996)
20. Moreno Saiz, V.M., Barros Guadalupe, J.: Application of kalman filtering for continuous real −21time tracking of power system harmonics. IEE Proc. Gener. Trans. Distrib. (C) **144**(1), 13–20 (1997)
21. Probabilistic Aspects Task Force of the harmonics Working Group Subcommittee: Time-varying harmonics: part I- characterizing measured data. IEEE Trans. Power Del. **13**(3), 938–944 (1998)
22. Exposito, A.G., Macias, J.A.R.: Fast harmonic computation for digital relaying. IEEE Trans. Power Del. **14**(1268), 1263 (1999)
23. Nagndui, E., Olivier, G., April, G.E.: Harmonics analysis in multipulse thyristor converters under unbalanced voltage supply using switching functions. Can. J. Elect. Comp. Eng. **24**(4), 137–147 (1999)
24. Soliman, S.A., Helal, I., Al-Kandari, A.M.: Fuzzy linear regression for measurement of harmonic components in a power system. Electr. Pow. Syst. Res. **50**, 99–105 (1999)
25. Bucci, G., Landi, C.: On- line digital measurement for the quality analysis of power systems under non-sinusoidal conditions. IEEE Trans. Instrum. Meas. **48**(4), 853–857 (1999)
26. Krim, F., Benbaouche, L., Chaoui, A.: A novel identification-based technique for harmonics estimation. Can. J. Elect. Comp. Eng. **24**(4), 149–154 (1999)
27. Lai, L.L., Chan, W.L., Tse, C.T., So, A.T.P.: Real- time frequency and harmonic evaluation using artificial neural network. IEEE Trans. Power Del. **14**(1), 52–59 (1999)
28. Lin, H.C., Lee, C.S.: Enhanced FFT-based parametric algorithm for simultaneous multiple harmonics analysis. IEE Proc. Gener. Trans. Distrib. **148**(3), 209–214 (2001)
29. Tsao, T.P., Wu, R.C., Ning, C.C.: The optimization of spectral analysis for signal harmonics. IEEE Trans. Power Del. **16**(2), 149–153 (2001)
30. Siarry, P., Berthiau, G., Haussy, J.: Enhanced simulated annealing for globally minimizing functions of many continuous variables. ACM Trans. Math. Softw. **23**(2), 209–222 (June 1997)
31. Yang, J.Z., Liu, C.W.: A precise calculation of power system frequency. IEEE Trans. Power Del. **16**(3), 361–366 (2001)
32. Dash, P.K., Pradhan, A.K., Panda, G.: Frequency estimation of distorted power system signal using extended complex kalman filter. IEEE Trans. Power Del. **14**(3), 761–766 (July 1999)
33. Sidhu, T.S.: Accurate measurement of power system frequency using a digital signal processing. IEEE Trans. Power Del. **14**(1), 75–81 (1999)
34. Sidhu, T.S., Sachdev, M.S.: An accurate technique for fast and accurate measurement of power system frequency. IEEE Trans. Power Del. **13**(1), 109–115 (1998)
35. Jose, A., Dela, O., Altave, H.J., Diaz, I.: A new digital filter for phasor computation, part I: theory. IEEE Trans. Power Syst. **13**(3), 1026–1031 (1998)
36. Jose, A., Dela, O., Altave, H.J., Diaz, I.: A new digital filter for phasor computation, part II: evaluation. IEEE Trans. Power Syst. **13**(3), 1032–1037 (1998)
37. Szafran, J., Rebizant, W.: Power system frequency estimation. IEE Proc. Gener. Trans. Disturb. **145**(5), 578–582 (1998)

References

38. Lobos, T., Rezmer, J.: Real-time determination of power system frequency. IEEE Trans. Instrum. Meas. **46**(4), 877–881 (1997)
39. Moore, P.J., Carranza, R.D., Johns, A.T.: Model system test on a new numeric method of power system frequency measurement. IEEE Trans. Power Del. **11**(2), 696–701 (1996)
40. Moore, P.J., Allmeling, J.H., Johns, A.T.: Frequency relaying based on instantaneous frequency measurement. IEEE Trans. Power Del. **11**(4), 1737–1742 (1996)
41. Terzija, V., Djuric, M., Kovacevic, B.: A new self-tuning algorithm for the frequency estimation of distorted signals. IEEE Trans. Power Del. **10**(4), 1779–1785 (1995)
42. Terzija, V., Djuric, M., Kovacevic, B.: Voltage phasor and local system frequency estimation using Newton-type algorithms. IEEE Trans. Power Del. **14**(3), 1368–1374 (1994)
43. Begovic, M.M., et al.: Frequency tracking in power network in the presence of harmonics. IEEE Trans. Power Del. **8**(2), 480–486 (1993)
44. Kezunovic, M., et al.: New digital signal processing algorithm for frequency measurement. IEEE Trans. Power Del. **7**(3), 1563–1572 (1992)
45. Soliman, S.A., Christensen, G.S.: Dynamic tracking of the steady state power system voltage magnitude and frequency using linear kalman filter: a variable frequency model. Electr. Mach. Pow. Syst. **20**, 593–611 (1992)
46. Soliman, S.A., Christensen, G.S.: Estimating of steady state voltage and frequency of power systems from digitized bus voltage samples. Electr. Mach. Pow. Syst. **19**, 555–567 (1991)
47. El-Hawary, M.E., Mostafa, M.A., Mansour, M.S., El-Naggar, K.M., Al-Arabaty, A.M.: A study of estimation techniques for frequency-relaying applications. Can. J. Elect. Comp. Eng. **21**(1), 9–20 (1996)
48. Chen, M.T.: Digital algorithms for measurement of voltage flicker. IEE Proc. Trans. Disturb. **144**(180), 175 (1997)
49. Michie, W.C., Cruden, A., Niewczas, P., Madden, W.I., McDonald, J.R. Gauduin, M.: Harmonic analysis of current waveforms using optical current sensor, IEEE Instrumentation and Measurement Technology Conference, pp.1863–1865. Budapest, Hungary (2001)
50. Soliman, S.A., et al.: Dynamic tracking of the steady state power system voltage magnitude and frequency using linear kalman filter: a variable frequency model. Electr. Mach. Pow. Syst. **20**, 593–611 (1992)
51. Begovic, M.M., et al.: Frequency tracking in power network in the presence of harmonics. IEEE Trans. Power Del. **8**(2), 480–486 (1993)
52. Akke, M.: frequency estimation by demodulation of two complex signals. IEEE Trans. Power Del. **12**(1), 157–163 (1997)
53. Soliman, S.A., Mantaway, A.H., El-Hawary, M.E.: Power system frequency estimation based on simulated annealing: part I: a constant frequency model. Arab. J. Sci. Eng. **28**(1B), 45–55 (2003)
54. Soliman, S.A., Alammari, R.A., Mantawy, A.H., El-Hawaary, M.E.: Power system frequency estimation based on simulated annealing: a variable frequency mode Scientia Iranica. Int. J. Sci. Tech. **11**(3), 218–224 (2004)
55. Dash, P.K., Pradhan, A.K., Panda, G.: Frequency estimation of distorted power system signal using extended complex kalman filter. IEEE Trans. Power Del. **14**(3), 761–766 (July 1999)
56. Jose, A., Dela, O., Altave, H.J., Diaze, I.: A new digital filter for phasor computation part II: evaluation. IEEE Trans. Power Syst. **13**(3), 1032–1037 (1998)
57. Kezunovic, M., et al.: New digital signal processing algorithm for frequency measurement. IEEE Trans. Power Del. **7**(3), 1563–1572 (1992)
58. Soliman, S.A., Christensen, G.S.: Estimating of steady state voltage and frequency of power systems from digitized bus voltage samples. Electr. Mach. Pow. Syst. **19**, 555–567 (1991)
59. El-Hawary, M.E., Mostafa, M.A., Mansour, M.S., El-Naggar, K.M., Al-Arabaty, A. M.: A study of estimation techniques for frequency-relaying applications, Can. J. Elect. Comp. Eng. **21**(1), 9–20 (1996)
60. Mantawy, A. H., Soliman, S. A., El-Hawary, M. E.: An innovative simulated annealing approach to the long-term Hydro scheduling problem. Int. J. Elec. Power & Energy Syst. **25**(1), 41–46 (2003)

Index

A
Active power transmission losses, 286, 292–293, 305, 312, 319, 322, 325
Actual recorded data, 21, 381, 389, 391, 396, 400, 405–407
Adaptive GA operators, 239
Advanced tabu search (ATS), 220–230
　for UCP, 223
Algebraic
　product, 68
　sum, 68
Algorithm
　for economic dispatch, 193–196
　of solution, 20, 105, 108, 164–165
All-thermal power system approaches, 96–145
Aspiration criteria, 55, 212
ATS. *See* Advanced tabu search

B
Basic
　concept of multi-objective analysis, 327–329
　fundamental of PSO, 74–78
　operation, 66–71

C
Combinatorial optimization, 16, 17, 48–50, 54, 56, 186, 191, 193, 196, 208, 213, 215, 246, 348, 373, 384, 396
Comparison
　of algorithms for the UCP, 230, 246
　of different objective functions, 292–294, 319–327
　between different tabu lists, 217–218
Complementation, 67
Computer simulation, 397–398
Constant frequency model, 396–397
Constraints, 1, 26, 84, 185, 281, 348, 383
Constraints handling (repair mechanism), 59–60, 233
Conventional
　all thermal power systems, 96–97
　techniques (classic methods), 3–7
Convexity and concavity, 66
Cooling schedule, 17, 21, 49, 51–54, 197, 205, 207, 248, 347, 348, 366, 369–371, 382, 384–386, 396, 402
Cost function, 17, 31–34, 47, 48, 53, 87, 102, 105, 107, 108, 112–146, 166, 170, 171, 194, 199, 200, 202, 290, 293, 348, 351, 353, 359, 360, 362, 364, 384, 386, 390, 393, 396–398, 400, 402, 405
Crew constraints, 190
Crossover, 8, 10–14, 59, 60, 73, 231, 233, 234, 237, 239, 253, 255, 263

D
Decrement, control parameter, 52, 53
Details of TSA, 56, 214, 373–376
Difference, 12–14, 21, 52, 68, 197, 238, 272, 281, 349
Differential evolutions, 13–14
Dot product, 43–44

E

Economic
  dispatch problem, 20, 83, 96–98, 111, 112, 120, 185–187, 193–196, 234, 239, 286, 291
  linear complementary, 187, 193, 194, 197, 207
Effects
  of harmonics, 398, 404
  of number of sample, 21, 388, 397–404
  of sampling frequency, 21, 388, 389, 398, 399, 403–404
Emission index minimization, 312–313, 317, 320–323, 326, 339
Equality constraints, 26, 33, 34, 47, 48
Evolutionary
  computation, 7–9, 12, 72–73
  programming, 8, 11–14, 299
  strategies and, 8, 11–12
  techniques, 3, 7
Exponential decaying frequency, 405–407

F

Feasible multilevel approach, 94–96
Final value, control parameter, 51–53, 370, 386
Fitness function, 58–59, 232–234, 236–237
Formulation, 1, 3, 9, 20, 84–88, 99–105, 113–124, 146–164, 185–274.
  See also Problem formulation
Functional optimization, 46
Fuzzy
  all thermal power systems, 112–128
  arithmetic, 68–70, 72, 121, 128–129
  economical dispatch, 145–181
  interval arithmetic, 72, 123–128, 148
  load, 98, 103, 105, 107–110, 120, 121, 144, 145, 150, 159, 161, 166, 168
  systems, 20, 60, 63, 98

G

GA. *See* Genetic algorithm
General
  genetic algorithm, 60, 234
  optimal load flow (OPF) formulations, 291–299
  optimal load flow (OPF) problem, 291–299
  particle swarm optimization (PSO) algorithm, 76–78
  simulated annealing algorithm, 50–51
  steps, 330
  tabu search algorithm (TSA), 56–57, 213–214
Generating
  an initial solution, 10, 20, 193
  non-dominated set, 329–332, 339
  techniques, 330–332
  trial solution, 185, 186, 191–192, 367, 368, 375–376
Generation
  constraints, 296
  fuel cost minimization, 297, 307–309, 311, 312, 323, 329, 339
Genetic algorithm (GA)
  operators, 58, 59, 232–234, 237, 253, 259
  part of GST algorithm, 261
  part of GT algorithm, 251–252
  for unit commitment, 231–245

H

Harmonics
  contaminated signal, 398–400
  problem formulation, 388–389
Heuristic search, 7–8
Hierarchical cluster technique, 333–337
Hybrid algorithm, unit commitment, 246
Hybrid of
  genetic algorithm, 251–255, 259–262
  simulated annealing and tabu search (GST), 259–263
  simulated annealing and tabu search (ST), 246–248
  tabu search (GT), 246–248
Hydrothermal-nuclear, 20, 84–96

I

IEEE 30 bus system, 299, 301–305, 307–310, 312–313, 316, 319, 322, 325, 339
Implementation of genetic algorithm, 232, 234–235
Inclusion, 62, 67
Inequality constraints, 26, 35, 36, 38, 94, 146, 194, 195, 292, 295–297, 351, 359–360, 383
Initial value of control, 51–53, 386
Inner product, 43–44, 46, 47.
  See also Dot Product
Intermediate
  term memory, 54, 208, 221–223
  term memory implementation, 223–225
Intersection, 67, 68
Interval arithmetic, 71, 72

Index 413

K
Kirk's cooling, 51, 53, 197

L
Length of the Markov chain, 52–54
Load demand constraints, 188
Long-term
  memory, 54, 208, 220, 222, 223, 225
  memory implementation, 225
LR-type fuzzy number, 70–71

M
Maximization of
  reactive power reserve margin, 293, 316, 326, 329
  security margin index, 317–319, 323–327, 329, 339
Minimization of
  active power transmission losses, 286, 292–294, 319, 329
  emission index, 294, 326
  generation fuel cost, 293, 307–309, 312
  reactive power transmission loss, 293–294
Minimum norm theorem, 20, 46–48, 356
Minimum up/down time, 60, 189–191, 199, 201, 203, 233
Multi-objective
  OPF, 299, 330
  OPF algorithm, 327, 329–331, 339
  OPF formulation, 329–330
Mutation
  operator (1), 238
  operator (2), 238–239

N
Noise-free signal, 397, 402
Nonlinear model, 357–366
Normality, 66
Norms, 42
Numerical
  examples, 198–207, 216, 239–244, 376–378
  results, 20, 218–220, 226–230, 248–250, 255–259, 263–268, 371, 372, 378
Numerical results of
  ATSA, 226–230
  GST algorithm, 263–268
  GT algorithm, 255–258

O
Objectives function, 20, 185
Optimal
  power flow, 21, 281–342
  single objective cases, 299–319
Optimal solution, minimum norm technique, 91–94
Optimization
  algorithms for OPF, 297–299
  procedure, 54, 87–91, 208
  techniques, 2–20, 23–78, 84, 96, 231, 285, 297–298, 301, 323

P
Parameters, 1, 32, 85, 198, 289, 368, 381
Particle swarm (PS), 14–16, 20, 21, 71–74, 77
Physical concept, simulating annealing, 49–50
Polynomial-time cooling schedule, 51–53, 197, 205, 248, 348, 370, 386
Pontryagin's maximum principle, 37–42
Power flow equations, 287–291
Practical application, 45, 356, 364, 370–371
Problem formulation, 1, 9, 84–86, 96–104, 113–123, 146–164, 291–299, 349–350, 386–389, 396–397, 402
Problem solution, minimum norm approach, 350–366
Problem statement, 185, 186, 207, 372
Production cost, 187, 193, 194, 199, 200, 202, 215, 282
PS. *See* Particle swarm

Q
Quadratic forms, 24–26, 29, 88, 149, 194, 366

R
Reactive power
  reserve margin maximization, 313–316
  transmission loss minimization, 319, 329
Rules for generating trial solution, 185, 186, 191–192, 207

S
SAA. *See* Simulated annealing algorithm
Security margin, index maximization, 294, 323–326, 329, 339
Signal with
  known frequency, 389–390
  unknown frequency, 390–393

Simulated annealing algorithm (SAA)
  part of GST algorithm, 263
  solving UCP, 196–207
Simulated examples, 84, 96, 105, 129, 165–167, 397, 402–404
Slack bus, 281, 288–291, 338–339
Solution
  algorithms, 20, 105, 108, 164–165
  chromosomes, 58, 232, 251, 263
  coding, 58, 232, 235–236
Spinning reserve constraints, 188–189
Start-up, 186–188, 190, 196, 199, 201, 203, 207, 215, 220, 228, 240, 249, 257, 263
Static optimization techniques, 26–37
Steady state frequency, 394–396
Step size
  adjustment, 384, 385
  vector adjustment, 369, 372, 376, 385
Stochastic, 6, 8, 12, 17, 60, 72–74, 246, 356
Stopping criteria, 55, 56, 75, 78, 213, 376
Strategic oscillation (SO)
  implementation, 226
Support of fuzzy set, 65–66
System
  constraints, 94, 112, 188–189, 290, 298, 338, 348, 372
  modeling, 185, 286, 350–351

T
Tableau size for economic dispatch, 196
Tabu list (TL)
  restrictions, 54–55, 209–212
  types, 216
  UCP, 216–217
Tabu search algorithm (TSA)
  part of GT algorithm, 253–254
  technique, 209, 214, 371–378
  UCP, 215
Tabu search (TS) method, 219, 230, 246, 251, 268, 373
Tabu search part of
  GST algorithm, 261–262
  ST algorithm, 247–248
Testing
  for harmonics, 389–393
  simulated annealing algorithm, 385
  for voltage flicker, 387–388
TL. *See* Tabu list
Transformations, 13, 45–46, 48, 84, 92, 331–332
Triangular and trapezoidal fuzzy numbers, 71
Triangular fuzzy numbers, 123–128
Triangular L–R representation of fuzzy numbers, 128–129
TSA. *See* Tabu search algorithm

U
Union, 68
Unit
  availability, 190
  constraints, 20, 188–190, 238
  initial status, 190

V
Variable frequency, 401–402, 405
Voltage controlled, 288
Voltage flicker, 386–388

W
Weighting method, 298, 329, 330, 332